Adrian Woolfson

was educated in London, Cambridge and Oxford. He is a Wellcome Trust Clinical Research Fellow in the Division of Protein and Nucleic Acid Chemistry at the Medical Research Council Laboratory of Molecular Biology in Cambridge and a Charles and Katherine Darwin Research Fellow in Molecular Biology at Darwin College, Cambridge.

More from the reviews:

'Be not afraid of the title! Adrian Woolfson loosens the meaning of the terms 'life' and 'genes', to provide us with new toys in the great playground of possibilities – where perhaps we shall come to understand the origins of life. It is serious fun.'

GRAHAM CAIRNS-SMITH, author of *Seven Clues to the Origin of Life*

'Incredibly versatile and ebullient . . . one of the most exciting reads in years.' ALAN KERSEY, *Cambridge Evening News*

'Gloriously playful antidote to the prevailing gloomily determinist notion that everybody and everything is inescapably genetically programmed. Woolfson wears his daunting erudition lightly and encourages us to think of DNA as, for instance, an infinitely flexible Lego set with all manner of enticing possibilities. Eye-opening and heartening.' *Scotsman*

'A welcome antidote to the naïve genetic determinism that is all too prevalent in popular science, and a pleasure to read. You have nothing to lose but your genes!'

IAN STEWART, author of *Does God Play Dice?* and *Life's Other Secret*

'There must have been a mix-up at the publishers. Some hapless office junior dropped the manuscripts for *Alice in Wonderland* and *A Textbook of Genetics* on the way to the photocopier and got them confused. The result is a book which jumps from the laws of thermodynamics straight into bizarre tales about the King of the Crocodiles who talks in rhyme. Woolfson alters the definition of life to reveal that our distant ancestors might have been just wisps of chemicals floating in an ancient ocean. His theories are very convincing. Anyone who has ever wondered how a planet of rocks and boiling seas could have given birth to Mozart and the Spice Girls will find the answer here. SARAH PLAYFORTH, *Manchester Evening News*

ADRIAN WOOLFSON

Life Without Genes

The History and Future of Genomes

Flamingo
An Imprint of HarperCollins*Publishers*

Flamingo

An Imprint of HarperCollins*Publishers*
77–85 Fulham Palace Road,
Hammersmith, London W6 8JB

www.**fire**and**water**.com

Published by Flamingo 2000
9 8 7 6 5 4 3 2 1

First published in Great Britain by
HarperCollins*Publishers* 2000

Author photograph by Anthony Oliver

ISBN 0 00 654874 1

Set in Janson by
Rowland Phototypesetting Ltd,
Bury St Edmunds, Suffolk

Printed and bound in Great Britain by
Clays Ltd, St Ives plc

to my parents
and Claire

Contents

	Preface	xi
	Acknowledgements	xvii
1	Adventures in Toy Space	1
2	The Structure of Gene Space	32
3	The Crocodile Holds Its Breath	54
4	A Visit to the Information Sea	74
5	The Origins of Life	103
6	Hunting for Intellectual Fossils	127
7	A World Without DNA	148
8	Analog Creatures	181
9	Life Without Genes	203
10	The Origins of Geneless Information	220
11	Patterns Without Programs	247
12	A Journey Through the Geneless Zoo	280
13	The Future of Life	325
	Bibliography	374

As it was in the beginning,
is now and ever shall be.

Preface

The message of this book is simple and can be expressed in a single sentence. Genes may not be necessary for life. Surprising as this seems, it may well be true. Although the creatures with which we are familiar owe their complexity to genes, the very first living things may not have used genes to generate and store their essential information. However, the geneless ancestors from which we might all have descended would have had only a very limited capacity to perpetuate their core information and evolve. They would consequently have been short-lived, incapable of generating extended lineages and able to attain just a rudimentary level of complexity. It was only following a technological innovation which enabled the structures of life to become digitally encoded within a genetic database, that this information could for the first time be transmitted faithfully from one generation to another. In so doing, genes were able to confer ancestral forms of life with a previously unattainable capacity for 'unlimited' heredity. This itself was followed by another innovation that allowed digital information encoded in one chemical technology to stand as a symbolic representation of another. Thus, although genes may not be necessary for life, the emergence of true complexity in our most ancient precursors depended on the unique ability of digitally encoded genes to separate the informational specification of living things from their material realization.

In order to examine these issues and sketch out a possible future for life, we will embark on a journey that will take us back 3.8 billion years or so to life's origin on Earth. This in turn will lead to an examination of the logical structure of genes and life itself. In so doing we will discover that living creatures may be viewed as gene kits that are not conceptually dissimilar to the model kits sold in toy shops. A full description of a living creature, however, encompasses dimensions of information that go beyond its genetic specification and include development, learning and culture. The information of an organism is therefore distributed across a number of different hierarchical levels. In considering what it may mean to have life without genes, we will help clarify the nature of both the genetic and other structures that underpin contemporary life.

The phrase 'life without genes' is itself intended to convey several meanings. First, genes may not be essential for all forms of life. The metabolic circuits of the very first living things may indeed have spontaneously self-assembled in the absence of digitally encoded genetic information and the 'organization of life' imprinted onto inanimate matter without the symbolic strategies of information representation employed by genes. Second, genes may not be adequate for life, as without the self-organizing factors that generate order and stability in the absence of the symbolic instructions of genes, the metabolic networks they regulate and encode are likely to be chaotic and unable to evolve. Third, genes do not necessarily have to be made from DNA, or indeed, translated into proteins. The essential information of living things might in principle just as well be encoded within some other substrate and the structures of life constructed from another type of non-proteinaceous molecular Lego. A logical process of life may thus be considered independently from the hardware in which it is implemented. An appreciation of this point focuses our attention both on the exact relationship between the formal and material aspects of living machines and on the constraints that any particular chemical technology imposes on them. It also means that it should, in principle, be possible to extract the logical essence from creatures and to regard them as substrate-independent 'information kits'. It should furthermore be possible to precisely define the mechanisms by which processes of life flow across their respective spaces of mathematical possibility. Fourthly and finally, all living and potentially living things may be viewed as having a timeless mathematical existence that is independent of their physical realization. Mathematical dodos, for example, continue to exist in a very real sense, despite the fact that their material counterparts have been extinct for many years.

It is hard to comprehend the extent to which our ever-increasing understanding of the genetic basis of biological existence will change the future of life both on Earth, and eventually elsewhere. When we have charted the genetic landscapes that have been explored by natural evolutionary processes and are in a position to fully appreciate the nature of the mechanisms responsible for generating and modifying living things, life will enter a new realm of history. It will no longer lie in the exclusive and capricious historical domain of chance and natural selection. It will instead be possible to design and construct new living things using ahistorical processes, in much the same way that we currently design and construct motor cars, traffic lights, helicopters and vacuum cleaners.

The essential information of all contemporary living things is encoded in genes. Indeed there is a sense in which a kangaroo is nothing more than a kangaroo gene kit, and a hippopotamus a sort of fully assembled hippopotamus gene kit. Much like model aeroplane kits, when the pieces are correctly manufactured and assembled, a corresponding 'model' creature is formed. Genes thus constitute an information database from which many or perhaps all the aspects of living things can be computed. Is it nevertheless possible that the information of the very first living things originated without genes? Might the denizens of Eden have been quite literally geneless?

Although modern genes are made from DNA, the first genes may have been fashioned from different building materials. The software that genes encode is consequently more fundamental than the hardware in which their programs are realized. We might, for example, choose to construct our spears, cooking pots and other implements from stone, iron, bronze, or aluminium. Each material has its own possibilities, limitations and constraints. The information of living creatures might similarly be represented in different materials, each with its own unique characteristics. Rather than being made from proteins, living things might in principle be constructed from breeze blocks, tin cans, or indeed any number of other technologies. The 'logic' of life is thus likely to be more 'essential' than any particular manifestation of that logic.

The dictates of nature and history are not, however, inviolable. Only a fraction of nature's infinite wealth of possibility has been explored and realized by natural search mechanisms. Sabre-toothed tigers, quaggas, penguins, panthers and all the other creatures that have been discovered and realized by history, represent only a fraction of the possible creatures which could have been realized and might one day still be created. Although to date all life on Earth has been designed and assembled by the indifferent process of evolution by natural selection coupled with historical contingency and the physical and chemical laws which underlie self-assembly and self-organizational processes, eventually mankind will be in a position to loosen the shackles of the past and to attain a degree of liberation from the chance perambulations and constraints of history. In the future, life will be designed ahistorically from first principles. It will not be necessary to restrict such enterprises to the historical DNA, RNA and protein technologies with which we are familiar, any more than it is necessary to restrict the design of a motor car to a particular form or construction material. Neither will such artificially designed organisms necessarily encode their information using genes, as the basic processes of life are unlikely to be exclusively dependent on either DNA or the spin of genes.

The universe is indifferent to the existence of any particular thing or event. The mathematical plane on which we currently stand is neither inevitable nor fixed. Just as we have come to reside in the region of mathematical space which we now occupy, we will inevitably one day melt off it. Familiar regions such as *Protein Sequence Space* and *DNA Space* which appeared to form the very cornerstone of our definition and existence, may become historical irrelevancies; ghost towns on an old and long-forgotten frontier. Nature itself is consequently nothing more than that with which we are familiar. If we are to give any value at all to words such as 'mankind', then it can only be arbitrary. One might indeed wonder how many genes could be pruned or removed from the human gene kit, before we would be obliged to erase the label *Homo sapiens* from the side of the box. It is hard to comprehend the extent and contents of the mathematical sea of all possible living things, however it is even more difficult to convince ourselves that we have any special significance in this infinite *Information Zoo* of all possible creatures.

This book is intended for anyone who has an interest in the history, structure and future of life. I have tried to reduce facts to a minimum and have concentrated instead on building a conceptual framework within which the nature of genes and geneless life can be understood. Although entitled *Life Without Genes*, it is as much a quest for an understanding of what genes actually are, as it is a journey to identify the hypothetical geneless antecedents of modern life. It assumes no previous knowledge of the subject and has been written with both the general reader and specialist in mind. It is possible that in trying to please both, I have ended up pleasing neither. If this is the case, then I offer you my sincere apologies. If on the other hand it is not, then I hope that you will find this exploration of life both interesting, entertaining and informative.

Although much of the subject matter is uncontroversial, some of the ideas may not be universally accepted as representing orthodoxy. Indeed, many workers in the origin of life field would not agree that life could have originated without genes, even in principle. This book should not consequently be taken as representing a mainstream synthesis of modern biology. Limitations of space have necessitated considerable economy with regard to what has been included and the detail in which each issue has been examined. The aim, more than anything, has been to leave the reader with a broad impression and sense of familiarity with the subject matter at hand. The rate at which information is being produced in this field is so high that one might reasonably question whether it is legitimate to write a book on this subject in anything other than an electronic format, which can be updated on a weekly, if not daily basis. My feeling however is that it is, and that the broad principles will survive most of the changes which new facts demand.

The fragmentary nature of the evidence for the specific composition and architecture of the earliest forms of life has led me to focus on the general principles that are likely to have underpinned the early history of life, rather than on particular candidate technologies. I have aimed to give a general flavour of the issues instead of offering definitive solutions. Rather than examining all of the major transitions which have defined the evolution of modern multicellular life, I have concentrated instead on the origins, structure and future of life. It has furthermore been necessary to be highly selective even within these well-circumscribed fields. Thus although I have examined the possibility that life began without genes, I have not explored related theories. These include those based on 'hypercyclically' organised collections of RNA molecules. My treatment of the subject matter should to this extent be considered partisan. Finally, the idea that genes may not have been necessary for the simplest manifestations of life should not be regarded as an experimentally proven fact, or even as a concrete hypothesis. It is instead a sketch for a hypothesis, which is nevertheless coherent and falsifiable.

I have chosen not to reference each point separately, but have indicated the source of the most important concepts at appropriate places within the text and have included a comprehensive bibliography. If more than two authors are associated with any particular study, I have tended to mention the names of only the first and the last. The central ideas in this book represent a synthesis of information from a wide variety of sources, but are based mainly on the work of the following: R. J. Bagley, D. Bray, S. Brenner, A. G. Cairns-Smith, C. Calvin, J. Chmielewski, J. E. Cohen, S. Conway Morris, K. Craik, B. Derrida, F. Dyson, M. Eigen, J. D. Farmer, R. Feldberg, W. Fontana, R. Ghadiri, H. Haken, R. W. Hamming, J. H. Holland, S. Kauffman, J. J. Kay, C. Knudsen, C. Langton, D. H. Lee, J. Maynard Smith, W. S. McCulloch, H. Morowitz, L. E. Orgel, G. Oster, N. H. Packard, A. Perelson, W. Pitts, H. Plotkin, I. Prigogine, S. Rasmussen, I. Rechenberg, R. Rosen, O. Rössler, E. D. Schneider, P. Schuster, E. Szathmáry, J. W. Szostak, D. Thompson, G. Wächtershäuser, G. Weisbuch, J. Wicken, S. Wright, S. Yao.

AW

Darwin College Cambridge, June 1999

Acknowledgements

I am and will continue to be greatly indebted to the following people: César Milstein, Barry Keverne, Herman Waldman, Nicholas Mackintosh, Miriam Rothschild, Anthony Dickinson, Chuck Yung Yu, Peter Salmon, Sue Henley, Benjamin Chain, Keith Peters, David Oliveira, Sheila Banks, Rachel and Joseph Charlaff, Gerald, Lynne and Alexander Woolfson, Karen, David and Daniel Woolfson, James Holt, John Grimley-Evans, Claire Sefton, Jane Wardle, Justin Stebbing, Henry Plotkin, John Ledingham, Chris Bulstrode, Geoffrey Lloyd, David Behrens, Anne-Louise Fisher, Naomi Rosenbaum, Martin Raff and Avrion Mitchison.

I am especially grateful to my agent Felicity Bryan for taking me on in the first place and for studiously overseeing the project with great skill, to her assistant Michelle Topham for her help and encouragement, to my copy editor Rebecca Porteous for her expert assistance, to my assistant editor Georgina Laycock for her help, enthusiasm and encouragement, and to Karen Spillard for helping to assemble the bibliography. My editor Philip Gwyn Jones showed much courage in taking on an unknown writer who had produced a rather rudimentary outline for a book which at that time was tentatively entitled *Beyond Biology*. I would like to thank him both for his confidence in my ability to complete this book, and for the great commitment, patience and enthusiasm that he has shown for this project since its inception.

I would also like to thank: the Wellcome Trust, the Medical Research Council, the Pre-Clinical Faculty of Medicine at King's College London, the Department of Psychology and the Department of Zoology at University College London, Gonville and Caius College

Cambridge, the Department of Zoology, the Sub-Department of Animal Behaviour and the Department of Experimental Psychology at Cambridge University, the Clinical School of Medicine at the John Radcliffe Hospital in Oxford, Balliol College Oxford, Darwin College Cambridge and the MRC Laboratory of Molecular Biology in Cambridge, for their help and support. The Wellcome Trust and Medical Research Council have provided me with generous financial support over the years, for which I am most grateful.

I am furthermore especially indebted to César Milstein, Barry Keverne, Gerald Woolfson, John Maynard Smith, Sheila Banks, Claire Sefton and Miriam Rothschild, for reading the entire book in draft and for all their helpful comments, thoughts, suggestions and discussions. I am, in addition, grateful to Kiyoshi Nagai for reading chapter 3, to David Grellscheid for reading chapter 4, to Michael Rodgers for reading chapters 1 through to 10 and to Wolf Reik, Alexander Woolfson and Myriam Altamirano for reading chapter 13. Needless to say, I remain entirely responsible for any mistakes that remain. I would particularly like to thank Sheila Banks and Claire Sefton for their great help, support and inspiration during the last few months of writing.

1

Adventures in Toy Space

If one were to walk into almost any well-stocked toy shop and visit the model section, the display of kits arrayed across the multiple towering shelves is likely to be quite substantial. I must admit to being uncertain as to which is currently the most popular. Perhaps dinosaur kits, flying saucer kits, or kits which may be assembled into various types of strange inter-galactic beings. When I was a schoolboy, almost every self-respecting friend had a large collection of plastic model aeroplanes. More enthusiastic individuals would paint their models with the thick, oily and pungent paint which came in small tins. Others would be content with attaching colourful stickers onto the wings, tailpiece and fuselage.

One exception to this rule, however, was my friend Conrad, who had only a modest collection of model aeroplanes which included just a single *Lancaster* bomber. He was, however, keen on collecting the plastic model heads of Indian chiefs which could be found inside the packets of certain breakfast cereals. It was unfortunate that Sitting Bull was discontinued long before he had managed to complete his impressive, but nevertheless imperfect, collection. Had he known this, his voracious and seemingly insatiable appetite for breakfast cereal might have been somewhat curtailed.

My personal favourites were the *Submarine Spitfire Mk1X*, *Hawker Hurricane*, *Messerschmitt 109* and a scale model of Lord Admiral Nelson's battleship *HMS Victory*. These and many other models could be found amongst the shelves of our local toy shop *Toys Toys Toys*, where we would spend hours gazing at the pictures on the boxes. Once home, we would construct a landscape that included toy soldiers, model aeroplanes, tanks, landing craft, battleships and support vehicles. Occasionally we would set a plane on fire and throw it from my bedroom window.

Although in some toy shops the models are arrayed somewhat haphazardly, owners generally tend to organize their displays so that kits of a similar kind, for instance aeroplanes, helicopters, tanks or ships, are clustered together. A secondary level of order is often found superimposed upon this primary level. World War II aeroplanes may, for example, be separated from those which are more contemporary. Kits may also be arranged according to their size, price, or country of origin.

The assembly of model kits was and still is very simple. The kit is selected on the basis of one's interests, recommendations from friends, television advertisements and the forcefulness with which the images on the packaging impinge on the imagination. Having purchased the kit and a tube of glue, all that is required is a little time, patience and an instruction manual. It is then just a matter of pulling the plastic pieces off the strips and organizing them into the correct pattern with the help of the accompanying instructions, a clothes peg and some glue.

The toy shops of today are however quite different from their predecessors and the small high street shops of the past are in the process of being superseded by toy hypermarkets. Instead of hunting for a model in several small and widely separated shops, it is often easier to pay a visit to the local toy hypermarket where an enormous variety of toys are housed under a single roof. The selection in these hypermarkets is in fact so large, that they are necessarily huge in order to accommodate them. Whereas in the past one would enjoy the advice of a friendly shop assistant, these days it is usual to proceed through a toy hypermarket accompanied by nothing other than a large shopping trolley.

In some cases, the choice of toys and model kits is so great that the hypermarket begins to approach what we might call completeness. In order to understand what I mean by completeness, we will briefly consider its antithesis. That is to say, what it means to be incomplete. This is not difficult as one only has to think of that annoying feeling when one walks into a shop to buy something only to find that they have either sold out, or never stocked the item in the first place.

Small shops have a limited amount of shelf space and so can only stock a limited selection of the items listed in the wholesaler's catalogue. A small shop may however increase its range by changing its stock seasonally. A fruit shop might, for example, sell strawberries and cherries in the summer, whilst replacing these with mangoes and passion fruits in the winter. Their actual stock at any given moment will however represent only a small fraction of the complete range of fruits and vegetables which are in principle available. But it is not only small shops that have incomplete stocks. Even the largest wholesalers are unlikely to stock rare items from Papua New Guinea or Madagascar. The complete theoretical repertoire of all known fruits thus greatly exceeds the size of both the actual stock of any particular shop and the available stock housed in the warehouses of wholesalers. In the case of small shops, there simply isn't enough space to store all of the items and the limited demand for specialized products does not usually justify their inclusion on the shelves. In the case of the largest wholesalers, although an almost complete repertoire is in principle attainable, its realisation is prevented both by market forces and other economic factors.

Let us now return to toy hypermarkets and examine the term completeness in more detail. There are several senses in which a toy hypermarket may be said to be complete. If, for example we were to visit the second largest imaginary toy hypermarket in Great Britain which we will call *Toy Land*, it might claim to stock every toy and model kit currently manufactured in the country. This toy hypermarket would in one sense be complete, as it would be possible to walk into *Toy Land* and emerge with any toy or model kit from the complete set of those currently made in Great Britain. It would only be the collector of model kits from Russia, or a child that wanted a model racing car from Sweden, who would be disappointed by their visit. So although complete at one level, *Toy Land* is incomplete at this more demanding level. The parents of these children had better perhaps take them to the largest imaginary hypermarket in the country, which we will call *Toys Are Great*: as it stocks every toy and model kit currently available in the world. The merchandise of this toy hypermarket constitutes a more comprehensive example of completeness as collectors of Russian and Swedish model kits alike would both leave with a full basket.

We will now however pay a visit to the largest toy hypermarket in the world. This imaginary hypermarket, which we will call *Toy World*, is truly enormous. It is popular and teeming with excited visitors as it too has a unique selling point. For *Toy World* stocks not only every toy currently available in the world, but every toy that has ever been manufactured. If I were to pay a visit to *Toy World*, I could walk along the aisles and pull out a brand new *Messerschmitt 109* model identical in every way to one I bought as a child. I would also be able to find toys made by the Romans, Greeks, Vikings and Victorians. The region containing spinning tops alone would be dizzyingly large.

Toy World is thus complete in a way that the other toy hypermarkets are not. This is because the depth of its completeness is enriched by a historical dimension. The completeness of *Toy World* and of all the other toy shops is, however, dependent on how one chooses to define the word 'toy'. One might, for example, choose to exclude toys made in antiquity because they were not mass-produced. There is also a sense in which something may transiently become a toy, if somebody should choose to relate to it in a toy-like manner. The completeness of any particular toy hypermarket must therefore be evaluated in the context of a very specific notion of what actually qualifies as a member of the set of toys that might in principle be found inside the shop.

We will shortly be completing our exploration of different types of toy shops. But before we abandon our current preoccupation with toy hypermarkets and model kits, I will ask for just a little more of your patience. For I would now like to discuss one final type of imaginary toy hypermarket, which will provide an example of the gold standard which, for want of a better word, I will call 'true' completeness. The truly complete toy hypermarket I would like to consider is called *Toy Space*.

Toy Space upstages all of the other toy hypermarkets and cannot be trumped by any other toy hypermarket. This is because it adds a third and hugely significant dimension to the concept of completeness. For *Toy Space* stocks not only the complete repertoire of all past and present toys, but also every possible toy that could ever be manufactured in the future. It thus constitutes a universal library of all possible toys. However, despite the attractions of this particular toy shop, it is unlikely that it will ever be constructed. This is because *Toy Space* is very much larger than *Toys Toys Toys*, *Toy Land*, *Toys are Great* or *Toy World*. In fact *Toy Space* is so large that if it could be built, it would cover all of the planet and would most likely reach high up into space. *Toy Space* is in fact the hyper-hypermarket of all possible toy hypermarkets and has an important property which distinguishes it from all other types of toy shop. This is that anybody could, in principle, at any time in the past, present or future, walk in with their shopping trolley and leave with any possible toy, or model kit, that they desire. Whether they could actually do so in practice however, is another matter.

Should children in the future happen to shrink to the size of a flea, this would not be a problem as they would find fathomless collections of toys one hundredth the size of a flea. If on the other hand they should happen to turn into giants, then this would also not be a problem as somewhere within the vast expanses of *Toy Space* they would find the full range of every possible imaginable toy appropriate for giants to play with. Should children in the future happen to have twelve hands and four heads, six eyes, and necks forty-two inches long, then this would also not be a problem, as they too will find appropriate toys and model kits. Toys for everyone and everything, at any time and in any place. Such is *Toy Space*.

It is easy to imagine each type of toy stacked tidily one upon the other in huge piles, arrayed up and down and in every direction by the *Toy Space* librarians. Hundreds of thousands of them, labouring continuously day and night like colonies of hyperactive ants. Boxing, cataloguing, ordering, arranging, shifting, packaging, labelling and relocating. Each toy and model kit neatly ordered according to its shape, size, colour and country of origin. And yet despite their vigorous work, there are still huge regions, mountains, valleys, plains and piles of disordered, random toys and model kits. Poking out at every angle and in every direction in a seemingly endless and multicoloured sea, they extend to the horizon and beyond. In many areas they are piled high up into the sky and the librarians sift through them layer by layer, like archaeologists excavating an ancient site. This vast and seemingly endless wasteland of disordered toys has not yet been sorted and contains a hotchpotch collection, many of which are quite dissimilar.

Those uncomfortable with heights are best advised to avoid shopping in *Toy Space*, as getting stuck in one of the many hundreds-of-thousands-of-miles-high *Toy Space* elevators might prove to be a most disconcerting experience. Parents similarly had better not lose their children in *Toy Space*. For once lost in the sea of toys, they would be difficult to retrieve. Beware also the dangers of falling toys, rocking horses larger than the Eiffel Tower, of whizzing plastic hoops, magic sets in which things disappear but never reappear, and of jack-in-the-boxes with coiled steel springs the size of France. Please also heed the signs: '*Do not walk inside this toy*', '*This toy will catapult you into the sky*', '*This toy may not be a toy*', '*Behaviour erratic and unpredictable*' or '*This part of the shop* is *a toy*'.

One might be forgiven for thinking that the owners of *Toy Space* are far wealthier than other toy shop owners. They do, after all, own the toy shop of all toy shops. Furthermore, rather than employing hundreds of thousands of designers to invent their toys, they use a random toy-making machine which has the ability to generate the entire contents of *Toy Space* without the intervention of a single designer. City analysts and potential shareholders might, however, choose to spend a little time studying the company prospectus before making investments. For like many good things, *Toy Space* has a catch. The problem being that shopping in *Toy Space* is, quite literally, a lethal business.

This inconvenient but somewhat important point arises from the fact that *Toy Space* is so large and the toys it contains so numerous that it would take very much more than a lifetime to travel even a small distance across it. An excursion into *Toy Space* might thus bring a whole new meaning to the phrase 'shop 'til you drop'! Even if you were able to extend your lifetime by several hundred years, it is arguable that you may not ever leave *Toy Space* alive once you had entered. This is because it does not in fact have an end. And as *Toy Space* contains absolutely nothing but toys, you would have to bring provisions such as food, water, bedding and other necessities with you on your excursion. A simple weekend shopping trip might come to resemble an expedition to the North Pole. If, by some miracle, you were able to find an exit door before you reached old age, it is unlikely that you would leave *Toy Space* with the toy or model kit that you had been looking for. And if by some good fortune you did, when you eventually arrived home, you might find that a vital piece of the kit was incorrectly shaped or missing, and that your model of *HMS Victory* had only a passing resemblance to the ship that the Lord Admiral Nelson knew so well. You might also find you were no longer interested in toys and that you would in fact have preferred a trip to *Gardening Centre Space*, *Bingo Space*, *Golf Space* or *Bowling Alley Space*. Such a state of affairs would clearly benefit neither the customers nor the shareholders, who would eventually feel the economic consequences of both the lack of cash in the tills and the costs associated with constructing the never-ending boundaries of *Toy Space* and the ongoing Relatives of All Past Customers vs *Toy Space* legal case. Perhaps we had better stick to old-style shopping after all!

My reason for discussing toy shops and toy models at such length is not because I have any particular fascination with toys. It is because I want to compare toy model kits with another sort of kit. This type of kit is far more complex than any toy model kit, as it is capable of generating living things. I will call this type of kit a gene model kit, or more simply in the words of Sydney Brenner, just a 'gene kit'. For the moment I will draw an exact analogy between a toy model kit and a gene kit. We may thus imagine a gene kit as a cardboard box filled with genes. On the front and sides of the box is a brightly coloured picture of the creature that might in principle be constructed if the information in the kit is used to instruct a biological manufacturing process. We could equally well call these kits 'DNA (deoxyribonucleic acid) kits'. This is because, with the exception of some viruses, the coded genetic instructions necessary for the construction of all known living things are written in the chemical language of DNA. Moreover it is not only genes that are needed to assemble living things, but also some of the bits of DNA which lie in between them.

Like toy model kits which contain both an instruction manual and the pieces of the kit itself, we may imagine that gene kits contain both a DNA instruction manual and a collection of chemical building-block 'pieces'. These may be stuck together by carefully following the instructions in the DNA manual. If executed correctly, this assembly procedure should result in the production of the creature that the manual specifies. In practice things are not quite this simple. But for the time being, we will assume that the information contained within a gene kit is all that is needed to construct a living thing.

When a gene kit is assembled according to the instructions inside the box, the end result is a living creature. However, unlike the instruction manuals of aeroplane kits which if carefully followed result in models which correspond to the picture on the box, the nature of the product resulting from the execution of the instructions stored within a DNA manual is far less certain. This is not to underestimate the importance of variability in the industrial production of 100,000 supposedly identical *Spitfire* kits. Machine error will inevitably lead to small changes in the components, and printing errors in the instruction manual might result in the construction of a *Spitfire* without a tailpiece, or with only a single wing. Such events are however rare, and imperfect models will usually be detected by a quality control team. Given an instruction manual, a safe pair of hands, a complete set of components and an appropriate environment, successful assembly of a toy model kit is more or less inevitable.

In contrast, the relative lack of inevitability when the information in a gene kit manual is used to instruct a biological assembly process, reflects the complexity associated with the construction of genetic 'model' creatures, or 'genetic toys'. The use of the word toy instead of organism is intended to reflect this fact, as the information needed to construct a living organism does not reside exclusively in its genes. Furthermore the information needed to successfully operate an organism may not reside exclusively within its genetic programs. The machinery for reading and interpreting the instructions stored within the DNA manual of sexually reproducing organisms, for example, is stored both in the egg, and in the mother's follicular cells which surround the growing embryo. These provide critical instructions which are not contained within the DNA manual of the developing organism. The information contained within the egg and the follicular cells is consequently just as important as that stored within the gene kit itself.

The implied equivalence between organisms and toys also emphasizes that, like toy model kits, living creatures are mechanisms with a defined organization. Living things may thus in principle be designed and manufactured by the implementation of an appropriate mechanical process. The differences between genetic toys and assembled living organisms will be explored in a later chapter. However I will take this opportunity to emphasize that a gene kit is a *specification* for a genetic toy, rather than a full *description* of a living organism. A complete description of an organism incorporates dimensions of information which go beyond the basic specifications encoded in genes and their associated regulatory DNA sequences. In addition to the information contained within the egg and follicular cells, this includes other types of developmental information and information that may be acquired by processes of learning and culture.

But what exactly is a gene kit and for that matter what is a gene? For the moment we will cut our cloth according to our needs and adopt a very simple model of a gene. However at a later point this expedient but somewhat impoverished model will be substituted for one which is both more complex and general. I will thus at present be happy to compare a gene to a string of four differently coloured beads. We will imagine that these are threaded one by one onto a long piece of wire to make a necklace. The process of threading differently coloured beads onto the wire generates the information which a gene encodes. The amount of information stored within a given gene depends on the necklace's length. But when using the term information, it is necessary to ask 'information for what or for whom?' We will see the full significance of this point later. For now, however, it is worth mentioning that only some of the information stored in DNA 'necklaces' encodes the information of genes. Other regions help regulate the pattern in which genes are switched on and off, or help translocate the genes through space when their information is being copied. Still others help package the sub-components of genes, or contain defunct genes. Surprisingly however, most of the sequences in DNA necklaces appear to encode irrelevant junk.

I am reminded at this point of a conversation I had many years ago in a village pub called the Wheat Sheaf to which we had adjourned following a college choir tour of Northamptonshire. The chaplain who had organized the tour had spent the day with a Pevsner guidebook in his hand leading us around various sites of architectural or other interest. These included a triangular building that had been built by the recusant Sir Thomas Tresham, who had been caught up in the Gunpowder Plot. Having been asked by the chaplain about my research and given him a brief explanation, I added, 'Do you know that in fact all you are is a string of genes?' Fortunately, on noticing his perplexed expression, I had the good sense to add 'but of course there is the spiritual dimension.' To which he smiled and replied, 'Yes, yes, we should not forget *that*!'

One can perhaps never fully expunge questions of metaphysics from living machines, as these are matters of belief and as such lie outside the domain of science. Indeed, the notion that creatures are machines which may be understood by invoking the physical and chemical mechanisms that underlie their function, is not necessarily incompatible with metaphysical beliefs of a general nature. The beauty of accounts couched exclusively in the language of genes and information, however, is that they are able to deliver a complete and consistent explanation of living creatures without invoking extraneous elements whose presence and influence is, by their very nature, impossible to prove or disprove.

Genetic 'necklaces' may differ either in the sequence of their beads, the total number of beads from which they are composed, or both. Eörs Szathmáry and John Maynard Smith have suggested that the total number of beads that encode meaningful information may be taken as a measure of the biological complexity of an organism. If we ignore the fact that only some DNA sequence information encodes functional genes, then the three and a half billion (3.5×10^9) or so genetic beads that are found in humans would pale into insignificance with the one hundred and thirty billion (130×10^9) of the flowering plant *Fritillaria*. However, perhaps fortunately for us, only about 0.02% of these encode genetic information. Although some of the other regions contain essential regulatory sequences, many of the other 99.08% of the beads in *Fritillaria* appear to lack a function. One cannot help wondering whether if this were not the case, flowering plants like *Fritillaria* might spend their Sunday afternoons pottering around human nurseries, looking for attractive human specimens to plant in their gardens. Men and women would be reduced to little more than ornamental slaves and we would be flattered if our plant masters condescended to speak even a single word to us.

The total number of genes contained in a kit, however, represents only one of the several tiers of complexity which define living things. These include the complexity of the genetic programs in which individual genes participate, and the non-genetic informational databases of learning and culture, which some gene kits are able to acquire. A true measure of the complexity of an organism should thus include all of the information that contributes to its successful operation, at every stage of its life cycle. Both genetic and extra-genetic.

We will now examine the structure of genetic necklaces. The first bead in a necklace might be a blue (**B**), followed by an orange (**O**), two greens and a red (**GGR**), five greens (**GGGGG**), two blues (**BB**) and so on. The first part of this necklace would thus read: **BOGGRGGGGGBB**. Another necklace may however begin with a red bead (**R**), followed by a green (**G**), two oranges (**OO**), seven blues (**BBBBBBB**) and a green (**G**). This necklace would read: **RGOOBBBBBBBG**. It is clear that if these necklaces were aligned, they would look quite different to one another.

In this bead-necklace model of genes, the coloured beads represent the four chemicals: **A** (adenine), **C** (cytosine), **T** (thymine) and **G** (guanine) which encode the information of genes and are known as nucleotide bases. In an exact analogy with bead necklaces, the difference between one gene and another lies in the sequence of its nucleotide bases. So whereas one gene might read: **AAAGGGCCTTTTAAACTTAGCCAT**, another might read: **TACTGGAACGCTTGTGAATGCCGC**. The total number of bases on the other hand reflects the *amount* of information that each gene encodes.

In higher (eukaryotic) organisms, genetic necklaces are housed within a compartment inside the cell known as the nucleus. However instead of rattling around the nucleus by themselves, genes and intervening DNA sequences are linked together to form genetic super-necklaces known as chromosomes. Punctuation marks within chromosomes mark the beginning and end of each gene so that, although joined together, each gene is clearly delineated from its companions. In addition to nuclear genes, a very small number of genes are stored in sub-compartments of the cell known as mitochondria.

With the exception of red blood cells (all of which lack genetic material), egg and sperm cells and B-cells of the immune system, every human cell has an identical set of forty-six chromosomes. There are two non-identical copies of each chromosome inside every cell, so that there are actually only twenty-three different types of chromosome. Species may differ in the total number of chromosomes that their cells contain. The cells of a kangaroo, for example, contain twelve chromosomes, whereas those of fruit flies have only eight. The DNA sequence of all the nuclear chromosomes and the mitochondrial 'mini-chromosome' that is stored in every nucleated cell of an organism is known as its genome. This contains all the genetic information needed for the construction and operation of an organism. A gene kit for a given creature thus contains its complete genome sequence. Not all genomes, however, are housed within a nucleus. The relatively simple genomes of unicellular organisms such as bacteria, for example, float freely around the cell. The process of determining the order in which individual nucleotide bases are connected together is known as sequencing. If we were to sequence the entire genome of a creature, we would end up with a unique string of **A**, **G**, **C** and **T** nucleotide bases. Give or take a few variations between individuals, this unique sequence fingerprint broadly characterizes all of the members of a species.

The only other detail I would like to discuss at present is how the information in genetic sequences is used to co-ordinate the construction of a living organism. It may not come as a great surprise that there are machines inside cells which are able to access the information stored in genes. This information is then translated into proteins, which are the principal blocks from which all known living things are built. The tiny machines that translate the genetic texts written in DNAese into the amino acidese language of proteins are called ribosomes. These machines are located outside the nucleus in the surrounding cellular space. Ribosomes have no key to the nucleus and are obliged to access the information of genes indirectly, by employing the services of chemical messengers known as messenger RNA molecules, or just mRNA molecules. These chemical go-betweens steal copies of genes and carry this information out of the nucleus to the ribosomes lurking outside. The main difference between DNA and mRNA sequences is that in mRNA the **T** (thymine) bases are substituted for a similar base called **U** (uracil). The words of the mRNAese chemical language are thus written in an **A**, **C**, **G** and **U** alphabet, as compared with the **A**, **C**, **G** and **T** alphabet of DNAese.

Ribosomes are universal translation machines that function like mini-computers. They are universal because they accept and respond to any program written in the **A**, **C**, **U** and **G** language of mRNA that is fed into them. This processing occurs regardless of the nature and origin of the input sequence. When fed with an mRNA program, the ribosome starts reading and translating at the beginning of the message and proceeds systematically to the end. The information encoded in the mRNA sequence is used to create a new type of necklace constructed from a different set of building-block materials. These are the twenty different types of amino acid from which proteins are made. Amino acid necklaces, assembled according to the **A**, **C**, **U** and **G** sequences of mRNA programs, are the one-dimensional elements of protein micro-structure from which the three-dimensional macro-structures of all known living things are constructed.

The digital mRNA program fed into a ribosome may be imagined as momentarily capturing control of its translation apparatus, which it then subverts to its own end. Ribosomes can be thought of as existing in one of twenty different states, each of which corresponds to one of twenty different amino acid adding behaviours. These may be sequentially selected from the machine's behavioural repertoire by the mRNA program which engages and is subsequently computed by the ribosome. Each represents the addition of a different amino acid to the necklace which is currently being synthesized.

In order to translate the **A**, **C**, **U** and **G** code of mRNAese into a language written in the amino acid alphabet of amino acidese, a translation principle is needed. It turns out that, with the exception of a small number of organisms that have acquired minor alterations to the code, an identical translation principle is employed by all living things. This genetic code is written in triplet sequences of **A**, **C**, **U**, and **G** bases. Thus a ribosome fed with an mRNA program that reads: **CCCGUACCGUAUUGCUACGUGAGG**, will parse the sequence and read it as eight lots of three symbols: **CCC / GUA / CCG /**

UAU / UGC / UAC / GUG / AGG. Each triplet combination of **A**, **C**, **U**, and **G** bases specifies one of the twenty different amino acids, or a punctuation START or STOP sign. These punctuation marks define both the beginning and end of each gene and the frame in which it is read.

The number of combinations of triplet sequences that can be constructed from four different bases is 4^3. There are thus 4 x 4 x 4 or 64 possible triplet combinations in total. Each of these is known as a codon. As the genetic code only needs to specify twenty different types of amino acid and two punctuation marks (START and STOP), there are in principle forty-two potentially spare codons. But the code is degenerate and all but two of the amino acids are specified by more than one codon. The amino acid leucine, for example, is encoded by six different synonymous triplets: **UUA**, **UUG**, **CUU**, **CUC**, **CUA** and **CUG**. Thus a ribosome encountering the codon **UUA** behaves in exactly the same way as it would when it encounters the codon **CUC**, and adds a leucine to the nascent chain. Three different codons **UAA**, **UAG** and **UGA** specify the punctuation STOP signal which delineates the end of genes. The beginning of all genes is however marked by a single invariant **ATG** START codon.

The important thing about gene kits is that they contain all the genes needed to construct and run the organism depicted on the side of their box. In addition to this, they also contain all of the other bits of DNA which are necessary to ensure that the organism functions correctly. When trying to understand any particular aspect of the design, control and functioning of an organism, the explanation should, at least in principle, lie within the gene kit box. There is indeed nowhere else to search for the fundamental elements of explanation, other than in the invisible molecular threads hidden within the kit, and in the dynamic patterns that are generated when the components are switched on and off in different cell types and at different times. The buck must stop with the genes inside the box. This genetic 'box of tricks' thus constitutes the minimal level of explanation necessary to specify the structure and function of a living thing. The contribution of any particular gene to the game of organism creation may be illuminated by systematically tinkering with the number and composition of the genes housed within a gene kit box.

I will not discuss the detailed wiring of gene kits here, as we will return to this topic later and because it is not at present necessary to do so. An electrical engineer is able to construct a radio set by connecting together a collection of capacitors, resistors, transistors and other components, about whose exact structure he or she may know very little. What is important to the engineer is not that a given capacitor or resistor is structured in a particular way, but rather that it behaves in a particular way when connected into a circuit with other components. Individual components thus represent minimal modules of function, while the circuits they are organized into constitute higher orders of modular function.

Groups of genes within a kit may similarly be imagined as being organized into self-contained modular circuits. Each of these may be assigned its own discreet sub-function, which is defined by the underlying genetic logic. There is thus a sense in which an organism emerges as a result of the combined agency of several assorted types of modular functions. Different biological architectures and behaviours may consequently be imagined as being generated by the introduction of novel modular functions into the basic set. Modules can also be modified, deleted, or rewired to their companions in different ways and the patterns in which they are switched on and off altered.

Organisms that appear similar might be expected to share large numbers of modules. The differences between a tiger and a leopard might, for example, result from the addition, deletion, modification or rewiring of only a small number of modules, while the modular 'distance' between a tiger and a porcupine is likely to be far greater. Despite the fact that different types of creatures are defined by their own private modules of function, there is likely to be a set of core modular functions which are shared by all living things.

We can now imagine a new kind of hypermarket, called the *Hypermarket of All Possible Modular Functions*. Each module is a black box and each function may be realized by any number of different technologies. The precise structure of the components inside the box need not, however, concern us. For as long as we are able to connect modules in an appropriate manner, we can construct a host of different organisms. To construct a baboon, for example, we need only walk through the hypermarket and collect the appropriate modules. These can then be assembled according to an appropriate set of instructions. A giraffe might similarly be imagined as having a genetic module for growing long necks, and a porcupine one for growing long spines. Rose bushes can furthermore be imagined as having modules which generate thorns, leaves or petals and if the black stripes of zebras were generated by the action of a stripe module, one might be able to produce 'zebra roses' by introducing the appropriate modules into the gene kit of a rose. This would be a certain winner at the Chelsea Flower Show.

The modular units of function whose collective actions define the structure and function of a given organism are topologically interconnected in a very specific manner. There is furthermore a sense in which the hypothetical module that controls the growth of a rose petal is more fundamental than that responsible for colouring the petals red or yellow. The colouring module may thus be considered subordinate to and nested hierarchically below the master module, which specifies the architecture of the petal. A colouring module cannot after all colour empty space. The more fundamental the function of a given module, the greater the impact that its damage, modification or removal is likely to have.

Although genetic circuits can be assigned discreet functions, when they are connected together new behaviours may emerge that are not reducible to the function of a single module. Complex phenomena such as consciousness and the patterns on a butterfly's wing are likely to fall into this category. The connectivity between genetic circuits is in fact so great that a genome may be thought of as a genetic network. The physical basis for this connectivity lies in the way in which genes are able to function both as generators of protein signals and as the target biochemical addresses where proteins exert their influence.

In recent years the gene kits of several simple and relatively complex organisms have been sequenced in their entirety. The first genome to be fully sequenced was that of the *phi-X174* virus which was completed by Fred Sanger in 1977. The small genome of *phi-X174* is only 5,386 bases long. Since then, the complete sequences of several other viral genomes have been determined. These include the very much larger *cytomegalovirus* which is 229,000 bases long, the *vaccinia* virus which is 192,000 bases long, the *variola* (smallpox) virus which is 186,000 bases long and the *molluscum contagiosum* virus which is 190,000 bases long. All of these gene kits can now be placed on the shelves of the imaginary *Virus Space* hypermarket. Many of the remaining empty shelves will no doubt be rapidly filled. Eventually all modern viruses and perhaps some extinct ones will be represented in the *Virus Space* hypermarket. Viruses are not, however, self-replicating entities, as they can only copy themselves by parasitizing the metabolic machinery of host organisms. A virus gene kit does not, therefore, constitute a specification for an autonomous creature.

In October 1995 the complete sequence of the bacterium *Mycoplasma genitalium* was published in the scientific journal *Science*. This was the result of a collaboration between three teams led by Craig Venter, Hamilton Smith and Clyde Hutchinson respectively. The project had, remarkably, taken less than eight months to complete. The race to obtain the first genome sequence of a self-replicating organism, however, was won earlier that year by Craig Venter's team with the 1,830,137-base-long gene kit of the bacterium *Haemophilus influenzae*.

The significance of the *Mycoplasma genitalium* sequence is that it is the shortest genome known to encode a self-replicating creature. The gene kit of *Mycoplasma genitalium* is in fact only 580,000 bases long. The *Mycoplasma genitalium* genome thus provides a unique insight into the minimal set of genes required to specify and direct the assembly of a living creature. It also raises the question of whether it is possible to define a minimal gene kit necessary for the construction of free-living creatures.

Mycoplasma genitalium lives as a parasite in the human respiratory and genital tracts and has a predicted total of only 470 genes. As it has the smallest known genome of any free-living organism, *Mycoplasma genitalium* may be taken as a standard against which all other kits may be compared. Any embellishment of the information encoded within the compact genome of this minimalistic organism requires an explanation. The human genome, for example, is around three and a half billion (3,500,000,000) bases long and is thought to contain around one hundred thousand (100,000) genes. These additions to the basic kit are responsible for making a human a human, as opposed to a *Mycoplasma genitalium*.

Modern techniques enable individual genes to be 'knocked out' of a kit. Knockout studies in yeast, worms, flies and mice suggest that only about one in three genes is essential, indicating that many genes could in principle be jettisoned. Arcady Mushegian and Eugene Koonin have compared the genomes of *Haemophilus influenzae* and *Mycoplasma genitalium* in order to identify genes shared between the two organisms. Of the 1703 genes in *Haemophilus influenzae* and the 470 in *Mycoplasma genitalium*, 233 were found in both organisms. Of these, 105 were found to be related or of recent origin and so were excluded, leaving a smaller pool of 128 shared genes. They concluded that these corresponded to the ancestral set used to assemble the earliest genetically encoded living creatures. The genomes of creatures both past and present may consequently be viewed as embellishments of this basic set.

The genome sequencing business is currently booming. New sequencing ventures are announced almost monthly and a host of projects, which include sequencing the tomato genome and that of cereals such as wheat and corn, are now underway in both academic institutions and private companies. The complete human gene kit will eventually be available on the shelves of genetic hypermarkets by the year 2001. Although the international project to sequence the human genome was initiated by the US government in 1990, under the banner of the Human Genome Project (HGP), the initiation of a controversial private venture headed by Craig Venter, led to a joint announcement of a draft version of the human genome on Tuesday 28th June, 2000.

Carl Woese has proposed that all life on Earth has descended from one of three genealogical lineages known as Archaea, Bacteria and Eukarya. In the last few years, representative gene kits from each of these have been sequenced. The 12,520,000-base gene kit of the yeast *Saccharomyces cerevisiae* was completed in 1996. The sequence of this kit is particularly significant, as the unicellular organism *Saccharomyces cerevisiae* belongs to the same eukaryotic lineage as man. The determination in 1998 of the complete sequence of the 97,000,000-base-long genome of the multicellular animal *Caenorhabditis elegans* was a major milestone, as this nematode worm is far more complex than a yeast or a bacterium. It should now be possible to obtain some insight into how the 19,099 genes of this creature differ from those found in the genomes of unicellular organisms.

The rate at which new sequences are being generated is so high, that any list of completed genomes is out of date almost as soon as it has been compiled. At the time of writing however, other bacterial gene kits which have been sequenced include: *Escherichia coli K-12* (4,639,221 bases), *Synechocystis PCC6803* (3,573,470 bases), *Helicobacter pylori* (1,667,867 bases), *Aquifex aeolicus* (1,551,335 bases), *Mycobacteria tuberculosis* (4,411,529 bases), *Treponema pallidum* (1,138,006 bases), *Chlamydia trachomatis* (1,042,519 bases), *Rickettsia prowazekii* (1,111,523 bases) and *Bacillus subtilis* (4,214,810 bases).

Only three Archaean creatures – *Methanococcus jannaschii* (12,052,000 bases), *Methanobacterium thermoautotrophicum deltaH* (1,751,377 bases) and *Archaeoglobus fulgidus* (2,178,400 bases) – have been sequenced to date. *Methanococcus jannaschii* is a strange science-fiction-like organism which lives at temperatures of up to 98°C, tolerates pressures of up to 200 atmospheres and is killed by oxygen. It was found in the Pacific Ocean at a depth of about three kilometres by a research submarine called *Alvin*.

Other genomes which are currently being sequenced include: the fruit fly *Drosophila melanogaster*, the pufferfish *Fugu rubripes*, the higher plant *Arabidopsis thaliana*, the crop rice *Oryza sativa*, the malarial parasite *Plasmodium falciparum* and the fungus *Aspergillus nidulans*. Many of these stand as ambassadors for the class of organisms that they represent. The tiny weed *Arabidopsis thaliana*, for example, serves as a model genome for over 250,000 different species of plants. The genomes of many other organisms of economic and pathological significance are also being sequenced. These include: *Neisseria meningitidis* (meningitis), *Mycobacterium leprae* (leprosy), *Neisseria gonorrhoeae* (gonorrhoea), *Streptococcus pneumoniae* (pneumonia) and *Vibrio cholerae* (cholera).

Once the generalized human genome has been sequenced, the emphasis will shift to sequencing the genomes of individuals with specific diseases, and to those of geographically and culturally isolated populations such as the inhabitants of Tristan da Cunha (who suffer from a high incidence of asthma) and the Onge tribes of India. The sequences of these genomes will help elucidate the genetic components of globally significant diseases. Indeed the genetic 'gold rush' is already underway, with bio-prospectors jostling one another to obtain access to valuable DNA collections and DNA 'mining' rights. Organizations representing indigenous peoples have condemned these initiatives, calling them 'biopiracy' and 'genetic imperialism'. Plans are nevertheless underway for a Human Genome Diversity Project (HGDP) which aims to survey mankind's genetic diversity.

The genes of all living species can be imagined as constituting a genetic landscape or 'genescape'. Much like the nineteenth-century colonialists, private companies and academic institutions are scrambling to acquire whatever genes they can. In some cases gene sequences are considered as inventions and intellectual property which can be protected with patents that confer them with potential commercial value. Eventually the construction of the *DNA Zoo* (modern section) hypermarket which contains the gene kits of all known living creatures, will be completed. With the help of technology that enables ancient DNA to be rescued, it might also be possible to construct a fragmentary *DNA Zoo* (ancient section) hypermarket which contains the partial and in some cases complete gene sequences of extinct organisms.

As each gene kit is sequenced, it can be compared with the sequences of other kits. The patterns of gene activity within a kit can also be studied. The new sciences of comparative genomics and bioinformatics use powerful computers to rationalize DNA sequence databases and are invaluable in helping, amongst other things, to identify genetic elements which have been preserved across evolutionary time. The evolution of individual genes can also be studied by comparing DNA sequences from different species and generating hypothetical lineages. Formal definitions of terms such as 'species' and 'life' may eventually have to be re-evaluated on the basis of molecular genetic considerations. The challenge in the future will be to find new ways of utilizing the information that DNA sequencing projects are generating.

Comparative genomics will also help identify redundancy within genes and genomes, that is to say regions which could be removed to generate more compact genomes. The pufferfish *Fugu*, for example, has approximately the same number of genes as humans, but has a genome which is around seven and a half times smaller. Although *Fugu* has less non-coding DNA than humans, the reduction in genome size is also achieved by reducing the size of its genes. It contains, for example, a compressed version of the gene thought to be responsible for Huntington's disease. Whereas in humans the gene is 180,000 bases long, in *Fugu* it is a mere 22,000 bases long. The structural organization of the gene however, is identical, demonstrating a high degree of conservation. Comparative genomics may thus provide insights into how genes could be miniaturized by an artificial process of 'genetic pruning'. Pufferfish genes have to some extent already had many of their irrelevancies excised by history and as such are 'discount' genes. Such discounted genes will help locate key regions of coding sequences, while discounted genomes will help identify essential non-coding sequences. As the genome of *Fugu* is much smaller than that of humans but has the same number of genes, it is potentially a good system for studying the function and regulation of human genes.

In humans, about 97% of the genome does not appear to encode genes. Various explanations have been put forward to explain the presence of this abundant non-coding DNA, the most popular being that most of it consists of irrelevant parasitic sequences. Some regions might however constitute a DNA workshop, housing superannuated sequences that could one day arise, phoenix-like, in their original or a modified form. Others may constitute a supply of useful genetic widgets that could be introduced into other genes. Using statistical techniques borrowed from linguists, the physicist Eugene Stanley has shown that junk DNA bears some resemblance to natural human languages. This suggests that junk may carry some sort of message fundamentally different to the information encoded in genes. There may consequently be 'something interesting lurking in the junk'. Perhaps a deeper code. A foreign and esoteric text, waiting to be deciphered.

At least some non-coding DNA sequences are involved in the control of gene expression and packaging. Gene expression is the process by which the information of genes is copied into mRNA and translated into proteins. Despite their considerable importance, these control regions are often difficult to localize. By comparing the non-coding regions of genomes from different species, comparative genomics may help define regulatory sequences which have been conserved between species. Comparative genomics and bioinformatics will also help determine the function of newly discovered genes. For example, if a sequence motif in a gene of known function is found in a gene of unknown function, it is likely that the genes have a broadly similar function. Comparative genomics also allows the genetic resource allocation of different organisms to be determined. That is the percentage of genes allotted to the various essential functions common to most organisms. For example whereas *Haemophilus influenzae* assigns sixty-eight (3.9%) of its genes to amino acid biosynthesis, in *Mycoplasma genitalium* the investment is quite different, with only a single gene (0.2%) being set aside for this purpose.

In addition to all of this, gene kits can be cracked open and systematically deconstructed. Genes may be added, modified, or removed from a kit and the overall effects of these manipulations observed. Studies involving random disruption of the *Saccharomyces cerevisiae* gene kit, for example, have demonstrated a remarkable degree of functional redundancy. In fact, under favourable conditions, up to 70% of the *Saccharomyces cerevisiae* genome may be jettisoned without interfering with its survival. It would appear, then, that organisms insure themselves against malfunction, or damage to their essential gene kit components by incorporating back-up systems that secure essential functions. *Mycoplasma genitalium*, as we have said, is an example of near-minimal life which, in the words of Barry Bloom, 'has learned to simplify its life by associating with its mammalian hosts'. It is indeed 'fascinating to learn what it could afford to shed and still survive'. We might similarly wonder how many genes could be shed from the pufferfish genome without it losing its essential 'pufferfish-ness', or from the gene kit of a fruit fly, without it losing its quintessential 'fly-ness'.

The answer to these questions depends to some extent on identifying the unique characteristics which define a species. Is it for example its external morphology, internal structure, behaviour, the dynamics of its cells and biochemical pathways, or a combination of all of these things? It is possible, for example, to imagine a pufferfish which looks like a pufferfish, but behaves like a fruit fly except for the fact that it swims, rather than flies. We might conversely try to imagine a fruit fly that thinks that it's a fish. Despite *looking* the part for the job, neither of these behaviourally inept creatures are likely to be very successful.

Although these more philosophical questions are harder to resolve, comparative genomics constitutes a powerful method for identifying the genes that can modify the properties of a minimal kit. If, for example, you were to compare the sequence of a pufferfish with that of a fly and then subtract the regions that are similar or identical, the remaining bits should embody the unique qualities of each species. Conversely, the regions which are similar constitute the invariant genetic background. One of the problems with this type of approach, however, is that it assumes that the removal of one or more genes from a kit will produce only local effects, without introducing disturbances into other genetic circuits.

The first studies of this sort are already underway. A comparison of the *Haemophilus influenzae* genome with *Escherichia coli*, for example, has identified 116 genes which are not found in *Escherichia coli*. These presumably encode some of the *Haemophilus*-ness that makes a *Haemophilus* a *Haemophilus*, as opposed to an *Escherichia coli*. Studies of the genomes of other organisms have identified 'orphan' genes, which bear no obvious resemblance to other known genes and for which no function has been attributed. These orphan genes are likely to be important in defining the unique features of different species.

We saw earlier that gene kits are sold in genetic hypermarkets and that these may exhibit various degrees of completeness. One might, for example, stock the gene kits of vertebrates, whereas others might specialize in the kits of insects, Amazonian parrots, trees, or fish. Let us now however walk past the entrance to the enormous *Sinogen* hypermarket, which stocks the gene kits of Chinese bacteria, and walk instead into the genetic hypermarket which houses the gene kits of animals found in Africa. Other branches of the same hypermarket might stock the gene kits of African flowers and birds, or of species, like the quagga, which once roamed the plains of Africa but are now extinct.

As I walk into the shop, I place a coin into the slot to release a shopping trolley and push it down the first aisle. My attention is immediately caught by a stack of gene kits piled high up to the ceiling.

Closer inspection indicates that the location of this stack is by no means random. The owner of this hypermarket clearly knows a potential best-seller when he sees one. Much taken by the portrayal of the king of the beasts on the front of each box, I put two kits in my trolley. They are after all on special offer. Situated next to the lion gene kits is a stack of zebra gene kits. And next to these is another stack of boxes, which at first glance also appear to be zebras. However I notice a placard hanging from the ceiling which reads 'zebras +1, *SPECIAL OFFER* – a free stripe in every pack!' And sure enough, when I count the number of stripes running across each type of zebra, the +1 does seem to have one extra. Walking past the hippopotamus, hyena and crocodile gene kits, I eventually arrive at the rhinoceros section. Another placard informs me that if I buy an eagle gene kit from aisle twenty-four, I will be entitled to a free rhinoceros +3. I look at the picture on the +3 box and find that it depicts a rhinoceros with four horns.

Although momentarily captivated by the flamingos, I pay scant attention to the other kits located in aisle thirty. Instead, I enter aisle thirty-one and in front of me now is an array of leopard kits. How odd that there are so many. And I realize that I am looking at what must surely be the prize display of this hypermarket. For there is not just one kind of leopard, but to name but a few, a leopard: +1, +7, –3, –6 and a leopard 0. It turns out that the +1 and +7 refer to extra spots and the –3 and –6 to a reduced number of spots. The leopard 0 is an unnatural-looking specimen which appears to have achieved the impossible and lost all its spots. Whoever said that a leopard couldn't change its spots?

By now I am curious about these variations of animals I thought I knew so well, and am relieved when I bump into the owner of the shop. He is a big man, with a large red face and short fat arms. He is wearing a khaki mohair suit and sports a pair of binoculars and a safari hat. Anticipating my question, he begins to speak. 'Before you say a word sir, the answer is *all* natural. I can guarantee you sir, *all* natural mutants. Admittedly it was not easy to find the 0 and –12. But the rest, just part of a normal jungle sir, a normal jungle. If you don't look sir, then quite frankly you don't find. At *C. J. Afribeasts* [for that was the name of the shop] we find!'

As he hurries away I think I hear him muttering something about tigers with extra stripes and spiders with sixteen legs. But fortunately his voice is obscured by the operator at the check-out till. She informs me that if I had only spent a little more, I would have been entitled to a free pelican 243.4. I resist the temptation of asking what the numbers refer to, but at this moment it becomes clear that any one gene kit can only encode a general organism, a statistical average of the natural variation in the gene pool of the population from which it was obtained. The genetic hypermarkets that stocked only one example of every species were nothing more than some type of sterile, postmodern Noah's Ark, and even Noah had had the sense to include two of every kind of creature.

No single kit can represent a species in anything other than a general sense, because the individuals that make up a species are genetically heterogeneous. As a result of this, a single gene kit can at best be only a statistical ambassador for its species. The variation is not, however, distributed evenly across the entire genome, and certain regions contain more variability than others. The more genetically homogeneous a species and the smaller the number of individuals from which it is comprised, the greater the degree of correspondence between a generalized kit and the species it represents. If the genomes of every member of a species were aligned side by side, the most common differences at each position could be determined. A genome representing the statistical average of all the genomes in the population could then be constructed. This 'wild type' genome represents the standard or average genome of the species. Any individual with a sequence that differs from that of the wild type is known as a mutant. Every species consists of a large spectrum of mutant sequences which cluster around the wild type.

Each slot on a chromosome may be filled by one of several alternative forms of the same gene, known as alleles. An allele is a version of a gene in which one or more mutations have been introduced into the wild type sequence. Any given organism may, however, fill a particular slot with only one allele from the gene pool. Each allele thus competes with its sisters for occupancy of the appropriate chromosomal slot. In sexually reproducing species, which have two copies of each type of chromosome, it is possible for an individual to have a different allele in the slot of each chromosome. If this is the case, then the individual is said to be heterozygous for the gene in question and every cell in the organism will possess two versions of the same gene. If on the other hand the same allele is found in each corresponding slot, then it is said to be homozygous.

A gene pool may be envisaged as a swimming pool filled with all the alternative forms of genes available in a population of organisms belonging to the same species. One can imagine gathering all of the members of a species together, plucking the genes from their chromosomes one by one and then in a bizarre rendition of a David Hockney painting, throwing them all into a sparkling turquoise Californian swimming pool. Alternatively, as each gene is plucked out, it could be sorted according to the slot from which it was taken. Each slot might then be allocated its own private swimming pool, into which genes from that slot and no other would be thrown. We might employ tens of thousands of Venice Beach bodybuilders to fish genes randomly from each pool using a specially designed 'gene pole' with a 'gene net' attached to its end. Once they had fished several examples of every gene from the private pools, their sequences would be faxed to an offshore creature-assembly factory. The machine in the factory would assemble new organisms according to the genetic information that had been fished out of the pools, and a corresponding database of essential non-coding DNA regulatory sequences. If the gene fishers were unbiased and their nets captured the alternative examples of genes in a truly random fashion, the organisms constructed by the factory would vary from one another in a host of different ways. If however the gene nets were more sticky for certain versions of genes than others, the creatures would be biased and more uniform in their design.

The advent of genome-sequencing projects and improved sequencing technologies enables us to consider what we might call a 'genetic superpool'. This is a hypothetical gene pool which contains the genes of all contemporary and perhaps some extinct organisms. One can imagine taking the genes of every living species from humans and whales, to ladybirds, greenflies, elephants, pelicans, penguins, sharks, monkeys, goldfish, snakes, plankton, viruses and yeasts, and putting them together in a huge genetic cauldron. Although nature does not usually allow the exchange of genetic material between species, artificial strategies for combining and manipulating genes need not be similarly constrained. The genetic superpool will provide a curious palette for the designers of the future, who will paint their genetic 'pictures' using genes. Instead of being viewed in conventional galleries, these new protean creatures with genetic components cobbled together from multiple sources will be 'installed' into specific environments. Really creative genetic artists will eventually realize that there is no reason to restrict themselves to the genetic material that has been offered up by history. They may instead prefer to generate new ahistorical genes and DNA regulatory elements. These artificial genes and regulatory elements might be test-driven using computer simulations of living organisms before eventually being tested in real organisms.

Having successfully completed my shopping excursion in *C. J. Afribeasts*, I finally arrive home. I put my leopard +2 gene kit down in front of me on the kitchen table and prepare to assemble it. However, I now begin to worry about what I will do with my +2 once I have constructed it. How will it fare in the English climate and what will I feed it on? Whilst pondering these and other matters, I notice some small print on the side of the box. I fetch a magnifying glass and begin to decipher the tiny words.

> *C. J. Afribeasts can only guarantee with a probability of 20% that, when constructed, this gene kit will generate a leopard –2 which corresponds to the one depicted on the front of the box. Furthermore, in the eventuality that your particular +2* ***does*** *physically resemble the creature on the front of the box, we cannot guarantee that it will* ***behave*** *like a normal leopard. We do not expect there to be any difference in behaviour between your mutant +2 and the conventional wild type beast. Should however your +2 behave abnormally or unusually, we request that you* ***DO NOT*** *return it to the store as* ***REFUNDS CANNOT BE GIVEN UNDER ANY CIRCUMSTANCES****. C. J. Afribeasts furthermore regrets that it* ***cannot take responsibility*** *for any non-leopard-like behaviour that the assembled +2 may exhibit. We* ***cannot be held responsible*** *for any damage to your +2 that may have occurred during storage or transportation, or which might occur due to lack of resistance to infections, other types of non-infectious diseases, or any other unforeseen*

circumstances related or unrelated to your leopard +2's lack of contact with other leopards. We ***cannot*** *furthermore* **be held responsible** *for any damage that your +2 might inflict on either* ***private*** *or* ***public property****, to you the* ***owner****, or indeed to* ***any other individuals*** *with whom it might have contact. Although we are confident that your +2 has a good chance of mating and producing offspring by conventional methods, we* ***cannot guarantee*** *that it will be able to do so.* ***Nor can we guarantee*** that it will be able to look after its offspring, in the event that it produces any. However, to minimise the likelihood of the above, C. J. Afribeasts are happy to say that ***all junk and non-coding sequences have been included*** *in the kit, along with full instructions of how they should be connected to the relevant genes. C. J. Afribeasts would like to take this opportunity to thank you for purchasing this product and looks forward to your continued custom in the future. Please note,* ***available soon in this series****, the leopard +8, tail-less +1 and the mark 7 'super fast' (guaranteed* ***at least 5km/hr faster than the wild type beast****).*

After some weeks I begin to get bored of the *C. J. Afribeasts* hypermarket. My interests broaden and I feel myself wanting to buy kits for all kinds of strange and curious creatures. The idea of a platypus and a five-legged pentapus come to mind one day. And then a fruit bat hanging upside down from a tree and a stick insect, like the ones they used to have at school. I even find myself becoming interested in yaks. Did you know, by the way, that for several months of the year Tibetan tribesmen survive on nothing but yak-milk products? My pre-occupation with polar bears and indeed every other type of bear lasts for a few weeks. And then the light filtering through the shutters in the sitting room reminds me of the pawn mutant paramecium, and the squids so beloved of my physiology professor. I decide one afternoon that I would like to build some of these creatures and realize immediately that there is only one place to go. And that is *Genes Are Great*, a hypermarket somewhere in the north near Newcastle, or was it Leeds? So I get into the car and sit on the motorway for several hours, humming a tune and listening to a conversation on Radio 4 about hedgehogs. 'Yes, I'd like to build one of them too,' I think to myself as I reach the junction and follow the signs to *Genes Are Great*. Finally I arrive, and find myself in a hypermarket far bigger than *C. J. Afribeasts*.

'This place really *is* huge,' I think to myself as the electric people-mover whirs me past the insect section, which seems to go on forever. The pictures of strange creepy-crawly insects stare at me like perverse gargoyles from the sides of their boxes and soon begin to irritate. I decide not to buy a stick insect kit after all. On and on past the dragonflies, ants and the ladybirds with all of their different spots. And then more bugs and beetles, and the giant wingless cricket-like insect, *Deinacrida heteracantha* from New Zealand, and their tree-dwelling cousin *Hemideina crassidens*. 'Will this ever end?' I ask myself, almost in despair. And soon it's lunchtime and I'm feeling kind of hungry. But all I can see is more and more insects, extending in every direction. The lady next to me starts to pull caterpillar kits off the shelves at such an alarming rate that I am forced to turn to her and shout, 'CAN YOU JUST SLOW DOWN A BIT!' I am obliged to resign myself to the low hum of the people-mover, as it drags me reluctantly past the spider kits and the scorpions. I begin to wish that I had never entered *Genes Are Great*, as it contains the gene kits of all living species. If only they would give you a map at the entrance! I make a mental note to send a letter of complaint to the manager, but my train of thought is distracted by something else. The sea of butterflies all around me dazzles my eyes and fragmented insect patterns perform a strange multicoloured dance in my head.

By the evening I am very tired. And yet I find that I am still whirring forward at a slow but steady rate. The lady next to me is fast asleep on a makeshift bed that she has constructed from her caterpillar boxes. I cannot help feeling that it is fortunate she dissipated her shopping energy before we reached the centipede and millipede section. Soon we are in the bear section. And finally my patience is rewarded, and I buy the last polar bear gene kit in the whole hypermarket. 'Well at least I got something!' I mutter to myself, but this does not seem to dull the monotony of the rows and rows of geometrical boxes.

Before long we are in the seahorse section. I find myself counting thirty-five species and start to repeat their names in order to pass the time. *Hippocampus ingens*, *Hippocampus erectus*, *Hippocampus zosterae*, *Hippocampus reidi*, *Hippocampus fuscus*, *Hippocampus hippocampus*, *Hippocampus comes*, *Hippocampus histrix*, *Hippocampus breviceps*, *Hippocampus bargibanti* and so on and so on. I am reminded of the nursery game of counting sheep. It does not seem as if this endless array of seahorses will send me to sleep. But I am wrong. Fairly soon my eyes begin to close and the names of seahorses mingle with those of a thousand other creatures.

When I wake up the next morning, I find that I am heading through the cactus section, which, according to the person in front of me, leads to the exit. Fairly soon the checkouts are in sight. Hundreds of thousands of them in fact. I pay for my gene kit and leave, thankful that I have avoided the *Tree World* and *Algae City* sections, but tired and very hungry. When I get to the car park I find that I have been given a ticket. And as I head in the direction of home, I decide that I had better find a new hobby.

I manage to do this, and for weeks do not so much as think about gene kits. But then a programme on television about dinosaur eggs arouses my interest in life long gone by, and soon my head is awash with dodos, sabre-toothed tigers, pterosaurs and mammoths. I take out *Jurassic Park* from the video shop and watch it several times. Could dinosaurs *really* run that fast? I read voraciously about meteorites and mass extinctions and strange, extinct Rocky Mountain creatures with names like *Wiwaxia* and *Dinomischus* and bodies with rows of elongated spines. I want to build a trilobite. I open a magazine and following an article on Cassius, a *Crocodylus porosus* that lives in a pond full of duckweed and is the largest crocodile in captivity, is ANOTHER article on dinosaurs. And this time it is about *Carcharodontosaurus*, which lived in the Sahara, had a skull at least several inches larger than that of *Tyrannosaurus rex* and could devour man-sized prey in a single bite.

And soon it becomes clear where I will have to go. I look up the address of *Fossilize and All the Other Bits that Didn't World*, but find to my dismay that it turns out to be just a corner of a far larger hypermarket called *Gene Space*. The hypermarket of all possible gene kits, the shop that stocks every type of DNA device that does, has or might ever in principle be able to walk, crawl, swim, float, fly, flutter, drift, run, hover, buzz or whir across the space of the universe. I look up the address of *Gene Space* in the telephone directory and find that it takes up several countries. In fact as I get into my car, construction work is still underway. Indeed the land masses of the Earth are slowly being covered over by the hypermarket, which now threatens the very existence of life *ex kit*. The pundits and mandarins are duly concerned and there are demonstrations in the streets of Moscow and talk of war in China. My car glides down the motorway, consuming miles in minutes. I drive and drive for what seems like ages, and suddenly I find myself in complete darkness. This seems a little strange as it is still daytime and only a few moments earlier I was surrounded by brilliant sunshine. I look in front of me, and far away in the distance I can just make out the outline of a huge grey building that appears to stretch for miles and miles in every direction and high up into the sky. Before long I realize that this is *Gene Space*. The shop of all shops, the hypermarket of all DNA kit hypermarkets.

2

The Structure of Gene Space

In my first dream, small wings sprout slowly from my sides. And like a character from Kafka, my metamorphosis proceeds. Until finally I am flying high. High in *Gene Space*. But what have I become? What indeed am I in the process of becoming? My feet change to hooves and my skin turns a sort of mottled, emerald green. The light is dazzling. And then suddenly I am shooting upwards and after a while, sideways. A thousand images fly past me. And when eventually I come to a halt, I am surrounded by a sea of frantic serpents which dance in every direction for as far as I can see. With my new insect eyes, the world seems imprisoned within an impossible, rigid geometry. Instead of a single image I now see eight hundred identical picture-frame windows. Each time I blink this number doubles, triples or quadruples. I try to orient myself, but everything seems back-to-front. My head is pounding and I am deafened by the mechanical buzzing of my wings.

I am woken, and as the images fade, I find myself attached to an endless piece of elastic. And now I am bouncing up and down and up and down. And the boxes, so numerous, tightly packed and brightly coloured, move around me like a merry-go-round. For a moment I am suspended in front of a wall of sabre-toothed tigers. Their leering grins are menacing and unfamiliar in the dim expanses of this limitless vault which extends above, below and all around me. In the next instant I am speeding through a sea of pelican-like birds and then through a landscape filled with creatures so strange, that I have never seen anything like them. And now I am in *Beetle Space*, which is doubtless only the tiniest corner of *Gene Space*. I find myself in an infinite sea of sparkling beetle shells. Curved like the bonnet of a

Volkswagen car and accompanied by ten thousand trillion beetle legs, all of which beat in unison. I am relieved to find that I have been spared the indignity of being transformed into one of these hapless creatures.

Only a moment later, yet another landscape. And now I am in a place called the *Land of the Mythical Beasts*. Soaring like a bird of prey and circling high in the air, I am looking down at an endless array of fauns. Flying still higher and higher, the air begins to thin. And then, without warning, I find myself tumbling through an opening which leads to the *Place of the Centaurs*. My vision becomes increasingly powerful, until I can see every minute detail of the creatures beneath me. Still further on I find myself in a valley inhabited by chimeric beasts. Lions with eagle-like claws and beaks, wolves with elephant tusks, and twisted fish with legs and horns that curve in unusual ways. At one point I see a signpost that reads: *This way to the Sea of Impossible Creatures*. But below me now is a sea of donkey-like beasts with humps like camels and long matted hair. Countless numbers of them stand in lines, like herds of wildebeest that have been snap-frozen in their tracks.

Whereas moments before I was overcome by a terrible terror, the feeling of unease that I am now experiencing seems incongruous. For the wall of shelves in front of me is packed with colourless, featureless boxes. I search each one for a name or picture, but am unable to find any. And then for what seems like hours, just more and more boxes. Dull and colourless with no writing or pictures. I wonder whether they will ever end, or whether this journey will continue for ever. If only I could find an exit. Surely everything, even *Gene Space*, has a beginning and an end?

It is now so dark that I can hardly see. And within this fearful abyss, I strongly feel the absence of life. Silence pervades this still, still place and soon I am groping with my hands on the wall, unable to find my way. I seem to travel for days like this. My rudimentary wings which have by now reappeared, buckle and crumple and are quite useless. And now I am falling through *Gene Space*. And as I fall, I become preoccupied with a single thought Organization. Is there any organization? My descent continues and at one point I am able to make out a sign which reads *Primeval Slime*. It is only now that I understand where I have arrived. Somewhere near the beginning. The beginning of life and the beginning of genes. And I reason that if I am near the beginning, then I might have a chance of escaping. But I seem to be getting smaller and smaller. So small in fact, that I am certain I will soon disappear.

In my second dream, which I later realize is a premonition, I imagine that I am in a glass capsule traversing a diagonal of *Gene Space*. However this is not *Gene Space* at all, but somewhere quite different. Inside the capsule are four men and a woman, all dressed for the sun. The lady wears a summer hat of the type seen at Ascot. The men, except for one who is dressed in a dinner suit and black tie, wear boaters and blazers. One sports a flannel jacket. I try to recognize the green of the blazer stripe, but its name eludes me. One man holds a strawberry in his hand. Another laughs uncontrollably. The woman is thinking about Wimbledon. And then she is playing badminton on the lawn of her country house and sipping a Pimms and lemonade.

'Should be a good day's shooting,' says the man in the flannel jacket, as the horses thunder across the stream and the fox makes a hasty retreat. The lady meanwhile has returned from the ladies' room. She is just in time for the second half. There is an announcement over the tannoy: 'Ladies and gentlemen, would you kindly take your seats, as the second half of the performance is about to begin.' The curtain rises and the soprano in the audience smiles as the second half of *The Pearl Fishers* begins. She thinks she may have had a little too much to drink, but she's in that sort of mood. And anyway she has after all been watching the second half since the beginning of time. She is happy that she has reapplied her lipstick.

'I always *do* look better with a bit of lippy,' she says to herself as the curtain rises. The man on her left coughs loudly and his wife rummages in her handbag. However despite the fact that she has been searching for fifteen billion years, she is still unable to find whatever it is that she is looking for. The man with the light blue blazer is lying in the sun. He holds a glass of sparkling white wine in his left hand. His father used to work for the East India Company. He started off in Pondicherry, but ended up in Chicago. The score is one hundred and eighteen for six. The sky is a brilliant blue and the elm trees cast long shadows on the green. He applauds as the new batsman walks across to the crease. That was a fine catch.

'Raaa!' he shouts to the spectators that have gathered in the bar. Meanwhile the pack is catching up with the fox and as the fastest hounds race ahead, the man in the flannel jacket imagines that he sees a prism. And as he watches it, the light splits into the coloured images of four men and a woman. One of the men imagines that he is sitting in a lecture hall. The professor raps the lectern with his stick and then points to the overhead projector. The machine whirs and hums and dust particles dance in the light, as the words 'the structure of moments' are beamed onto the screen. And then I am back in the auditorium and am just in time to see a lady returning from the ladies' room. Her crimson mouth is striking. I listen to the orchestra tune their instruments and then there is an announcement over the tannoy: 'Ladies and gentlemen, would you kindly take your seats, as the second half of the performance is about to begin.'

In my third and shortest dream, I pick a gene kit off one of the shelves and shake it hard. It feels light and curiously empty. As I open it, a single gene falls out. My wings by now have vanished and resolving immediately to adopt a more conventional mode of transport, I step into one of the glass elevators. My legs lose contact with the ground and soon I am floating. As I shoot upwards in *Gene Space*, I cannot say where I am. I can give no exact position. But as before, the innumerable shelves in this sub-archive of logical impossibility are filled with hundreds and thousands of pictureless boxes. And given that only recently I was in an area signposted '*Primeval Slime*', I infer that I am still there. It would appear that the librarians of *Gene Space* are an orderly bunch and have arranged their collections so that each gene kit bears a particular relationship to its neighbours.

My first and third dreams offer a tantalizing glimpse of *Gene Space*. The second dream, however, provides an inkling of a space far stranger even than *Gene Space*, as this extraordinary hypermarket houses not only the complete set of all possible creatures, but also the complete set of all possible events. In the *Space of All Possible Events*, collections of mathematical specifications for events are stored like gene kits. Each box contains a different version of events and each set of possible events has a timeless existence.

I would now like to examine the anatomy of *Gene Space*. This will help us to understand both our movements through it and the processes which will enable us to navigate its vast uncharted expanses. Although we have referred to the space of all possible gene kits as *Gene Space*, it is perhaps more accurate to call it *DNA Sequence Space*, or just *DNA Space*. This is because most of the sequences in the space do not encode genes. We will nevertheless for the moment stick to the *Gene Space* terminology that we have adopted.

If asked to list the differences between a koala bear and an ostrich, one might note that koala bears are small, furry creatures which live up gum trees and are found in Australia, whereas ostriches are large, clumsy creatures that have black and white feathers and are found in Africa. One might, however, choose to provide a more precise answer that identifies the points at which their DNA sequences differ. These differences could be determined by two simple procedures.

The first is to count the number of bases in the complete set of ostrich chromosomes and then compare this with the number in the koala chromosomes. We would then be able to say that a koala genome has this many or that many bases more or less than an ostrich. The second is to take the ostrich chromosomes and sequence them. Starting with the first, you could move each base one position to the left, as one would the beads of an abacus. Proceeding in this way to the end of the necklace, you could jot down the identity of each base that you encounter. At the end you would be left with a piece of paper which reads something like: **TTGACGC** and so on. If you were of an inventive nature, you might construct a machine to perform this task for you. On feeding in the ostrich sequence, the machine would automatically pull off one base and after recording its value, would move on to the next. The koala sequence could be fed into the machine in the same way, and once the sequences of the two creatures had been determined, it would be a trivial task to align and compare them.

Let us now imagine that a koala is specified by a tiny six-base-long genome which reads: **TTCACG**, whereas ostriches on the other hand are specified by an eight-base-long genome which reads: **TTGACGAT**. If these sequences are aligned, the differences between the ostrich and koala genomes can be clearly defined. In this example, ostriches have two extra bases (an **A** and a **T** at positions 7 and 8 respectively), and the **G** at position 3 in the ostrich genome is replaced by a **C** in the koala genome.

The value of placing our explanation at the level of DNA sequence differences lies not just in its precision and economy, but also in that it enables us to go one step further. For once we have determined the differences between an ostrich and a koala, it is not difficult to see that we have also generated a formal procedure for changing one into the other. To change a koala into an ostrich for example, all we need do is replace the **C** at position 3 in the koala sequence with a **G**, and then add an **A** at position 7 and a **T** at position 8. It is possible to imagine a simple universal gene-transformation machine which can replace any particular base in a sequence with any other. It can also add or remove any number of bases to or from a sequence. Given a machine of this sort, we could write a simple program to transform a koala into an ostrich. The machine is able to accept any DNA sequence from *Gene Space* as its input. It will also execute as few or as many manipulations as the program demands. Given sufficient time and an unlimited supply of bases, the machine could transform any DNA sequence in *Gene Space* into any other. We will call this the principal of universal interconvertibility.

The term 'formal procedure' refers to the instructions we would need to feed into the machine in order to realize the transformation. These might be of the form: (1) **Go To** position 4 on the sequence. (2) **Erase** the **T** and **Replace** it with an **A**. (3) **Proceed** to position 12. (4) **Erase** the **G** and replace it with a **C**. (5) **Go To** position 87. (6) **Insert** 4 **G**'s. (7) **Stop**. A set of instructions of this sort is known as an algorithm.

In our original example, the procedure required to convert a koala DNA sequence into one that specifies an ostrich is a koala to ostrich conversion algorithm (KOCA). We can imagine some of the other types of algorithm that could be realized by our machine, for example: a rhinoceros to armadillo conversion algorithm (RACA), a giraffe to turtle conversion algorithm (GTCA), a dandelion to penguin conversion algorithm (DPCA) or a stickleback to shark conversion algorithm (SSCA). We need not, however, limit ourselves to creatures with which we are familiar. We may wish to convert a firefly into a flamingoceros (FFCA), which we can imagine is a hypothetical cross between a flamingo and a rhinoceros. It is not inevitable that a DNA specification for such a creature actually exists in *Gene Space*. However when the relationship between sequences and morphology is better understood, it will be possible to search *Gene Space* for flamingoceros-like specifications and then 'run' candidate genomes that we have fished out in order to see what they produce.

There is a sense then, in which every kit in *Gene Space* is connected to every other kit by a huge web of DNA conversion algorithms. Movement through *Gene Space* may thus be accomplished by the implementation of an appropriate algorithm. Our universal gene-transformation machine is thus an engine which will help us to explore *Gene Space*. A suitably powerful machine would have an enormous repertoire of conversion algorithms in its memory banks. We might for example choose to feed in the DNA sequence of a porcupine, whilst typing the word 'pig'. After some whining and whirring of the machine's mechanical cogs which implement the required additions, deletions and substitutions, the old gene kit box would be destroyed and a brand new box emblazoned with a pink pig would take its place.

Gene Space contains DNA sequences of every possible length and composition. These may be imagined as being arranged so that similar sequences are clustered together. DNA sequences might be as short as a single base, or infinitely long. In practice though, several factors are likely to limit their length. At the smaller end of the really huge sequence range, the problem of excessive length can be avoided by cutting chromosomes into smaller fragments. This appears to be nature's strategy, as organisms with large genomes have multiple chromosomes. At the larger end of the really huge DNA sequence range, however, several new constraints come into play. The volume of the nucleus for example sets a limit to the amount of DNA that it can accommodate. The resources that a cell is able to allocate to DNA repair and surveillance are also limited. Beyond a certain genome size, the efficiency of DNA repair might consequently be compromised. It may however be possible to engineer the size of a nucleus, so that it is able to accommodate larger volumes of DNA. The efficiency of DNA repair might also be artificially increased.

The number of different combinations that can be generated for a DNA sequence of given length is easily calculated. For sequences of length one base, there are 4^1 (4) combinations (**T** or **C** or **A** or **G**) and for sequences of length two bases, 4^2 (16) combinations. There are similarly 4^3 (64) combinations for sequences three bases long and 4^4 (256) for sequences that are four bases long. If cells only had a storage capacity of four bases, the maximum DNA sequence length would be four bases. The total number of realizable sequences would thus be: 4 (all sequences of length one base) + 16 (all sequences of length two bases) + 64 (all sequences of length three bases) + 256 (all sequences of length four bases). This gives a total of 340 different sequences. Real cells, however, are able to accommodate much longer sequences and as the constraint on length is relaxed, the number of possible sequences becomes astronomical. *Mycoplasma genitalium*, for example, clocks in with a genome which is 580,000 bases long. If we were to generate the complete library of all possible sequences of this length, it would contain $4^{580,000}$ items. It is unclear how many of these genomes would compute a living organism, or indeed would resemble *Mycoplasma genitalium*. It is nevertheless evident that the set of sequences which specify all living and extinct organisms represents only a tiny fraction of the possibilities that *Gene Space* has to offer. History has realized only a perfunctory example of what might have been, or might one day still be.

To give an indication of the vastness of the numbers involved, I will draw your attention to a calculation made by Manfred Eigen, who considered a DNA sequence consisting of just 1000 bases. The number of possible combinations of the four different bases in this sequence is 4^{1000} (10^{602}). Eigen makes the sobering point that the material content of the entire universe corresponds, weight for weight, to fewer than 10^{75} sequences of length 1000 bases. There is thus insufficient matter in the universe to construct even a tiny portion of this library.

Humans are specified by a genome which is about 3.5 billion bases long. In the context of Eigen's calculation, the extraordinary size ($4^{3,500,000,000}$) of the complete collection of all possible DNA sequences of human genome length should be self-evident. I use the phrase 'human genome length' rather than 'human genomes', as many genomes of human genome length will not specify humans. If we were to imagine, conservatively, that human genomes differ in size by only plus or minus a single base, then the total number of possible genomes of human genome length increases to the astronomically large number $4^{3,500,000,000} + 4^{3,499,999,999} + 4^{3,500,000,001}$.

The question of how many bases can be added or removed from the human gene kit without having to remove the label *Homo sapiens* from the box is filled with semantic complexities. Many of these relate to our minimal expectations of what we feel humans should be capable of thinking, perceiving and feeling, and how they should behave. Others relate to criteria which colour our view of what we are prepared to regard as acceptable manifestations of internal and external human form. Still others relate to which particular bases are changed. It might for example be possible to remove several million bases from apparently non-essential 'junk' regions of the kit, without making any difference to the essential information that it encodes. Indeed if 97% of the human genome consists of junk, it might be possible to remove 3.395 billion bases without adverse consequences. However, the removal of a single base from an essential gene or regulatory region may have devastating consequences. At the end of the day, the question is unlikely to have a clear cut answer. But once the analysis of many different types of genome has been completed, it should be easier to make distinctions between species on a more logical basis.

Regardless of the limit we arbitrarily set for the largest and smallest possible human genome, the number of bases that can be added or subtracted without introducing appreciable changes into the nature of the creature that the DNA sequence encodes is likely to run into at least the tens of millions. If the combinatorial possibilities of this additional collection of genomes is added to the calculation, then the size of the repertoire of all possible human genomes reaches even giddier heights. If we assume that the maximum sustainable population on Earth at any time is in the tens of billions, then if the average human life-span remains constant and the realization of all DNA sequences is equally probable, many billions of years would have to elapse before even the tiniest fraction of the collection of all possible human genomes could be realized. But the universe has existed for only 15 billion years, and life on Earth for only 3.6 to 3.85 billion. The processes by which the sea of potential DNA sequence information is explored and realized are thus limited both by time and space.

It should by now be clear that *Gene Space* is big. It is in fact so big and the number of DNA sequences it houses so immense, that it would appear not only impractical to store genes inside the boxes, but physically impossible. So instead of storing the genes themselves, we will imagine that the *Gene Space* librarians have decided to represent the information of genes symbolically and what they actually store in each box is a DNA conversion algorithm. They have also devised a unique way of providing customers with the kits that they require.

Each customer is supplied with a free universal gene-transformation machine, a supply of nucleotide bases and the DNA sequence of the newt *Triturus*, whose genome is around 19 billion bases long. If for example you wanted to build the lungfish *Protopterus*, you would purchase the appropriate gene kit box but instead of housing a *Protopterus* genome, it would contain a *Triturus* to *Protopterus* conversion algorithm (TPCA). As you left the hypermarket you would pick up your free machine, a supply of bases and a *Triturus* DNA sequence. Once you arrived home, constructing your *Protopterus* would be a simple matter. The *Triturus* sequence would be placed in the machine and the TPCA then fed into the input slot. After filling the machine with bases, you could go for a short walk whilst the machine performed its task. The genome of *Protopterus* is 140 billion bases long so, in addition to editing the *Triturus* sequence, the machine would need to add about 121 billion bases to it. Every transformation is characterized by a unique pattern of additions, deletions and modifications. The process of running a *Triturus* to *Mycoplasma genitalium* conversion algorithm (TMCA), for example, would result in the removal of all but 580,000 bases from the *Triturus* sequence.

In the interests of profit maximization, the number of beads dispensed to each customer would be proportional to the DNA sequence distance between *Triturus* and the target genome. The sequence distance is equivalent to the number of modifications that need to be made to the *Triturus* sequence in order to transform it into a DNA sequence which specifies the target organism. It is also a measure of how many bits of information are needed to implement the transformation. The DNA sequence distance between two gene kits in *Gene Space* can thus be expressed as an information distance.

Whereas it might take billions of modifications to turn a *Triturus* into an ant, it might take only a few million alterations to convert it into a frog. So it makes sense to have a number of different reference kits that could be dispensed to customers along with their conversion algorithms. If, for example, you wanted to build a centipede, you would be better off starting with a caterpillar gene kit than a polar bear reference standard. If on the other hand you wanted to build a baboon, it would be faster and more economical to start with a human standard than a daffodil. Similar types of creature are likely to be clustered together within *Gene Space*, so that the DNA sequence distances between them are small compared with the distances separating them from other creatures. Distances in *Gene Space* reflect both the logical similarities between creatures and the time and effort needed to journey from one region to another.

If we were to limit the contents of *Gene Space* to creatures which are currently living or once existed, the information distances between kits would acquire a new significance, because they would enable us to determine the likely historical relationships between all of the creatures in the space. A complete table of the information distances between each pair of organisms would help us to construct lineages that indicate the likely relatedness of each type of creature.

Conversion algorithms constitute a sort of *A to Z* road map of *Gene Space*, which enables travel from one creature to another without getting hopelessly lost in the infinite sea of alternative destinations. Only a tiny number though are likely to terminate in a gene kit which specifies a viable organism. Natural evolutionary strategies for navigating *Gene Space* demand that each DNA sequence location passed *en route* to a target destination must specify viable creatures that can endure the idiosyncrasies of the world in which they find themselves. Artificial transformation processes like those implemented by the universal gene-transformation machine are not subject to this constraint, as intermediate sequence loci can be skipped over without being realized. Artificial engines for exploring *Gene Space* may consequently travel across any route with impunity.

Unlike artificial devices, which exercise the principle of maximal economy and select the shortest possible routes to their destination, natural processes of navigation must choose their routes blindly. The availability of routes are determined by chance mutation events and environmental circumstances that allow some of them to be explored.

Once a pathway has been selected, the number of pathways that are subsequently available for exploration is reduced. Natural jaunts through *Gene Space* are consequently unlikely to be as extensive as their artificial counterparts.

The *Gene Space* librarians eventually realize that designing their information storage system around the DNA sequence of *Triturus* and a host of other reference standards is both inefficient and impractical. It is more economical to dispense with the notion of conversion algorithms altogether and to replace them with what we will call a construction algorithm. Construction algorithms are also run on universal gene-transformation machines. Given a sufficient supply of bases, a universal machine will assemble any DNA sequence from scratch, according to the dictates of the construction algorithm with which it is fed. The construction algorithm itself mirrors the logic of genes, and employs four symbols that are combined in a unique sequence. These symbols are the letters **A**, **C**, **T**, and **G**, which correspond to the nucleotide bases in which the information of DNA is encoded. When the letters **A C T C G A** of a construction algorithm form the input to a universal machine, the genetic sequence **ACTCGA** is assembled.

Although the genetic sequence and algorithm might appear identical, there is an important difference between them. Whereas an algorithm specifying the construction of a genetic sequence represents pure information that exists independently of any particular material representation, the same is not true of genes. Once the abstract algorithmic information has been committed to the natural **A**, **C**, **T**, and **G** language of genes, it is also committed to being translated into amino acidese. But couching the logic in DNAese and the realization of the logic in amino acidese places its own constraints on the type of information, that is to say the genetic and amino acid 'utterances', that can be expressed. Thus although there is a sense in which a message encoded in the nucleotide base language of DNAese presumes the existence of amino acids and the chemistry which enables them to be joined into proteins, genes and the collections of interacting genes within a genome may be considered as embodying a higher order logic.

Is it possible that this logical structure or organization could be examined independently from the particular construction materials in which it is realized? If so then the 'logical form' of living things should be more fundamental than any particular manifestation of that form, and may perhaps be realizable in different technologies. Some of these may be quite foreign to the historical DNA and amino acid building materials with which contemporary life is familiar. DNA and amino acids are thus not necessarily the only, or the best of all possible building materials. Although their considerable versatility should not be underestimated, the historical search for possible life technologies was almost certainly far from exhaustive.

The wiring diagrams of organisms that specify the ways in which the genes in each kit are interconnected and co-ordinated in time by being switched on and off in a programmed manner should alone be sufficient to give an intelligent Martian robot an insight into how life on Earth is generated. Although it may not be possible to directly infer the existence of DNA, RNA and protein molecules from the logical form of the programs that they encode, the abstracted logic should suggest the kind of technologies which might be appropriate for realizing the covert biological structures represented in a given wiring diagram.

Ludwig Wittgenstein suggested that language limits the expression of thoughts. There is similarly a sense in which the information encoded in genetic construction algorithms may be considered as being to genes (the first language) and amino acids (the second language) as thoughts in a language-free form are to the mother language in which they are articulated and any foreign language into which they are translated. However, unlike the more abstract symbols of written languages, which are purely symbolic, DNA and amino acid languages are embedded within the physico-chemical world. This imposes several constraints, both on the nature of the information that they are able to express, and on the manner in which the biological information articulated in these languages is able to change across time. One might predict that there are large numbers of potential utterances which, although having a legitimate expression in the abstract logical form of algorithms, are nevertheless inexpressible within the constraints imposed by the DNAese and amino acidese languages of contemporary life.

One of the most significant of these constraints is imposed by the twenty different amino acids from which natural proteins are made. The physical and chemical properties of amino acids and the structures they form determine the repertoire of three-dimensional shapes or folds that protein molecules are able to adopt. All complex structures of life must be crafted using different components from the finite library of structural possibility that this limited repertoire of amino acids allows. Cyrus Chothia has predicted that modern proteins cluster to around one thousand different structurally related families, although the number may be as high as eight thousand. However, the number of all possible types of protein families housed within the shelves of *Realisable Protein Fold Space* hugely exceeds the size of the contemporary repertoire. Nature thus utilizes far fewer protein folds in her construction processes than are theoretically possible and modern life operates with a very restricted set. Furthermore the folds which are used, represent a very biased set that are unlikely to represent the diversity which might, in principle, have been utilized by history.

The contents of *Realizable Protein Fold Space* themselves probably represent only a fraction of the items stored within *Theoretical Protein Fold Space*. But many of these are forbidden by physical and/or chemical constraints which prevent their native forms from being realized in the world as we know it. This is not to say that some of these folds might not become plausible if artificial techniques were used to facilitate their formation, or if some or all of the laws of physics and chemistry did not apply. Other folds may be plausible on a physico-chemical basis but nevertheless unrealizable as a consequence of their being inaccessible to natural evolutionary processes.

Even allowing for these constraints, the huge discrepancy between the actual number of protein folds and the theoretical number of allowable folds suggests that the number of structural protein tools available to contemporary life has been limited, at least in part, by some factor. The most likely limiting principle is the cursory historical manner in which nature has explored *Protein Sequence Space* and thus *Realizable Protein Fold Space*. Although the modern repertoire of protein folds has been subjected to multiple rounds of selection over several billion years, it has also been shaped by the types of random events which characterize historical processes. In structural terms, traditional life-construction technologies are stuck in a historical rut, albeit one which has to date been very successful.

We have speculated that there are algorithms within the hypothetical library of all possible construction algorithms *(Construction Algorithm Space)*, which the DNA and amino acid languages are unable to articulate. This might result from the physico-chemical implausibility of the protein fold which the algorithm specifies. The information contained in these algorithms must thus be passed over in silence. The full richness of the potential tapestry of algorithmic information housed within *Algorithm Construction Space* is thus realized as only an incomplete and loosely woven collection of historical threads.

I would now like to return to the question of the pictureless boxes that appeared in my first and third dreams. If one were to draw a map of *Gene Space*, it would look like a vast and complex network of three-dimensional intersecting grids. Each gene kit would be housed within a different grid and allotted a unique address within the grid network. However, there are so many gene kits that what we really need is a multidimensional space to display them all in an appropriate manner. In order to facilitate the task of partitioning *Gene Space* into crystalline packets of order so that the map might be of use to prospective *Gene Space* travellers, a meeting is scheduled with the *Gene Space* librarians. It comes as a great relief when they divulge that the kits have not been stacked randomly but are organized in as orderly a manner as possible. Each kit is thus arrayed with kits of a similar sequence length and positioned next to the cluster of gene kits which are their nearest relatives. The small information distance between individual gene kits within such a cluster makes interconversions between such sequences a trivial matter.

At this point I find myself at the entrance to the *Gene Space* hypermarket. In the first aisle I find four gene kits stacked next to each other. Each of these lacks a picture on its box, and contains only a single DNA base. Situated next to these are sixteen other kits, each of which contain two bases and also lack pictures. As I walk up, down, forwards, sideways or diagonally from the entrance, I encounter gene kits with genomes of a similar or identical length. If I walk far enough, I find kits with larger numbers of bases. Any increase in the number of bases in a box results in a corresponding increase in the cluster of related gene kits. Whatever direction I walk in, the outcome is initially much the same; different variations on a DNA sequence theme, more and more bases in each box and a consistent lack of pictures on the boxes. Eventually, after travelling a great distance, I begin to find clusters of boxes with pictures on them. Some of these form islands surrounded by a sea of pictureless boxes. Others, like a seam of coal in a mine, lead to a procession of different picture-boxes. I resolve to keep a tally of each kit that I encounter and record whether the box has a picture on its side (a picture-box) or not (a plain box). This information is then incorporated into my map. If I encounter a plain box, the corresponding grid on the map is filled in with a white crayon. If on the other hand I encounter a picture-box, the corresponding grid is coloured black.

Despite the physical constraints which limit the size of the DNA sequences that can actually be realized, *Gene Space* is infinitely large. The holy grail of a definitive map will thus elude even the best cartographer. An incomplete map, however, is better than no map. Traipsing around *Gene Space* on foot proves to be an exhausting and cumbersome task. And I am relieved when the librarians provide me with a space capsule that enables me to travel at the speed of light. Like a bee systematically visiting and inspecting the cells of a huge honeycomb, I am now able to visit every grid position within my incomplete three-dimensional map. The majority of the creatures I encounter on my journey are quite unfamiliar. Giraffes the size of the Post Office Tower and countless other creatures that lie outside my realm of experience. As I zip around *Gene Space* I record the value of each box I encounter. That is to say whether it is a picture-box or not. As before, this information is jotted down on a piece of paper and on my return, I begin the tiresome task of colouring in the grids on my map. Black for a picture-box and white for a plain box.

After many hours of colouring, I complete my incomplete three-dimensional grid map of *Gene Space*. It is very large and in order to fully appreciate the complex pattern of black and white grids that I have created, I build a huge perspex cube which is divided into vast numbers of cells. Each cell represents a grid on my map, which in turn represents a gene kit box. I proceed to colour every cell in the cube black or white according to the colour-coded information in my map. As the cube is transparent, I am now able to examine the whole pattern of black and white cells across *Gene Space* and what I observe is very interesting. Distributed across a sea of white cells is an interconnected network of black cells which spread across the cube and form a tortuous, three-dimensional pattern. These branching pathways remind me of a complex system of crisscrossing, tarmac, cycle pathways. The width and depth of the lines varies from place to place, but they are all connected. Some of them are dead ends, whereas others spiral and ramify upwards, downwards, across, forwards and outwards, splitting off into hundreds of thousands of daughter lines as they do so.

I now imagine that I am cycling along one of these pathways. Unlike normal cycling however, in the perspex cube I am able to cycle up, down, diagonally and sideways, as well as forwards and backwards. Starting at any position in the cube, I can cycle to any black cell simply by consulting my *Gene Space* map, selecting the appropriate route and sticking to the path. Some routes are relatively direct, whereas others are more circuitous. If I cycle too far to the right or left of the path however, I find myself at the edge of the *Sea of Whiteness*. There are no short cuts across this forbidden sea, so it is consequently impossible to veer off the pathway. Occasionally I take a wrong turn and find myself cycling up a blind alley, which leads to the *Sea of Whiteness*.

Picture-boxes are gene kits that specify the construction of a potential organism, whereas plain boxes do not. The black pathways that represent continuous trails of picture-boxes surrounded by a sea of white picture-less boxes depict 'life-lines'. These pathways of potential life lead through the sterile *Sea of Whiteness* forming an extended three-dimensional ragged-edged peninsula of possible life which is only just discernible amidst the vast expanses of the *Sea of Whiteness*. Interconnected, privileged picture-box sites of logical order, which have the potential to harness the power of life. Plain boxes, however, fall short of this. The extent to which they do so may be objectively determined by counting the number of cells that need to be traversed in order to reach the nearest picture-box, by the shortest possible route. This is equivalent to the information distance between the plain box and the nearest picture-box. In a later chapter we will examine nature's historical exploration of some of the black crayoned grids which ramify through *Gene Space*. For the moment however, we will address a few associated issues.

The first of these relates to the abruptness with which the first picture-box was encountered on our exploratory ramble through the *Sea of Whiteness*. The significance of the first picture-boxes we find is that they represent potential transition points from non-life to life. For this reason, they raise the theoretical and historical question of how a gene kit that specifies a potentially living thing could suddenly emerge from a lifeless sea of pictureless boxes. How, indeed, does a genetic specification for a living thing come into existence in the first place? What are their precursors and where might they be found? This relates to the general question of how living things can be generated from non-living elements and also to the historical process by which life originated on Earth.

What I would like to suggest is that the abrupt and sudden transition from white to black boxes, that is to say the apparently sudden appearance of life in *Gene Space*, is nothing more than an artefact. This stems from the fact that we have, until now, limited our study of life to gene kits made exclusively from DNA and housed in *Gene Space*. Indeed, we have assumed that the information of living things must be encoded within genes and that these must be made from DNA. This is not however, necessarily a reasonable assumption. It has been suggested, for example, that the first organisms may have encoded their essential information within crystals of clay. As *Gene Space* only houses gene kits made from DNA, the precursors of the first picture-boxes will only be found in *Gene Space* if they are indeed gene kits and made from DNA. Every item must be hunted down in its own appropriate type of space. We would not, after all, look for toys in a fruit shop, or for garden furniture in a book shop.

The very first picture-boxes that we encounter in *Gene Space* may, however, be connected to invisible threads that form another network of pathways. These are analogous to the life-lines of *Gene Space*, but traverse its edges and terminate in another space located somewhere beyond *Gene Space*, which we will call *Chemistry Space*. This space houses the complete set of all possible chemical reactions. Because genes are chemicals and the addition or subtraction of a base to or from a sequence involves a chemical transformation, *Gene Space* must be a subregion of *Chemistry Space*. Using the perspex cube model, *Gene Space* is a small cube located within the larger cube of *Chemistry Space* and it is here that we must search for the earliest incarnations of life. The proto-organisms that are the potential precursors, and in some cases the actual historical antecedents, of the simplest picture-boxes in *Gene Space*. We might choose to colour the grids which correspond to these candidate precursors organisms green.

We will eventually move on to unravel the history of life's procession through *Gene Space* and *Chemistry Space*. But before proceeding, I would like to address one final point. In the previous discussion, we assumed that it is possible to determine which boxes merit pictures at all. But how, when they were constructing the library of all possible gene kits, could the *Gene Space* librarians have known which boxes were able to encode potential organisms and thus required pictures? One might also wonder how they were able to paint pictures on the boxes at all, without actually constructing the kits so that they could see what the organisms looked like. History can be of some use here, as the librarians of *Gene Space* have kept a record of all the gene kit boxes that have been opened. This gives a full account of the construction algorithms which have been run, tried and tested. The librarians know that given the right circumstances, these kits have the potential to generate living organisms. But what about the boxes that history forgot? The picture-boxes that were never opened and which might never be opened? How do we know that their construction algorithms will generate living creatures if they are one day tested? The answer to these questions is by no means straightforward, but I will offer the following.

It might one day be possible to define a formal logic of life, a design logic common to all living things. As the number of complete genome sequences in our database grows, bioinformatic techniques will help define essential sets of genes shared by organisms of a certain type. A search for these genes constitutes one strategy for screening *Gene Space* for candidate picture-boxes. However, it is unreasonable to assume that all the potential organisms within *Gene Space* will utilize the same broad set. Life might equally well be constructed using different sets of genes.

In the future it will be possible to finely detail the pattern of gene activity in organisms. There may however be some way of predicting this pattern, or at least the set of likely patterns, from the genome itself. The patterns of living and potentially living things will almost certainly be different to those of creatures that are unrealizable. If presented with the genome and inferred pattern of gene activity of a putative living creature, we should be able to discern whether it is likely to specify a living thing. Creatures which utilize a different inventory of genes may, however, generate patterns quite different to those of known organisms.

Once the logical structure of a creature has been established, it should provide some insight into its likely form. This would enable us to paint a general picture on the front of each picture-box. However, until a creature has been run and tested, such pictures must remain tentative. We might find that it becomes necessary to erase and redraw many of the pictures once an organism has been run. In some cases we may have to erase the picture permanently. There is also another problem to confront. For, when deciding whether a box should be assigned a picture, is it enough that the creature can be constructed and is able to survive for a few hours or days, or must it be viable for a longer duration and able to leave offspring? Furthermore, in what environment should each creature be tested? Each kit should ideally be tested in every possible environment, although the genome itself might give some clues as to which is likely to be optimal.

It is, however, one thing to talk about *Gene Space* and an infinite sea of gene kits, and another to acknowledge that none of these contains the means for its own execution. In order to implement the instructions written in the DNA programs that are contained within each gene kit, translation machinery is required. In the case of sexually reproducing species, this is contained, as I have already mentioned, both within the egg in which each gene kit is stored and in the induction signals delivered by the follicular cells that surround the growing embryo. But *Gene Space* does not contain eggs or follicular induction signals. It will therefore be necessary to infer yet another set of crisscrossing invisible threads, which on this occasion lead to a subregion of *Chemistry Space* that we will call *Egg Space* and *Follicular Induction Signal Pattern Space*. This is the space of all possible machines that are able to read instructions written in DNAese and kick-start a gene kit into life. Each of the kits in *Gene Space* are thus connected by an invisible thread to every egg in *Egg Space* and to every different pattern of induction signals in *Follicular Induction Signal Pattern Space*. Only some of these pairings will however result in a meaningful translation of the gene kit's information. So when we record whether a given box deserves a picture, we must test the kit repeatedly so that each potential pairing with different eggs and follicular induction signals can be evaluated. Moreover, the information stored in the egg and the patterns of follicular induction signals are not the only sources of essential 'extra-genetic' information. The membrane which surrounds the single cell that is formed as a result of the fusion of a sperm and an egg, is also derived from the egg. The organization of this pre-existing cell membrane thus templates the structure of all subsequent cell membranes. We can thus imagine yet another space called the *Space of All Possible Cell Membranes*. Presumably only a small number of these would be compatible with normal cellular function.

Given these complications, we must proceed with caution and continue to use the term 'potentially living organism' rather than the more presumptuous 'living organism'. At the end of the day, we might at best be able to say that a given gene kit has a certain probability of producing a living organism. But there will always be some degree of uncertainty, and it is likely that the only definitive test is to allow a gene kit to run in the context of a particular environment, egg and set of follicular induction signals. The logic of the potential organism and the machine responsible for decrypting its logic could be test-driven in real time and space, or using a computer simulation. Until such a time, however, we are obliged to place a question mark on every unopened box and to qualify each picture with phrases such as 'is likely to produce' this or that creature, or something to that effect.

In this computational model, living creatures are envisaged as being computed by their genetic sequences. The genes in sexually reproducing organisms are the program, and the egg and follicular induction signals the hardware on which it is run. Kurt Gödel has demonstrated however that there are some logical propositions which are undecidable; structures that can never be shown to be true or false using computational procedures. If we were to proceed through *Gene Space* and feed the genetic program of each potential organism systematically into a computer, we should expect there to be some programs for which the computer is unable to give a definitive 'yes this program, if run, would generate a living thing', or 'no this program would not generate a living thing'. Some potentially living organisms are thus likely to be non-computable and may, consequently, be placed in a sub-space which we might call the *Undecidable Possible DNA Zoo*. In the case of these gene kits, the corresponding grids would have to be coloured grey instead of a definitive white or black. The only way to resolve the uncertainty in these cases would be to run the organism and see what happens. It should be remembered, however, that a picture-box which generates life today will not necessarily do so tomorrow. Gene kits are not constructed in a vacuum, and must instead to be born into a particular part of a particular world. Even then, the picture on the box can never constitute an exact representation of the logic of the kit, as in order for this to be the case, the environment would have to be specified to an infinite degree of precision.

3

The Crocodile Holds Its Breath

Have you ever wondered how crocodiles manage to stay underwater for such a long time? The Nile crocodile *Crocodylus niloticus* can in fact remain underwater for more than an hour without having to come up to breathe. If you are ever unfortunate enough to get caught in the powerful jaws of one of these creatures, you might discover that crocodiles tend to kill their larger prey by drowning them, instead of using their usual and marginally more merciful method of gripping their victim by its legs and then rotating themselves rapidly so that it is torn apart. This predilection for drowning rather than dismembering presumably reduces the risk of them losing their meal as they open their jaws to deliver the *coup de grâce*.

Incidentally, should you ever get caught by a crocodile and happen to get eaten head first, I have heard (albeit from a quite unreliable source) that it is possible to outwit the unsuspecting beast and beat it at its own game. All you apparently need do is reach forward and pull as hard as you can on its epiglottis, which is the little piece of pink tissue that hangs down at the back of its throat and covers the opening to its lungs. If implemented correctly, this manoeuvre will make the crocodile drown. The adducter muscles, which snap a crocodile's jaws shut with a suddenness and force not easily forgotten, are said to be more powerful than the abductor muscles which open its jaws. So if you manage to disentangle yourself from the crocodile's jaws and avoid the second inevitable snap, it should in principle be possible to hold its jaws firmly shut. Fortunately I have not had the opportunity to test either of these survival methods, so am unable to guarantee their efficacy.

A recent edition of the *Australian Medical Journal* helpfully advises that if visiting a region inhabited by crocodiles, activities to avoid include: paddling light craft, swimming, wading in shallow water, cleaning fish at the water's edge and camping close to the river bank. It also provides a degree of tangential credibility to the defence tactics suggested above, by invoking the general success of strategies involving retaliation. These are presented in the form of testimonies from three surviving victims in the Northern Territory of Australia. Each claimed that they were able to expedite their escape from a crocodile's jaws by gouging the beast in its eyes and nostrils. Three other victims were released following a prolonged tug-of-war and two of these received assistance from an obliging friend. It is reassuring to know that, in the Northern Territory at least, the Conservation Commission employs a policy of relocating 'problem crocodiles' that are found in tourist areas to alternative regions of the countryside.

On a good day, I estimate that I can hold my breath for about three minutes. However, instead of guessing, I have just tested myself and found that I can only manage a rather disappointing forty-six seconds. Perhaps I had better pay some more visits to the health club. Professional divers on the other hand, are able to swim to a depth of one hundred metres in a single-breath dive which may last up to about four minutes. The best divers rely on techniques such as yoga to reduce their metabolic rate and thus their oxygen consumption.

The reason humans and many other creatures can hold their breath at all is because of a protein called haemoglobin which is able to bind and release oxygen. Haemoglobin is stored and transported within red blood cells, each of which is packed with haemoglobin proteins. When oxygen is in short supply, haemoglobin is able to relinquish some or all of its cargo of four oxygen molecules. But when exposed to the high levels of oxygen found in the lungs, each haemoglobin protein loads up with a new supply of oxygen. Each haemoglobin protein thus functions as a tiny oxygen-collecting and -distributing machine. It might appear paradoxical that haemoglobin is able both to load and subsequently relinquish its supply of oxygen. The solution to this enigma was provided by Max Perutz who elucidated the three-dimensional structures of oxygen-loaded and empty haemoglobin molecules. These demonstrated how haemoglobin's unique geometry influences its behaviour and provided the first insight into the design of a complex protein machine.

Haemoglobin is made from four amino acid chains of two different types. These fold into four separate sub-units, each of which has a similar shape. The sub-units are closely intertwined, like a three-dimensional jigsaw puzzle. Each contains an oxygen binding site known as a heme and can accommodate a single oxygen molecule. Experiments have shown that the four hemes are able to communicate and co-operate with one another. Although reluctant to fill the first heme, once this has been filled the 'appetite' of haemoglobin for further oxygen molecules increases dramatically. Conversely, although reluctant to relinquish its first oxygen molecule, once it has done so haemoglobin readily gives away the three remaining oxygen molecules. Max Perutz has noted the similarity between this state of affairs and the parable which concludes 'whosoever hath, to him shall be given, and whosoever hath not, from him shall be taken even that which he seemeth to have.'

The three-dimensional structures of oxygen-loaded and empty haemoglobin molecules suggest a remarkable explanation for this phenomenon. It appears that haemoglobin is able to change its shape and, like a circus contortionist, click back and forth between two different three-dimensional forms. When adopting the relaxed (R) conformation that it has in the lungs, haemoglobin is able to load oxygen. However when it reaches metabolizing tissues, it switches to a tense (T) conformation. This facilitates oxygen unloading by lowering the haemoglobin's affinity for oxygen by up to twenty-six times. The ease of transition between the two conformations is influenced by metabolic products that accumulate in the blood. These chemical modulators include: bisphosphoglycerate, chloride ions, acids and carbon dioxide. The accumulation of acid in metabolizing tissues stabilizes the 'T' conformation, whereas in the lungs, where the blood is less acidic, the 'R' conformation is stabilized.

Humans and all other creatures that rely on haemoglobin as their principal reservoir of oxygen molecules are able to hold their breath for as long as it takes to deplete their oxygen supply. Some creatures however, have evolved ingenious strategies for oxygen conservation. King penguins reduce their oxygen consumption by allowing their abdomen to fall to temperatures as low as 11°C. This suppresses their metabolic rate and enables them to stay underwater for up to seven and a half minutes. Whales, seals and many other diving mammals, on the other hand, supplement their supply of haemoglobin with a large supply of an oxygen-binding protein called myoglobin, which in the words of John Kendrew, is a sort of 'junior relative' of haemoglobin. It has only one amino acid chain instead of four and consequently only a single heme. It is stored in muscle tissue and acts as a 'spare tank' of oxygen molecules which is filled by oxygen transported from the lungs, bound to haemoglobin.

When a whale dives, it rapidly depletes its supply of haemoglobin-bound oxygen and switches to its myoglobin-bound reserve tank which functions as a chemical lung and releases oxygen to metabolizing tissues. Crocodiles have nearly a hundred times less myoglobin than whales. There is no obvious reason why crocodiles should not have explored the possibilities of a myoglobin-like oxygen storage system. Evolution has, however, provided them with an equally effective but quite different mechanism for maximizing their oxygen delivery capacity. Crocodile haemoglobin incorporates two unique design features. First, it is exquisitely sensitive to the bicarbonate produced when carbon dioxide dissolves in water. Bicarbonate accumulates in the crocodile's blood when it is underwater and binds to a regulatory site which lowers the haemoglobin's oxygen affinity. Second, even in the absence of bicarbonate, crocodile haemoglobin is primed to deliver oxygen as it has a much lower oxygen affinity than either human, whale, seal, or king penguin haemoglobin.

The implications of the bicarbonate effect for oxygen delivery are best understood by considering populations of haemoglobin molecules rather than individual proteins. As we have seen, haemoglobin molecules have four hemes, each of which may bind a single oxygen molecule. In the 'T' state all four hemes may be empty, whereas in the 'R' state all four may be full. Some haemoglobin proteins will be in a state of transition between the 'T' and the 'R' states and will have only two or three oxygen molecules bound. The size of the pool of haemoglobin molecules in both crocodile and man is vast. But let us imagine that it contains only ten haemoglobin molecules, with a combined total of forty hemes. If all forty hemes are occupied, the population is 100% saturated. If none of them are occupied, the population is 0% saturated. In man, the oxygen saturation of populations of haemoglobin molecules varies from 97% to 75%. Translated into our imaginary pool of ten haemoglobins, this means that a maximum of only ten hemes can be empty at any one time. The other thirty oxygen molecules remain bound and thus unavailable.

The bicarbonate effect enables crocodiles to gain access to some of these normally inaccessible oxygen molecules. This is achieved by introducing bias into the population of haemoglobin molecules, so that the probability of encountering a 'T' state in a given molecule is much increased. In this way the population of haemoglobin molecules is allowed to fall to saturation levels far below the minimum value permitted in man. This oxygen-scavenging mechanism, coupled with the reduced oxygen affinity of crocodile haemoglobin, enables crocodiles to hold their breath for up to one hour, which is around sixty times longer than the average human and about fifteen times longer than a king penguin.

Noburu Komiyama and Kiyoshi Nagai have compared the haemoglobin genes of humans and crocodiles and identified a discreet region of the crocodile haemoglobin gene that is responsible for the bicarbonate effect. Having identified the positions at which the genes differed, they 'humanized' the crocodile gene within this control region. If at one position the crocodile sequence read **AAGACA** compared with **AACACA** in the human gene, the **G** in the crocodile sequence would be changed to a **C**. This process was repeated until only thirty-six differences remained. The hybrid gene encodes a chimeric protein which is part human and part crocodile, which they called haemoglobin 'scuba'. This responded to bicarbonate in an almost identical manner to that of wild type crocodile haemoglobin. The ability of crocodiles to remain underwater for extended durations thus appears to be dependent on the introduction of only thirty-six or fewer changes into the human haemoglobin sequence. The bicarbonate binding site is located at the interface of the four haemoglobin sub-units. These slide across one another and are responsible for switching the protein between its two conformations. Bicarbonate functions as a molecular clamp which stabilizes the 'T' state and thus the oxygen-delivering mode of the haemoglobin mini-machine.

A few changes at key positions in the haemoglobin gene sequence have enabled crocodiles to gain access to a secret underwater world. Had things been different, we might have joined our crocodilian friends in their underwater perambulations, perhaps living side by side with them in the swamps, lakes, rivers, estuaries and coasts of Africa, Asia, Northern Australia, Mexico, Central America, the West Indies and South America. In the future, human haemoglobin genes may be routinely 'crocodilized', so that we too will be able to hold our breath for one hour. The haemoglobin of any creature could in fact be engineered to behave like crocodile haemoglobin by altering as few as 4.2% of the 861 or so bases that it takes to encode a haemoglobin protein. Crocodilized humans might thus be joined by crocodilized cats and bats. But we need not restrict ourselves to borrowing sequences from haemoglobin genes. Fashion houses might one day sell a wide selection of designer genes. For why should one buy a crocodile-skin coat when it is possible to transform one's own skin into attractive green scales? Why, furthermore, should one be content with a two-dimensional mouth, when one could up-grade to a snout that comes complete with an impressive array of razor teeth?

The selection of crocodiles alive today is quite varied. The Nile crocodile *Crocodilus niloticus* and the estuarine *Crocodilus porosus*, for example, may grow up to twenty feet long. The dwarf crocodile *Osteolaemus tetraspis*, on the other hand, is a mere six feet in length. Whereas the American crocodile *Crocodilus acutus* has a long, thin, tapering snout, that of *Caiman crocodilus* is shorter and more rounded. The twenty-one different species of living crocodiles nevertheless represent only a trifling sample of the complete collection of all of the possible crocodile gene kits, which are located within the *Crocodile Space* subregion of *Gene Space*. Many of these are quite different from their modern counterparts. But despite the depth, richness and extent of this sea of crocodilian possibility, only a very small number of these kits will ever see the light of day.

The archaeological records attest to some of the crocodile kits which have been selected from *Gene Space* in the past. Kits which were assembled and tested, only eventually to be disassembled as the result of a process of extinction and returned to the crowded shelves from which they came. The Protosuchians were a group of crocodiles that inhabited the swamps of the Upper Triassic around one hundred and ninety to two hundred million years ago. These ancient beasts had exceptionally short snouts, which are atypical of modern crocodiles. Their behaviour, which unfortunately does not fossilize and can only be inferred from their structure, is also likely to have been different from their modern counterparts. The Protosuchian *Protosuchus*, for example, was almost certainly a terrestrial crocodile. Other extinct crocodiles include the Mesosuchians of the Jurassic and Cretaceous, which lived around sixty-five to one hundred and ninety million years ago. One of these, called *Bernissartia*, lived in Belgium and England and was only two feet long. The Eusuchians, which appeared in the Upper Jurassic, include *Deinosuchus* which translates as 'terrible crocodile'. It is also known as *Phobosuchus* which means 'horror crocodile'. *Deinosuchus* lived in the swamps of Texas toward the end of the Cretaceous and is likely to have reached a length of around fifty feet. But *Phobosuchus* was not the only gigantic crocodile kit to have been visited by history. *Rhamphosuchus*, which is known from a single Pliocene jawbone found in India, is likely to have been at least as big as *Deinosuchus*. Another Eusuchian crocodile, called *Pristichampsus*, lived in the Eocene and like *Protosuchus* was also a terrestrial crocodile. Unlike modern crocodiles, *Pristichampsus* had long legs which were suited for running and hooves instead of claws.

But packed alongside the small collection of crocodile gene kit boxes that have been touched by the hand of history, experienced the thrill of life and tasted blood, scattered somewhere deep inside *Crocodile Space* is the much larger collection of crocodile picture-boxes that time has passed over. Who knows what secrets lurk in these dusty, unopened kits? Crocodiles the size of tadpoles, winged crocodiles, tree-climbing crocodiles, or crocodiles with elephant tusks and tiger stripes. Crocodiles that lack tails, hop like kangaroos and howl like hyenas, or crocodiles with a hundred insect legs. Hairy crocodiles, blue crocodiles, two-headed crocodiles and crocodiles with silver scales like fish. A walk through *Crocodile Space* would certainly not be boring.

History has performed only a rudimentary and incomplete search of this space. Indeed huge numbers of swamps, rivers, lakes, coasts and estuaries, coupled with vast expanses of time, would be needed to test drive even a fraction of the potential crocodiles that are housed within the vaults of *Crocodile Space*. Although the majority will never be assembled, their genetic specifications have a timeless mathematical existence which is independent of their realization. If, for example, we had not found the fossilized remains of *Protosuchus*, it is unlikely that we would ever have known that a *Protosuchus*-like creature could exist. However as we do know that it existed and was specified by a *Protosuchus* gene kit, we could in principle infer the contents of the kit and construct a new *Protosuchus* from scratch. Although no *Protosuchia* grace the Earth today, the potential for recreating one is always 'out there', its information stored as an inviolable mathematical specification. It is unlikely, however, that all of its information was encoded genetically. In order to recreate a credible historical *Protosuchus*, we would have to extend our excursion to include information spaces that lie far beyond *Gene Space*.

During our brief visit to *Crocodile Space*, the pictures on the gene kit boxes were taken at face value and all of the possible crocodiles were treated as if they had only external form. If two picture-boxes shared the same picture, they were taken to be equivalent. Our discussion of crocodile haemoglobin, however, illustrates that small changes to the internal structure of an organism can have consequences that are every bit as profound as those resulting from changes to their external structure. Although the picture on a gene kit box is a representation of the external form of the organism that the kit specifies, it tells us nothing about its internal form. The picture symbol on the front of each box is flat in this dimension. A crocodile with human haemoglobin would after all look identical to one with crocodile haemoglobin or haemoglobin scuba. This flatness in the dimension of internal structure becomes apparent when we walk through the *Crocodile Space* hypermarket and find it far less interesting than we had expected. This is not due to the lack of variety, as the crocodile picture-boxes are every bit as entertaining as we imagined them to be, but is instead a consequence of the huge distances that we have to travel before noticing any changes in the picture on each box. On entering the *Deinosuchus* locality of *Crocodile Space* for example, the aisles contain billions of kits with identical picture-boxes.

The thrifty tour operator whose brochures offer 'chance of a lifetime' trips into *Crocodile Space* is quick to realize that visitors are rapidly bored by the apparent repetitions. Who, after all, would pay to enter an art gallery in which every picture was identical? Following a meeting with the directors of *Crocodile Space PLC*, a survey is commissioned which demonstrates that the number of boxes with different pictures constitutes only a small fraction of the total number. It is suggested that *Crocodile Space* be pruned and the apparent repeats destroyed; but this is rejected on account of the capital expenditure it would necessitate. Being of an enterprising disposition, the operator finds his own solution to the problem, which involves busing visitors from one region of *Crocodile Space* to another to avoid areas of repetition. Using his privileged access to the surveyors' map of *Crocodile Space*, only he can guarantee that more than ten thousand different types of crocodile will be viewed on any given *Crocodile Space* safari. On the old-style trips, you would be lucky if you viewed more than four different types on a single outing. Following an advertising campaign, the new venture is launched. After viewing a single example of a given type of crocodile, the shutters in the bus wind down, the lights are extinguished and the visitors are treated to the latest movie releases. On arriving at a new locality, the shutters wind up so that the next type of crocodile can be viewed, videoed and photographed. After a few minutes they wind down again to minimise interruptions to the movie. For those of an academic inclination, the bus also contains a small lecture theatre in which a professor speculates about how each of the potential crocodiles might have behaved if they had been realized by the capricious machinations of history.

In actual fact, as has already been intimated, the apparently identical picture-boxes are non-identical at the level of their DNA sequences and represent variants on a multitude of internal morphological themes that have no effect on the external appearance of the crocodile. In the pictorial locality of the Nile crocodile *Crocodylus niloticus* for example, the spectrum of gene kits with identical pictures might include a crocodile with a human haemoglobin gene. As it lacks a crocodile haemoglobin gene, the humanized crocodile is unlikely to be able to hold its breath for more than four minutes and might consequently have trouble drowning its prospective victims. Another kit might specify a humanized crocodile which lacks a crocodile haemoglobin gene, but has a myoglobin supply equivalent to that of a humpback whale. Although this crocodile might be able to stay under water for an hour, it is unlikely that it would be able to stay submersed for as long as a crocodile whose kit includes both the crocodile haemoglobin gene and genes that increase its myoglobin supply. Another crocodile kit might contain a haemoglobin gene which encodes a protein that is inept at oxygen binding and delivery. These crocodiles might only be able to hold their breath for a few seconds. Had these creatures existed, they might have been called 'gasping' crocodiles. Despite the considerable differences in their lifestyles and abilities to survive, all these potential crocodiles are likely to look identical.

We have for the first time then, been given reason to doubt whether the picture on a gene kit box is a true representation of the organism it specifies. Regardless of its external form, a crocodile's survival depends on the fact that its internal structure mirrors an important aspect of its relationship with water. Without the ability to drown large prey, crocodiles would often go hungry and might occasionally starve. Significant modifications may thus be made to the internal structure of an organism without affecting its external morphology. Two fossils which appear identical might consequently represent very different creatures. Changes to the internal structure of an organism are not, however, limited to haemoglobin genes. All regions of a genome both coding and non-coding, are mutable, although most mutations have no effect on either internal or external structure. In order to determine the full effects of a mutation, it would be necessary to produce an inventory of every aspect of the mutant organism. This would include amongst other things: its external morphology, cellular and sub-cellular morphology, anatomy, physiology, biochemistry and pathology.

We will have further reason to question the ability of the picture on a gene kit box to stand as a symbol for the organism it specifies. For the moment, however, we will consider another dimension of internal structure which is not reflected in external structure. In order to do this we will imagine a line of potential crocodiles in the *Crocodylus niloticus* locality of *Crocodile Space*. Despite looking identical to its companions and having a normal capacity for underwater diving, one of these crocodiles is very different to the others as it has a major problem. The problem being that it is scared of water, and that it is, in short, a hydrophobic crocodile. For the purposes of our argument we will assume that the reluctance of this crocodile to go near water results from a single mutation in a hypothetical water-liking gene. Although we might find the crocodile's hydrophobia endearing, for the crocodile it is fatal. This is because there is a sense in which a hydrophobic crocodile is hardly a crocodile at all. For as we have already seen, modern crocodilian existence is largely defined by its relationship to water. Although crocodiles enjoy eating fish, land animals constitute a significant part of their daily menu. The behavioural programs associated with hunting thus incorporate the tacit assumption that crocodiles like, or at the very least, are indifferent to water.

One favourite hunting strategy of crocodiles is to float impassively at the water's surface so as to be almost indistinguishable from a drifting log. They then wait patiently for something to swim, fly or walk by. At other times they might stand motionless at the edge of a water hole, waiting for their prey to come down and drink. If their hunting trip is successful, they might use their tail to knock an unsuspecting beast into the water and then drag it beneath the surface to a watery demise. A hydrophobic crocodile would not be able to hunt using either of these strategies. And if it didn't die of starvation it would certainly not leave offspring, as for crocodiles mating is an activity confined exclusively to water.

The process of assembling a protein is much like constructing a microscopic building, and the introduction of a mutation into a gene analogous to changing a detail in an architect's plan. The significance of the change, however, depends on its location. Let us imagine that we are constructing the Eiffel Tower. A change to the part of the plan pertaining to a detail at the top of the tower is unlikely to influence the overall nature of the final structure. However, a modification to the part that specifies the foundations or one of the weight-bearing arches would result in the tower falling down or at the very least being lop-sided. Proteins are very sensitive to structural changes. Sequences of (and individual) amino acids located at key strategic positions are crucially important in giving a protein the unique structure that defines its function. Other amino acid sequences, or constellations of amino acids cluster together in three-dimensional space, form important recognition sites. Proteins constitute the interface between the information of genes and the biological world. They are the universal structural currency of life on Earth. The structure of proteins determines their function, and a single change to a gene can profoundly influence the higher dimensions of protein structure.

Proteins are constructed from one-dimensional sequences of amino acids that are specified by the one-dimensional sequences of genes. Although in principle amino acid sequences can orient themselves in a number of different ways, they in fact adopt only two fundamentally different types of basic pattern. These two-dimensional patterns are known as alpha helices and beta sheets. Individual alpha helices and beta sheets are connected to one another by short segments of primary amino acid sequence. The numbers of these two types of structural elements defines a protein's secondary structure, and the combinatorial manner in which they are interlinked, its topology. If the topology of proteins was completely random, they would be expected to have mixtures of both alpha helices and beta sheets. However, most proteins appear to fall into one of four combinations of: (1) mainly alpha helices, (2) mainly beta sheets, (3) alternating mixtures of both alpha helices and beta sheets, or (4) segregated mixtures of alpha helices and beta sheets.

The primary sequence contains all the information necessary to specify the higher-dimensional patterns of protein structure. Each amino acid appears to have a general predilection for appearing in certain types of secondary structure. This intrinsic preference may, however, be modified by the environment in which it finds itself. The amino acids valine, isoleucine, threonine, phenylalanine, tryrosine and tryptophan are, for example, frequently found within beta sheets and thus in this sense 'born to be beta'. Alanine, aspartate, glycine and proline, on the other hand, are rarely found in beta sheets. Nevertheless, evidence suggests that only a relatively small number of the amino acids in the primary sequence of a protein make a significant contribution to the generation of its higher order patterns. Indeed, the position of many amino acids appears to be entirely random. These features of primary sequences allow them to be considered as random sequences that have been edited in critical regions to generate the core structural 'fold'. This can then be 'decorated' by additional changes, some of which may generate motifs that have a regulatory function. Lynne Regan and her colleagues at Yale have shown that it is possible to transform a protein composed principally of beta sheets into one that contains four alpha helices by altering only 50% of the amino acids in its primary sequence. Structural studies have also provided examples of proteins whose primary sequence identities are no greater than those of randomly selected sequences, but which nevertheless adopt near identical higher-dimensional patterns.

Combinations of beta sheets and alpha helical structural elements may be bent, stapled and folded back upon one another to generate such higher-dimensional patterns. These three-dimensional architectures are known as a protein's tertiary structure. In some cases, amino acid chains exhibit structural heterogeneity and oscillate between two or more alternative tertiary structures. Some proteins are composites of more than one three-dimensionally folded amino acid sequence. Haemoglobin, for example, is assembled from four discreet sub-units. The manner in which each of these slots into place with the tertiary structures of its companions generates yet another level of complexity, known as quarternary structure.

Some proteins have identical tertiary structure architectures but differ in their topology. The existence of topological variants suggests that proteins have been subjected to a process of topological optimization. While they may have identical distributions in three-dimensional space, different topological organizations of the alpha and beta structural elements are likely to influence the efficiency with which the protein can fold and in so doing realize its higher order spatial patterns. Evolution may have selected topological variants on the basis of their relative folding efficiency. It is also likely to have selected topologies that impose minimal constraints on potential architectures that might be generated in the future. This topological plasticity would help ensure that the repertoire of potential folds is kept to a maximum.

In the absence of chemical modifiers of structure, and given sufficient information about the environment in which the protein is likely to find itself, a protein's higher-dimensional patterns should in principle be inferable or 'computable' from its primary sequence. Experts in structural prediction thus find that many of their predicted structures have an excellent correspondence with experimentally determined structures. Powerful algorithms now exist which are able to accurately predict the nature of the fold that a primary sequence is likely to generate. These types of algorithms, and the as yet only partially understood rules of protein folding that they incorporate, should eventually allow protein folds to be both accurately predicted and designed from first principles. In the case of some proteins, however, the assistance of specialized chaperone proteins and foldase enzymes are necessary to facilitate correct folding. Others will only fold correctly in the context of proteins with which they form a permanent complex.

We have seen how the higher order patterns of haemoglobin molecules enable them to function as tiny machines. It is the quarternary pattern however that makes the principal contribution to the complexity necessary to generate the functional anatomy of the machine. The manner in which the four tertiary structural sub-units of haemoglobin rotate and slide across one another to generate different quarternary structures, has a critical role in influencing the way in which oxygen molecules are able to associate and disassociate from the four binding sites located in crevices near the exterior of the protein. So how are crocodiles able to tune and modify the operation of their haemoglobin oxygen-transporting machinery and to squeeze some extra performance out of it?

The mutations in the crocodile haemoglobin gene change just twelve amino acids located at the interface between two of its sub-units and generate a bicarbonate binding site. As the crocodile goes underwater with its prey clasped tightly in its jaws, bicarbonate ions accumulate in its blood. When bicarbonate ions bind to the regulatory site, they induce a change in its quarternary structure which enables more oxygen to be delivered than normal. This mechanism allows individual haemoglobin molecules to sense and respond to the chemical texture of the outside world. The bicarbonate binding site thus functions as an oxygen sensor which can shift haemoglobin machines from a normal mode to an oxygen stress mode of operation. Action-at-a-distance effects of this kind between two spatially distinct sites in a protein, for example the bicarbonate binding site and the four hemes or any individual heme with all of the others, are known as allosteric interactions. The introduction of an allosteric effect into the haemoglobin of the Nile crocodile has allowed it to take great behavioural strides into previously forbidden territory. This was made possible as a result of only a small number of modifications to its haemoglobin gene. The introduction of the allosteric effect by such parsimonious means illustrates nature's ability to find elegant and economical solutions to design problems that might otherwise seem intractable.

The success of naturally occurring biological structures is based upon their versatility and evolvability. It would appear that proteins are poised for change and have been selected, amongst other things, for their protean structural possibilities and chameleon-like properties. Historical success thus invokes structural components that are optimized not just for current functions, but which also have the ability to be modified in the future. Nature appears to accomplish such transformations with relative ease, allowing creatures to take large strides through the *Space of All Possible Places to Eke Out a Living* which we will call *Niche Space*, by taking just tiptoe steps through *Gene Space* and the corresponding *Amino Acid Sequence Space* and *Gene Regulation Space*.

The economy of nature's allosteric solution to the problem of modifying haemoglobin in order to furnish it with a new property, may be affirmed by considering an alternative strategy that might have been used to solve the same problem. This involves the artificial engineering of haemoglobin, that is modifying its structure by 'instruction' rather than selection; redesigning it from first principles, instead of evolving a modified structure by constructing a repertoire of random variants and selecting only those best adapted for survival. Whereas selection of variation on a historical theme benefits from knowledge gleaned from millions of years of previous experience, redesign from first principles does not. The success of a design solution may be judged on the basis of a number of criteria that include: the number of genetic modifications needed for its implementation, the speed with which the change can be realized, the extent to which the solution satisfies a complex array of different requirements, and its relative efficiency and flexibility.

Let us now cast ourselves into the fictional kingdom of the king of the terrestrial crocodiles. The king, Pristichampsus, is bored. Bored of the land and the sensation of hard earth under his hooves, and bored of his terrestrial friends. What he really wants is to be able to swim like a fish. Of course he doesn't want to give up life on land completely, but he is growing increasingly restless. How he longs to investigate the depths of the rivers, to chase sparkling catfish and, in the summer months, to immerse himself in the cool, murky water of the swamps. Passing the time with his favourite sports of hunting and maiming, he might spend his remaining years in this sub-aquatic idyll. Sitting on his throne and dressed in purple, he summons the court architect Crotopus, who is also known as 'thumping foot'. Crotopus is his chief genetic engineer, an artificer and designer of considerable repute.

Humbled in the presence of the king, Crotopus averts his eyes respectfully as he is addressed.

'I'm short of money and short of time,
But would like so very much to dine
On those sparkling little silver fish
Which would doubtless make a perfect dish.'

The architect looks concerned. 'In short,' the king continues,

'Three days you have and modest sums,
In which to fulfil my greatest wish, to be a sort of crocofish.
Suffice to say that if you fail,
Your teeth and scales I shall impale!'

Crotopus backs away, growling a subservient crocodile growl, his hooves slipping clumsily on the red carpet. He has almost left the throne room, when the king stands up and adds rather menacingly, 'Above all be efficient.' Crotopus bows a crocodile bow, scuttles backwards through the arched entrance and disappears into the depths of the palace.

So now Crotopus's paradox. The king wants to swim, the king wants to catch fish, the king wants to *be* a sort of crocodile fish. But the king is in a hurry and is frugal with money. He sets to work at his drawing board. Fully comprehending the enormity of his task and given the constraints of time and finances, he realizes that his only chance is to focus down onto a single component of the king's metabolic machinery. A single protein encoded by a single gene. Haemoglobin comes to mind as an obvious candidate, but Crotopus knows that the king is capricious. Although content to catch river catfish today, he might easily decide tomorrow that catfish are boring and that what he really wants to do is to hunt the deep seas. If Crotopus focuses his design on catfish hunting, one hour of underwater time will suffice. Deep-sea fishing on the other hand, is an entirely different thing, and without the keen sight and speed of a penguin, it could require up to three hours of underwater time. Then there is the problem of the king's ambiguous use of language – the preference of this crocodilian Coleridge for esoteric couplets instead of clear, straightforward instructions. Furthermore engineering the king's haemoglobin constitutes just one of several alternative design solutions, each of which must be considered on the basis of its relative economy, ease of implementation and possible adverse consequences. Knowing the king's foibles and being a competent and efficient creature, Crotopus lists the properties that his new protein will have to incorporate.

Given the constraints that we have outlined, he uncharacteristically decides to forgo aesthetic and artistic considerations. So although he has made his name as someone who is both daring and contemporary, able to break with tradition and can be relied upon to make a statement, he must put such issues aside for now. He must be pragmatic and elegant, letting utility and efficiency be his principal guides. The number of changes he can make is limited, as the modification, addition or deletion of any base in a gene has associated costs. Most importantly, however, the new protein must be able to store and deliver much more oxygen than the crocodile haemoglobin that he has in front of him.

Unfortunately, however, a seemingly expedient and economical short term solution might in the long run prove uneconomical. If Crotopus were able to 'soup up' the function of wild type crocodile haemoglobin, he might be able to squeeze out just enough function to enable the king to live in his aquatic idyll. If this could be done for a modest sum, than so much the better. In a cynical rendition of a politician tinkering with the constraints inherited from a predecessor and searching avariciously for short-term advantage, Crotopus might in this way win the immediate favour of the king. But in so doing, he might inadvertently compromise the ease with which the king could engineer himself in the future. For what if the king did develop an interest in deep-sea fishing? It is possible that no amount of tinkering could get the three hours of performance out of haemoglobin which would be necessary for the king to pursue his deep-sea fishing hobby.

So although tinkering might be expedient in the short term, the tinkerability of haemoglobin might itself be finite. Once it had reached its limits, Crotopus or one of his successors might be forced to go back and redesign an oxygen-transporting protein from scratch. This would be both expensive and extremely tiresome. Especially if the task were to fall onto his shoulders. If his design was seen to lead to a dead end, it would also damage his reputation. Crotopus did not want to be remembered by history as the designer who did not have the insight to realize that haemoglobin was an old-fashioned and inept piece of machinery. No, he was Crotopus. The visionary, the innovator, the designer who set fashions rather than following them. A big designer with big designs. And Crotopian designs were built to last.

The design details of 'Crotoglobin' came to Crotopus only two days later, appropriately enough in the form of a dream. In his dream he saw a snake biting its tail, and when he awoke, he realized that each Crotoglobin sub-unit was going to be C shaped, with the oxygens fitting neatly into its centre. Yes I did say oxygens, the plural. For each Crotoglobin sub-unit was able to accommodate not just one oxygen molecule like myoglobin or haemoglobin, but two. As Crotoglobin was composed of seven independent sub-units instead of four, it had the capacity to deliver fourteen molecules of oxygen. One for each of the letters in the king's name. 'A big protein for a big designer,' Crotopus chuckled to himself as he threw the plans across his desk. 'And capable of delivering enough oxygen for at least four hours of underwater diving!' And the master stroke, the defining feature of his genius? Instead of switching allosterically between a mere two alternative 'T' and 'R' architectures like haemoglobin, Crotoglobin incorporated a third allosteric switch to the 'Crotopus' or 'C' state. On switching from the 'T' to the 'C' conformation, Crotoglobin would be able to deliver all fourteen oxygen molecules to needy tissues.

The journey from his rooms to the palace was a memorable one. Even the birds seemed to sing out with joy at the beauty of his new creation. Indeed the whole world appeared to be crying out in boundless adulation. 'Crotopus, the greatest designer the world has ever known!' Crotopus watched as the countryside disappeared past the carriage window. He sank back in his velvet chair and closed his eyes. He scoffed with disdain at the tinkerers of the world, the short-sighted designers who stuck to protocol. The fiddlers and adjusters that stole their solutions from history. How easy it would have been to purloin the myoglobin solution from seals or to construct a haemoglobin/myoglobin hybrid. But to assemble new form in one bold step, using nothing more than the brilliance and profundity of abstract thought itself; *that* could only come from a deep and detailed understanding of the way in which nature is constructed. A communion with the very nuts and bolts of existence. This type of insight touched only very few and passed over so many. To think that he Crotopus, the humble son of a fisherman, had been blessed with such abilities!

The coach stopped, the palace gates swung open and Crotopus dismounted. Dressed in green velvet with a hat of mauve and red, he swaggered into the throne room and grinned a crocodile grin. His head swam with images of promotion, lavish gifts and adulation. How the king would adore him. Enter Sir Crotopus! How the crowds would cheer and raise their glasses with clamorous praise. To Crotopus, the greatest architect that the world has ever known! He stopped and bowed a crocodile bow, his hooves thumping heavily on the marbled floor.

'Your Majesty,' he said in his most eloquent voice, his teeth crunching shut and his small yellow eyes darting around the room as he spoke. 'Your most *gracious* Majesty.' The king paced up and down by the throne and seemed preoccupied. Crotopus raised his voice. 'I come with some good news. In fact with some *very* good news'. The king looked up impatiently, muttered to himself and then scribbled on a piece of paper. 'Sir, Your Highness,' continued Crotopus in his most subservient growl, 'may I present to you for the first time in crocodilian history, Your Majesty, Crotoglobin, named you will notice your Royal Highness sir, after the inventor. That of course, Your

Highness, being none other than I, your very own humble, obedient and faithful servant, Crotopus. It is, of course, a completely new design. And if you could spare just a few moments, Your Majesty, I could explain how it works.' But the king by this time was fast asleep and, as it turned out, did not live long enough to ever hear the full design details of Crotoglobin.

The news of the king's poor health arrived early the next day. He was, it was rumoured, pale, breathless and suffering from terrible headaches. Two days later he was seeing flashing lights and complaining of weakness in his tail and by the third day he was dead. It was the queen Psamathe who first suggested that the king's untimely demise and the replacement of his haemoglobin gene with that of Crotoglobin might in some way be linked. This was later confirmed by the court physicians, who diagnosed sudden and severe hypertension as the factor precipitating his death. Tests later revealed that unlike haemoglobin, Crotoglobin was unable to bind nitric oxide, a chemical which is crucially important in the regulation of blood pressure. In focusing on and optimizing only the oxygen-transporting properties of haemoglobin, Crotopus had forgotten that haemoglobin subserves more than one function. Poor Pristichampsus. He was not destined to become a crocofish after all.

Things moved swiftly from then, and Queen Psamathe condemned Crotopus to death. The trial of Crotopus by the Grand Crocodile Court was a sorry affair. The judge Linus looked solemn as he said, 'Will the accused please stand.' Shouts and cries of 'Justice for the king!' and 'Crotopus has killed the king!' filled the courtroom. In his summing up, the judge said only that 'nature was evidently a better architect than Crotopus'. He added that Crotopus had 'broken the golden braid of history' and 'tampered with the imprint' of crocodilian existence. Throngs lined the streets to watch as the wretched Crotopus, his hooves thumping even more clumsily than usual, was led away in chains. He was left on top of a mountain, and by the morning had been torn to pieces by wolves. In trying to modify the king, he had inadvertently engineered his own destruction.

4

A Visit to the Information Sea

In the beginning there was mathematical possibility. At the very inception of the universe fifteen billion years ago, a deep infinite-dimensional sea emerged from nothingness. Its colourless waters, green and turquoise blue, glistened in the non-existent light of a non-existent sun and were cooled by the non-existent moonlight of a non-existent moon. And in its deepest depths, order crystallized from chaos. A strange sea though, this *Information Sea*. Strange because it was devoid of location, was watery, but at the same time unable to make one wet. A sea that one could spend a lifetime searching for, but which could never be found. A sea that was everywhere simultaneously and yet at the same time nowhere. A sea stripped naked, a sea devoid of form. A strange sea indeed.

Search to the end of the world, investigate the deepest caves, the thickest forest, the furthest lands, the distant planets. But elusive like the tiger, padding softly and stealthily across the jungle floor, it defies capture. And yet how odd that in a sea so unassuming, a sea so obscure that it might not even in principle be found, lies the potential for everything. For within this sea lies not just you and me, not just this day or yesterday, but everyone and everything and all the days that have been, might, or ever will be. Battles and armies, victors and vanquished, the sweetest tune and the bitterest tear. Creatures so strange they can hardly be imagined and events so bizarre they defy comprehension.

The *Information Sea* is a hypothetical sea. An abstract sea. And yet its mathematical landscapes are as real as any of the vistas we might chance upon in the Swiss Alps, on the canals of Venice, or in the streets of Paris. They are as real as Piccadilly Circus or Trafalgar

Square and the mathematical fangs of its fiercest creatures as deadly as the venom of the most poisonous snake. In a later chapter, I will demonstrate the physical reality and tangibility of these abstract mathematical landscapes by systematically exploring some regions of their infinite expanses. This will not, however, be achieved using natural, historical processes of exploration, but instead by utilising artificial, ahistorical processes of landscape construction and discovery. But the actual existence of the *Information Sea*, real or imagined as it may be, is not the current focus of our attention. It would indeed be a shame to allow any disagreements over the exact status of its existence, to distract us from matters of more importance. We will consequently for the greater part of what follows, confine ourselves to examining the value of the concept of an *Information Sea*.

However before proceeding further, I should explain what I mean by the term *Information Sea*. In an earlier chapter we discussed a hypothetical space, which we called *Gene Space*, *DNA Sequence Space* or just *DNA Space*. The value of the *DNA Space* concept is that it allows us to sketch out the complete set of genetic specifications for every possible type of living creature whose essential information is encoded in DNA. It was emphasized at the time that the DNA micro-instruction specification for the construction of a potential organism does not constitute a full description of that organism. In fact, although an organism assembly program written in DNAese encodes the essential informational foundations of an organism, a considerable portion of the information of complex organisms is furnished at extragenetic levels. These include development, learning, and in some cases, culture. If we wanted to formulate a complete description of a complex organism, we would need to specify the information stored at each appropriate level, both genetic and extragenetic. Although the language in which each component of the total information of an organism is encoded is different, the raw information itself constitutes a sort of *lingua franca* in which it can be expressed. The complete informational form of an organism may thus be precisely described, defined and accommodated within a multidimensional information space. The information contained within *DNA Space* may itself be imagined as slotting neatly into this space, but occupying only a vanishingly small fraction of its infinite expanses.

An information space of this sort would furnish a complete description of all potentially living and unrealizable creatures to an infinite degree of precision. It would contain, amongst many other things, anomalous creatures with plausible architectural designs but nonsensical and incompatible behaviours.

But what if we wanted to increase the resolution of our description still further, taking it in fact to its very limits? What, for example, if we required our description of genes and their non-coding regulatory elements to include not just the DNA sequences that encode their information, but details of the structures of the atoms from which each base is constructed as well? A truly complete description of an organism would contain information so detailed that, if fed into a hypothetical atom-assembly machine, a perfect facsimile of the organism could be assembled. In the case of humans, the description might amongst others, include the contents of their stream of consciousness and the cultural information they had acquired during their lifetime. This might itself include a knowledge of the latest football scores, the poetry of Keats, recent developments in television soap operas, a knowledge of popular tunes, how to operate a mobile telephone, a taste for beer, Indian take-aways and an appreciation of brightly coloured, convertible motor cars. It might also include the information contained within the books on their bookshelves, or the entire collection of records, compact discs, tapes and books in their local library which they are able to access by virtue of the fact that they possess a library card.

But for the description to be truly complete, it would also need to include an account of things such as the clothes they were wearing, their hairstyles and exhaustive details of their surroundings. The limit that one sets to the resolution of a description is arbitrary. One might argue that a complete description of an organism should include an infinitely detailed description of the ecosystem in which it is embedded. Alternatively, one might argue that a genetic specification is sufficient. We might even decide that we can do without a description of a particular organism and settle for a generalized account of the whole species instead. The resolution of a description must consequently be tailored to the specific purpose for which it is required.

But how are we to catalogue all of this information? How can one store a person's knowledge of the rules of tennis alongside the information of their genomic sequence, their love of seafood pasta and the fact that at the exact moment that they are being described they are reading by candlelight and listening to Elgar's First Symphony? The answer, as before, is to store each element of the overall description in the universal language of substrate-independent information. Once the private languages of DNA, tennis rules, a person's love of seafood pasta and the information of the candlelight and music have been translated into the mathematical language of information, these diverse types of description can be stored within the same space. Just as all possible DNA sequences are housed within *DNA Space*, so all possible bits of information are housed within an information space that we have called the *Information Sea* and which accommodates every element of an infinitely detailed description of the state of the world at any moment in the past, present or future. An infinite repertoire of descriptions of this sort would contain the information necessary to recreate any moment in history. It would also furnish the information needed to generate every alternative version of history and indeed every alternative version of the present and future. Any specific fragment of actual or possible history is described by a unique 'fingerprint' of mathematical descriptions.

The *Information Sea* is thus the space of all mathematical spaces, a hypothetical information space which contains the complete collection of all the infinite libraries of description that document every possible state of the universe to the highest degree of resolution. The mathematical descriptions within these libraries define all possibility. We should note, once again, that an acknowledgement that such infinite repertoires of mathematical descriptions might exist in no way implies that it is possible for them to be detailed. The fact that they cannot be fully detailed does not, nevertheless, negate their existence. Similarly, although the mathematical libraries of description within the *Information Sea* harbour information that might, in principle, be used to generate exact recreations of historical events, in practice the successive mathematical descriptions that define processes of historical change leave no trace and cannot be inferred in any deterministic way from an initial description. Thus, although one might imagine generating a plausible representation of a particular moment in history, it would be impossible to determine whether this had any correspondence with what had actually happened.

Irrespective of the particular position that one adopts with regard to the objective existence of the *Information Sea*, it seems fairly uncontroversial to suggest that there are at least some senses in which it might be useful to regard the phenomena of the material world *as if* they reflected an indifferent exploration of the infinite collection of mathematical possibilities whose objective, but in some cases unknowable, descriptions are housed within the *Information Sea*. The *Information Sea* may be taken as being something akin to a gene pool. However instead of a sparkling turquoise-blue Californian swimming pool filled with genes, we have an infinite ocean replete with raw mathematical information. There is a sense, then, in which any organism and its actual or alternative life histories can be formally described by a set of dynamic ripples through the mathematical substance of the *Information Sea*.

The *Information Sea* is not necessarily defined and underpinned by a mathematical structure. Indeed a complete description of the material world may require elements that lie beyond mathematical description. The universe may not ultimately have a mathematical texture, in which case an *Information Sea* with a mathematical composition would be an imperfect sea. Mathematics itself may constitute no more than a model or logical picture of an underlying objective reality. There might be languages of representation that transcend mathematics and capture facts that fall through holes in the fabric of mathematics. However, in the absence of any candidate language of description that might replace mathematics, we will assume that the *Information Sea* does have a mathematical structure and that the mathematical edifices within it represent not merely a model of possibility, but the very flesh and bones of possibility itself. The *Information Sea* might thus equally well be known as *Mathematical Space*.

So how might this repository of all possible mathematical descriptions be structured? The *Information Sea* may be imagined as being composed of two interlocking parts; an infinite *Sea of Facts* and a *Logical Space* in which the facts are arrayed. A fact may be viewed as an atomic piece of information about the world. For example, 'the sky is blue', 'the cat is black', or 'this atom is in this position'. The relationship between atomic facts is detailed in logical space. By embedding the *Sea of Facts* firmly within *Logical Space*, all possible sets of logical relationships between all possible facts are precisely specified. Any particular state of any possible world is represented by a particular pattern of facts within *Logical Space*. The dynamic pattern of facts which describes me sitting at my desk at any particular instant today will be different from that which describes me sitting at the same desk tomorrow, yesterday, or on any other day. If we were able to peer down at the *Information Sea* from a great height and to colour each fact with one colour or another, the patterns generated on each occasion would be readily distinguishable. Life, its history and all of its alternative and as yet unrealized possibilities may thus be represented by an infinite library of 'fact movies', each of which may be imagined as being accommodated within the vast expanses of the *Information Sea*.

In this model of the *Information Sea*, facts are taken to constitute the basic atomic units of description. As these facts are atomic, they may not be split into any smaller sub-facts. Each fact thus describes one of the essential pieces of information needed to specify a detail of the material world. Any fact may contribute to an ensemble of facts which specifies the construction of an object that could in principle be represented in a physical world of some sort or other. This might be the world with which we are familiar, but it might equally well be one with which we are not. A world, for example, more like Neptune or Venus, or in which the laws of physics with which we are familiar no longer apply.

The representation of a fact or set of facts in a material world of some sort may be thought of as generating a 'picture', or symbol of that fact or set of facts. A symbol, much like the underlying fact or facts that it represents, embodies information embedded in logic. A set of facts arrayed within *Logical Space* describes both the factual form of the target structure and the logical relationships between each fact, or their 'logical form'. What use after all would a pile of bones be, if we were unable to assemble them into a logical whole? The pieces of information needed to specify a set of disjointed bones are facts, whereas the instructions necessary for their articulation embody their logic. The combination of both facts and logic are necessary for the assembly of a symbolic structure in the material world. Everything that has, can, or ever will exist, in this or in any other world, may be fully described by the complete collection of relevant facts and the corresponding set of logical interconnections.

The *Information Sea* remains unchanged as the result of furnishing any particular piece of information in any particular way. We have stated that it is a mathematical space which contains all possible information and all of the possible logical inter-relationships that may exist between atomic facts. A collection of logically interconnected facts may be referred to as an 'information cluster'. Logically interconnected clusters of information clusters may, similarly, be referred to as 'information superclusters', or molecular facts. A molecular fact may be broken down into its constituent atomic facts. Because no fact may exist independently, there is a sense in which every atomic fact within the *Information Sea* has some type of relationship with every other fact. Each piece of information is thus simultaneously a member of several distinct but overlapping molecular facts. But despite the high potential for inter-connectivity between all possible facts, the strength of any particular connection may vary greatly.

Let us now throw ourselves into a fictional *Lego Toyland*. The *Lego Toyland* universe consists of a Lego kit containing 1000 bricks of assorted colours and a building base on which any Lego model can be constructed. In this 'toy universe', the position of each Lego brick is an atomic fact. The relationship of each brick to its 999 companions defines the region of *Logical Space* in which it is embedded. When a model has been constructed using some or all of the bricks in the kit, those which form part of the model constitute an information cluster; that is to say a set of logically interconnected atomic facts. If we were to construct two separate models on the same board, the information cluster represented by both would form part of a molecular fact.

The extent of the space of this toy universe is defined vertically by the maximum possible height achieved by stacking all 1000 of the bricks on top of one another. The horizontal extent of the space is defined by the maximal possible horizontal distance achieved by arranging all 1000 bricks end to end in a line. There are a huge but nevertheless finite number of ways in which the 1000 bricks may be arranged. Each alternative configuration could be precisely described by numbering each brick from 1 to 1000 and then recording its spatial coordinates within the *Lego Toyland* universe. The set of all of these sub-collections of mathematical coordinates constitutes a complete library of descriptions that document every spatial possibility within *Lego Toyland*. If supplied with a list of all these descriptions, we could set about trying to build every possible Lego model described by these abstract mathematical specifications. In so doing we would discover two things. First, that the majority of descriptions did not specify the construction of anything that even vaguely resembled a building or any other recognizable object and second, that a large number of the descriptions could not be constructed, even in principle, as the structures that they specified would fall apart under the influence of gravity. Rather than testing every single description, we might attempt to extract invariant features from the descriptions encoding impossible or irrelevant structures. These could then be used to screen the entire library of Lego model construction specifications, in order to purge it of all the descriptions which do not specify a plausible object. The probability of building a viable model would then be greatly increased, and irrelevant and implausible descriptions could be thrown into the waste-paper basket.

The *Information Sea* may be imagined as being a sort of *Lego Toyland* writ large. Clearly, however, the resolution of the description of *Lego Toyland* just discussed was low. The facts which described the ,position of each brick were essential in helping us define the precise state of *Lego Toyland* at any moment. But these facts were not strictly speaking atomic facts, as the position of a given Lego brick may be refined into a more detailed description which documents the position of every molecule within the brick. A still more detailed description might record the position of every atom within each brick, and a yet higher resolution description might document the position of every sub-atomic particle within each brick. A complete description must, by definition, be written in the language of such a set of atomic facts. It is by no means certain, however, that it is possible, even in principle, to formulate an objective description of such a state of affairs.

This uncertainty arises from the puzzling fact that the laws of physics that apply to large-scale objects like cricket balls, Lego bricks or tennis rackets do not appear to apply to objects the size of sub-atomic particles. When the Lego bricks in *Lego Toyland* represent large-scale objects of the macro-world, we are able to assign an exact position to every block, and to measure that position. But when the Lego bricks stand as ambassadors for the sub-atomic particles of the micro-world, a very strange thing happens. We are no longer able to say that a given brick is in this position, or that position, and must instead accept the bewildering fact that each brick simultaneously occupies every position within the *Lego Toyland* universe. The description of the position of a given brick is now defined by the superposition of all its spatial positions. These are weighted so that the block is over-represented in some regions and under-represented in others. We thus have to come to terms with the unfamiliar and somewhat absurd notion of Lego bricks whose matter does not localize at discreet positions, but is instead distributed across the whole *Lego Toyland* universe.

This strange weighted superposition of the brick's alternative spatial locations is described by what in quantum mechanics is known as its 'wavefunction', or quantum state. Curiously, the quantum state appears to be momentarily destroyed if the system is measured or observed. If we do not take the lid off our *Lego Toyland* universe and try to make a measurement of the position of the bricks, the quantum state of the system remains intact. However the moment we open the lid and observe the system in order to measure its quantum state, the wavefunction 'crashes' and the quantum state is destroyed. Once this has happened, the bricks no longer behave like micro-objects and behave instead like macro-objects. The matter of each brick condenses down in space and realizes only one of the multiple alternative positions from which the quantum description is comprised. The moment the measurement or observation ceases, each brick begins to spread out, reverting once again to its objective but immeasurable quantum state.

The *Information Sea* is thus a quantum-mechanical sea, composed from infinite repertoires of entangled quantum descriptions. Although defying description, they appear nevertheless to be completely objective and thus characterizable by the measurements that we might in principle perform. It is possible that the ultimate description of reality is not written in the language of quantum wavefunctions. Indeed, quantum wavefunctions might themselves constitute incomplete pictures of a still more fundamental level of description. But until such a time as an alternative or modified quantum theory is proposed and substantiated, we will have to accept the disconcerting but nevertheless experimentally verifiable picture of reality that quantum theory presents.

There is a sense in which the first step towards life on Earth may be traced back to the inception of representation, the origin of the first symbol. That is to say that the first steps towards life occurred when the first atomic and molecular facts located within the *Information Sea*, were realized in the physical universe. Real space, as opposed to abstract space. We can date this event to the origin of the universe and all of the matter in it. It was at that moment, with the creation of real space, real matter and time, that the possibility for representation came into being. One might be tempted to speculate whether the origin of the *Information Sea* preceded the origin of the universe. The potential for the creation of the universe might itself be located within the infinite mathematical libraries of the *Information Sea*. But such a notion appears unreasonable and it seems more appropriate to imagine space, matter, time and the *Information Sea* as having come into existence simultaneously. However by even attempting to address this type of question, one runs the risk of giving the impression that the *Information Sea* is anything more than an expedient concept.

So why bother with the notion of an *Information Sea* when this infinite information space must by definition remain abstract, intangible and unknowable? One of its values lies in the fact that the structures and processes which define life may be imagined to behave *as if* there was an *Information Sea*. Any potentially living creature and the events in which it has, does or might participate can thus be thought of as being fully described by alternative sets of interconnected mathematical patterns arrayed within the *Information Sea*. The *Information Sea* is thus the ultimate refuge of description and provides the logical foundations upon which all symbolic representation is based. It is the place where the logical and factual form of all possible existence may be systematically stripped down and reduced to naked mathematical description. From a living organism to a hot summer's day, the position of a sub-atomic particle, or the structure of a kettle, the complete but inaccessible quantum-mechanical descriptions of all may be found within its fathomless depths. The *Information Sea* is thus a crucible in which the information of any structural element of the material world may be melted down. Description furnished and articulated in the form of pure mathematical information is thus the universal language that defines all existence. Indeed all existence is defined by flickering patterns of alternative descriptions within the infinite dimensions of the *Information Sea*.

The collection of all the local patterns across time and multidimensional space generates a higher order pattern, which constitutes the 'information signature' of whatever is being described. The more bits of information that are needed to describe an aspect of the material world, the greater the number of ramifications into different pools or sets of alternative molecular facts and the more complex the higher order pattern is likely to be. If one were able to fly high above the *Information Sea* and peer down simultaneously at all of its multiple dimensions, each thing, collection of things, event, or collection of events would be recognizable by the unique trace that it makes in this non-observable informational shadowland of form. However, because we are forbidden even to open, let alone fly high inside the spaceless box which contains the *Information Sea*, the finer details of these traces will remain forever invisible.

The *Information Sea* presents an interesting paradox. On the one hand it houses the infinitely extensive collection of raw mathematical descriptions that define all possibility. On the other, it appears so small that it occupies no space at all. So how is it that something without size is able to accommodate the potential for everything? If we were to use contemporary methods of binary information storage, we would need vast amounts of memory to store even the minutest fraction of the information needed to fully describe a complex quantum system such as a cat, St Paul's Cathedral, a football match and the systems with which they are entangled. And this assumes that it is even possible to measure a quantum state without destroying it. The apparent paradox is, however, a straw man, as there is no paradox. None of the information in the *Information Sea* needs to be stored, as none of it actually exists. All that does exist is the mathematical potential for this information. Stored as mere potential, the entirety of every possible description of every possible thing may be telescoped onto a point smaller than a pin head. A point so small that it has no size and doesn't in fact exist at all. How wonderful and ironic that the largest and most complex thing in the universe occupies no space whatsoever and that the place in which one searches for the information of all things, cannot itself be found!

The reason why the *Information Sea* cannot be found and does not exist is because it is located outside time. It is only as a consequence of accessing information by the process of representation that a small part of the *Information Sea* is realized inside time and indirectly becomes visible indirectly. Only a symbol for a piece of information may exist within time. But the destruction of the symbol is not synonymous with the destruction of the potential for that information. This remains unchanged irrespective of the creation or destruction of a strand of time which may occur as a result of the generation or destruction of a symbol of that information.

This has interesting consequences for the concept of time travel, for if the *Information Sea* is a truly complete sea which contains the information necessary to describe the quantum states that define all possibility, then all history and all possible alternative histories must be located somewhere within it. Lurking deep within the *Information Sea*, for example, lie the wavefunctions that fully describe the Battle of Hastings in 1066, exactly as it was. There are also multiple alternative functions which describe the infinite set of battles that were not realized. In one of these William may have been killed by an arrow which pierced his eye, and Harold may have ridden off as the triumphant victor. In another, instead of arriving and setting fire to his ships, William the 'almost but didn't quite make it' Conqueror might have changed his mind about invading England and sailed back to Normandy. Housed along with these alternative versions of the Battle of Hastings, one might also find the full set of histories of the world, many corresponding to alternative outcomes of the key events. What games history might have played!

There is a sense then, in which time travel can be reduced to a search problem; the challenge being to retrieve correct information structures from the infinite library of all possible facts. But in order to travel within time, one would first have to travel outside time by traversing the timeless mathematical landscapes of the *Information Sea*. Once a given set of candidate quantum mechanical descriptions had been selected, it might then be possible to generate symbols of those states without having to observe the symbols themselves. It would not, however, be possible to verify whether the set of quantum specifications that had been pulled from the quantum hat of the *Information Sea* were historically authentic versions of the events that they purported to represent.

By authentic versions, I mean historically correct. There has, one presumes, only been one Battle of Hastings. A complete quantum description of this event would contain all the atomic facts necessary for its exact recreation. These would include descriptions of the firing patterns of the neurons in every soldier's brain, descriptions of the molecular activity of the sarcomeres in every soldier's muscles and descriptions of the position and nature of every blade of grass on the battlefield and of every grain of sand on the beach. It would also include the movement of every leaf on every tree, the position and type of every cloud and so on. Two versions of the same battle might differ solely in the fact that one particular leaf on one particular tree moved a fraction more or less at a given moment. However inconsequential this difference might appear to be, only a single description can be historically correct.

Given a machine which could systematically search a region of the *Information Sea* for plausible versions of the Battle of Hastings, and another able to translate this information into symbolic representations, we could, in principle, travel back to the battle by running the machine's output into a piece of empty space. By recreating the information of a historical event, we may attain the same result as if we had actually travelled back in time. We might alternatively choose to supply the appropriate sensory information directly to our brains, so as to enable us to 'experience' the Battle of Hastings, without actually having to be there. It is most unlikely that we will ever be able to retrieve the actual Battle of Hastings, even though its complete quantum description must reside somewhere within the *Information Sea*. We might, nevertheless, be able to find a version of the Battle of Hastings which, despite historical flaws, appeared sufficiently close to the actual battle to be of interest. An imperfect form of time travel then, limited by the power of the search algorithm that we employ to differentiate between the alternative candidate information structures, and our intrinsic inability to objectively evaluate the historical plausibility of the candidates.

The realization of a fact is synonymous with the inception of its representation. This occurs when a symbol in the real world acts as an ambassador for its underlying mathematical description. Every symbol or 'picture' in the real world maps onto a corresponding description in the *Information Sea*. But although the picture in real space and the information cluster that it represents share logical form, we should not mistake one for the other as there may be information clusters for which no symbolic representation is possible. The constraints on the realization of these facts arise as a consequence of both the nature of the historical biological languages in which their information is articulated and as a result of the natural historical process by which the *Information Sea* has been explored. We may imagine that the collection of these forbidden or unrealizable information clusters is housed in a subregion of the *Information Sea*, called the *Secret Sea*.

The facts within the *Information Sea* are neutral with respect to both the nature of the symbols that may represent them and the context in which they might be represented. For example although the *Information Sea* holds a description of an apple falling from a tree and hitting Newton on the head, it contains an equally good description of an apple falling off the tree and shooting upwards into space. The logic of information structures does not take gravity as a premise. The *Information Sea* is able to furnish descriptions of a world both with and without gravity equally well. It also holds descriptions of potential organisms that would not be plausible on Earth. Strange creatures that might flourish in outer space where gravitational forces are negligible, but which become absurd when considered in the context of gravity. Contextual circumstances thus play an important role in determining the plausibility of potential representations, which are to some extent defined by their logical relationship with the surrounding information clusters in which they are embedded. The

Information Sea itself has no direct experience of the material world. Its logic lies beyond existence, as its structure owes nothing to its protracted fifteen-billion-year dialogue with the material world.

One might at this point wonder how the information in the *Information Sea* is distributed. Are the facts embedded in *Logical Space* randomly arrayed, or ordered? Is it possible to break the *Information Sea* down into regions of thematic order, and are these hypothetical regions themselves organized so that their subregions are correlated? Is each region discreet, or are the borders between each locality continuous? One might also ask whether it is possible to draw a limit to the extent of the *Information Sea*. That is to say, once we have entered its mathematical expanses, is it possible, even in principle, to implement an exhaustive search of its contents?

As the *Information Sea* is a repository of potential information and exists outside the world of space and symbols, the presence or absence of order in its organization is inconsequential. The distance between each pair of facts is the same, irrespective of how different they are. It is helpful, though, to imagine that it is organized so that rather than being randomly distributed throughout the sea, similar kinds of facts embedded in similar types of logical structures are grouped together. The *Information Sea* may consequently be imagined as being subdivided into regions in which similar types of information are housed. Rather than this being just a coarse grouping, the information is finely sorted, so that there is a smooth progression from one information cluster to the next. As the *Information Sea* contains the mathematical descriptions of all possible things, anything in the material world may be taken as just one example of a mathematical repertoire of related items. Each of these may be found in an appropriate region of the *Information Sea*.

A repertoire of related things may be just a small subset of a higher order set of things. The description of a Manx cat might, for example, be found in *Manx Cat Space*. But this space is itself just a small region of a far larger space called *Cat Space*. *Cat Space* however is but a small region of *Cat Family Space*, which is in turn a trifling region of the *Space of All Mammalian Species*. This occupies just a tiny region of the *Space of All Living Things*, which is but a small region of the *Space of All Possible Living Things*. One is reminded of Russian dolls, each of which fits neatly into the other. But unlike a Russian doll which belongs to only a single hierarchical set of larger and smaller dolls, each item in the *Information Sea* may be simultaneously a member of an infinite number of interconnected spaces.

The notion of a mathematical space which is partitioned into regions of relative homogeneity that contain collections of related things, raises the question of how one might define a particular type of thing. Perhaps there are formal logical criteria which might be used to unambiguously assign an information cluster to one region of the *Information Sea* rather than another. If this is the case, then what happens at the edges of a region? Is the transition gradual and continuous, or characterized by a discreet jump? Are there grey areas between different types of mathematical structures, where descriptions of potential things are trapped in a logical no-man's-land?

To explore these questions, we will now visit the region of the *Information Sea* which houses the set of information clusters that have the potential to be realized in the form of apples. It is aptly named *Apple Space* and is itself but a corner of the far larger *Fruit Space* hypermarket. Rather than directly exploring the extensive mathematical orchards of formless information contained in this region, we will instead borrow apple symbols from the real world. These will be used to represent each of the information clusters stored within *Apple Space*. As *Apple Space* holds the complete set of all possible alternative descriptions of apples, not all of the symbols that we borrow will necessarily look, smell and taste like the apples with which we are familiar. We should remain aware that the inclusion criteria and thus the limits that we impose upon *Apple Space* are to some extent arbitrarily determined.

On trips to the local department store with my grandmother when I was a child, it was impossible to return home without a chocolate animal, or a red apple. The chocolate animal would usually be devoured within seconds of being purchased, although occasionally I kept it in my pocket and ate it on the top deck of the number thirteen bus as we travelled home. The sting in the tail was the apple that usually followed. It was not so much that I didn't like red apples, but rather that chocolate animals tasted so much better. I never counted how many such trips we went on, nor how many apples my grandmother gave me. But let us now imagine that I enter *Apple Space* in search of those apples. Not similar apples, but the very same apples, still warm from the touch of my grandmother's hand. I know that those apples are there, because they *must* be there. And I resolve to find them and eat them for a second time. To feast upon them, gorging myself until I can eat no more.

Fortunately *Apple Space* appears to be partitioned into multiple subspaces. Each of these is well signposted and several of the signs indicate subregions of possible interest. There is, for example, the *Space of Every Red Apple Ever Eaten On a Number Thirteen Bus*. This sub-space must certainly contain at least some of the apples I am looking for. Nevertheless I soon stumble upon a much larger sub-space, which is the *Space of Every Red Apple Ever Eaten by a Young Boy Within Twenty-Four Hours of Having Eaten a Chocolate Animal*. I consider myself exceptionally lucky, however, when I discover an even more excellent hunting ground. This is the *Space of Every Red Apple Ever Sold in John Barnes*. Since my grandmother never shopped anywhere else, all of my apples must lie somewhere within this sub-space. By good chance, when I enter this sub-space further signposts direct me to several sub-sub-spaces of this sub-space of *Apple Space*. One of these is the sub-subregion which contains the *Space of Every Red Apple Ever Purchased at John Barnes by a Grandmother for her Grandson*. I enter this region and eventually find a signpost which directs me to a sub-sub-sub-space which is the *Space of All Red Apples Ever Purchased at John Barnes by a Grandmother called Rachel for her Grandson Adrian*. This region contains several hundred or so apples and I conclude that these are the very apples that were given to me by my grandmother when I was a child. How satisfying to know that they are still there! How curious, indeed, to think that I may eat them for a second time. But unfortunately, as was the case with our search for the historical Battle of Hastings, there is in reality no obvious way in which we could evaluate the historical validity of any candidate apples that we find. Thus, although these apples could in principle be found, they are in practice likely to be irretrievable.

These sub-sub-spaces of *Apple Space* are the higher dimensions of *Apple Space*. Each contains a set of clusters in which the same information is embedded in a host of different logical and informational contexts. The number of dimensions is limited only by the number of ways in which it is possible to partition a mathematical space. This sentence, for example, is fully described in the *Space of All Possible Versions of Potential Books Whose Titles Contain the Word 'Life'*. An identical description of this sentence may also be found in the *Space of All Possible Sentences Beginning with the Word 'An'*, or in the *Space of All Possible Sentences Ever Written in Cambridge*, or in the *Space of All Possible Sentences Ever Tapped into a Word Processor Late on a Tuesday Night*. In this sense then, the *Information Sea* is highly degenerate. If the information in one dimension were destroyed, it could be retrieved from a logically related dimension. When you stand in one region of the *Information Sea*, you thus stand simultaneously in every other dimension to which it is logically connected. Any given information cluster is thus a wormhole that connects to an infinitude of logically interconnected spaces and sub-spaces. As each fact is distributed across multiple dimensions, in order for the mathematical potential of a symbol to be fully erased, the relevant information clusters would have to be deleted from a potentially infinite number of dimensions.

We will now imagine that *Apple Space* lacks any sort of organization. The consequences of this can be illustrated by setting ourselves the task of entering *Apple Space* to search for the reddest of all possible apples. The only constraints on our search are that we must complete it within a period of twenty-four hours and may pick only one apple. As we enter *Apple Space* we are surrounded by a sea of apples which stretches to the horizon in every direction. Red apples, green apples, yellow apples, brown apples, ripe apples, rotten apples, spotted apples, bruised apples, big apples, small apples, cooking apples, irregular apples, and mottled apples. They are all there. Some taste sour, some are sweet, some have pips, whereas others don't. Some are as large as the World Trade Centre and others as small as a ladybird. Occasionally we spot a red apple in a sea of green and are led to believe that we have located a seam of redness, only to be disappointed when we find that it leads to another sea of yellow. Once we pick up an apple we are committed to bringing it back, but if we choose not to, it is unlikely that we will encounter it again. And if we are unfortunate enough to slip into an inappropriate dimension of *Apple Space* for example the *Space of All Possible Green Apples*, we will not encounter a single red apple, let alone the reddest of all possible apples. We might even find that we have fallen into some obscure region of *Fruit Space* and are surrounded by a sea of strangely coloured and unfamiliar fruits that defy description.

After much searching, we finally select a marvellously red apple. How fortunate that we didn't pick up any of those we saw earlier. Yet like someone stuck in the maze at Hampton Court, what we don't know is that just a few rows of apples separates us from an even redder apple. The reddest apple of all, however, is located many orchards away in a distant outpost of *Apple Space*. But we didn't have even the remotest chance of finding it. Actually in a completely random *Apple Space*, we would be lucky to find an apple that contained even the faintest hint of red. In this chaotic *Apple Space*, we are forced to make compromises, opting for local solutions rather than global optima. Given the low probability of finding any red apples at all, we would be foolish to ignore an apple that even remotely resembled what we were searching for. Another problem with uncorrelated landscapes is that even if, by some remarkable chance, we did stumble upon what we thought was the global optimum, this could never be verified. Any solution to a problem would always be overshadowed by uncertainty. Might there be a better solution lying elsewhere within the vast sea of possibility?

But in correlated spaces, where similar types of things are located together, searching for a specific type of item is much easier. Once you hit upon the correct region, you will be led to similar types of the same thing. In a correlated *Apple Space*, all green apples would be clustered together in *Green Apple Space* and all yellow apples in *Yellow Apple Space*, as would every other type of apple in its own appropriate sub-space. The more highly correlated a given landscape, the greater the number of dimensions it will have. In *Red Apple Space*, the task of locating the reddest of all possible apples would be much facilitated. The most difficult aspect of the task would be to locate the first tendril of redness. Once this had been done, locating the reddest apples would involve following the seam of redness until you arrived at the reddest apple. The higher dimensions of this space would however correlate the collection of apples even further, into say *Light Red Apple Space* and *Dark Red Apple Space*. We will see later that the correlated nature of some sub-regions of mathematical space enables them to function as suitable substrates for natural evolutionary processes of exploration.

Now that we have given some consideration to the organization and content of the *Information Sea*, let us consider its extent. Instead of asking whether it would be possible to walk to the end of the *Information Sea*, we will first examine the simpler question of whether it would be possible to walk to the end of a given region. The *Space of Every Red Apple Ever Sold in John Barnes* is one such region and contains information embedded not only in *Logical Space*, but also in time. Time functions both to constrain and prescribe the nature of the possibility allowed within this space, giving it a clearly defined beginning and end. The only limits are set by the precision with which it is defined. One might, for example, wonder whether it contains apples that were initially purchased from John Barnes, but were returned because they were bruised or past their sell-by date? Might it also include apples that were stolen rather than purchased, or those surreptitiously eaten by hungry shop assistants? We could also return to the region which holds all of the possible descriptions of alternative Battles of Hastings and wonder how it might be organized. Did the battle begin when the first Norman soldier set foot on England, or at the moment the invasion was conceived? And did it end the moment the arrow hit Harold in the eye, or when William the Conqueror rode triumphantly into London? Any attempt to impose an artificial structure on this event must be arbitrary. And what about *Apple Space*? Does it have an edge, or do the apples just keep on getting larger and larger, smaller and smaller, or infinitely variable?

If we were able to reach the end of the *Information Sea*, would we be able to satisfy ourselves that it was finite? In the fourth century BC Plato's friend Archytas produced a compelling argument for the infinity of space, which directly challenged the Aristotelian notion of finitude. In a thought experiment, he imagined that he had reached the edge of the universe. On arriving there, he wondered whether it might be possible to poke his stick out beyond the edge and concluded that it would be absurd if he were unable to do this. However if he were able to poke his stick out, then what lay outside must be either a body or a place. Thus Archytas would go on in the same way forever, reaching a new limit each time and asking the same question. The best challenge to Archytas in antiquity came from Alexander of Aphrodisias, who stated that once he had reached the edge of the universe, Archytas would be prevented from poking out his stick not by some obstacle bordering the universe, but simply because there was nothing outside. How, after all, is it possible to poke a stick into nothing and how can a thing come to exist in something which does not itself exist? A contemporary challenge to Archytas is that space could be finite without having an edge. Despite these and other objections to the possibility that spaces might be infinite, it is interesting to reflect briefly on the consequences this would have for our search for the reddest of all possible apples within *Apple Space*. The notion that it is possible, even in principle, to find the reddest of all possible apples implies that the search space is finite. But if the space were infinite, it would not be possible even in principle to look for the reddest of all possible apples, as it would not be possible to perform an exhaustive search.

Let us now, like Archytas, imagine ourselves at the edge of the *Information Sea*. As it occupies no space, no boundaries can limit its extent. For the sake of the present argument, we will crack the *Information Sea* into its two component sub-spaces, the *Sea of Facts* and *Logical Space*. Since the *Information Sea* consists of one embedded within the other, there is a sense in which *Logical Space* could limit the extent of the *Sea of Facts* if, for example, there were facts for which no logical relationships existed. If we were to adopt the Aristotelian point of view that space is finite, we could argue that the matter contained within the universe is also finite. It would follow then that the set of all possible mathematical descriptions of all possible things in the universe must also be finite. This argument is of course nonsensical because as we mentioned before, the *Information Sea* knows nothing of the real world. It merely contains the potential for realizing symbols. It is hard to imagine that the *Information Sea* might contain descriptions of things that cannot be realized in the known universe, but this may nevertheless be the case.

What if the *Information Sea* were to have edges beyond which no stick can be poked? This would imply that there are mathematical constraints that limit the extent of possibility and that existence has a well-prescribed boundary past which there is true nothingness.

Beyond these mathematical cliffs, lies a sheer drop into a hypothetical void where there is no potential for information or logic and in which no symbolization can occur, for a symbol by definition must stand as a representation of something, though one might still argue that anything which lies beyond the edges of mathematical space may stand as a symbol for the void itself. If, however, there are no limits to the *Information Sea*, then we can never fall off its infinite mathematical planes and possibility itself is limitless.

It should by now be clear that for every type of thing that has the potential to exist, one can imagine at least one type of appropriate space in which all of the possible examples of that thing might be housed. All symbols may be imagined as behaving as if they were exploring a prescribed region of a mathematical sub-space. It should also be apparent that the complete library of all possible examples of a given thing vastly exceeds the time and space available to realize them. The Earth originated about 4.6 billion years ago and by around 3.9 billion years ago had cooled sufficiently for the first rocks to form. It is a sobering thought that the stage of the entire Earth coupled with the 3.9 billion years which have elapsed since the formation of the first rocks would allow for the realisation of only a fraction of the infinite sea of descriptions housed within the region of the *Information Sea* that describes all of the possible alternative versions of the Battle of Hastings. Having used up all of the available time and space with different renditions of the Battle of Hastings, there would be little room for anything else. How tedious history might have been!

We have so far only hinted that there might be some logical structures within *Apple Space* that cannot be represented in the material world. We called these forbidden regions of the *Information Sea* the *Secret Sea*, because their unrealizable contents are hidden from expression in the real world and must consequently be passed over in silence. If such regions do exist, then we might expect some of the potential apples in *Apple Space* to disappear when we touch them. These forbidden fruits cannot be eaten as they are unrealizable in the real world, and the information clusters that specify their construction lack symbols. It is as if entire regions of *Apple Space* were fenced off from us, holding apples which we can sense the presence of and which are tantalisingly close, but which must remain forever out of reach.

A world beyond representation and history, defined by information and logic but devoid of form, the *Information Sea* is a peaceful world. A world where all possibility nestles together in an interconnected unity. A world where every thing is part of every other thing. It is only when elements of the *Information Sea* are realized by history, that alternative representations are forced to compete in the tortured dance of symbolism and became locked in a struggle for perpetuation, that the game of life begins. It is only when they have made the transition from the shadowlands of the *Information Sea* to the real world that the information in genes behaves *as if* it wishes to perpetuate itself, asserting itself in the real world of wet logic, as opposed to abstract logic.

History only exists because the number of things that are possible greatly exceeds the resources of time and space available for their representation. Herein lies the tension that defines the essence of the process of life. Life is one huge resource allocation problem. But how is this tension resolved in the real world? Clearly a search and selection principle is needed. So life becomes synonymous with the process by which the mathematical space of all possible things is searched. This applies equally well to events as to individual things. Life is thus a contest for the perpetuation of representations in real time and space.

The advantage of this information-based model of life is that it makes it possible to fashion a complete description of an organism that incorporates all of the non-equivalent dimensions across which its information is distributed. When these are all translated into the universal language of information, they can be mapped onto the same space. An interesting consequence of viewing life as an exploration of a mathematical space, is that nothing can be invented or designed, as the potential for everything already pre-exists. Life is thus defined by the process of symbol creation and the exploration of potential symbols within the *Information Sea*, the place from which we all came, and the place to which we must all return.

I would now like to focus on the *Information Zoo* region of the *Information Sea* which contains information clusters able to generate living things. As we enter the *Information Zoo* it turns out to be just another type of hypermarket, but instead of stocking gene kits, its shelves are packed with information kits. It might thus equally well be known as the *Information Kit Hypermarket*. Unlike gene kits which are flat in the dimensions of developmental information, learning and culture, information kits contain all the information needed to provide a complete description of an organism. In contrast to the kits stored in *Gene Space*, if one were to take an information kit from the shelves of the *Information Kit Hypermarket* and assemble the organism it describes, it would be an exact facsimile of the multidimensional creature that the picture on the box represents.

The *Information Zoo* is partitioned into an infinite number of sub-spaces, each of which contains information kits of a similar type. One sub-space might, for example, contain an infinite sea of seemingly identical Mozarts. If, however, one were to select forty different Mozart kits randomly from disparate regions of *Mozart Space* and then construct them, we would unearth some interesting surprises. Illiterate Mozarts and Mozarts that spoke no German, Mozarts inept at musical composition and any number of other unfamiliar Mozarts, each of whom have little in common with the historical Mozart, despite the fact that they are morphologically indistinguishable. But in order to find Mozarts which differed in such obvious ways, we would first have to wade through huge seas of apparently identical Mozarts which differ from one another in more subtle ways and are consequently for all intents and purposes indistinguishable. These would include Mozarts which had identical external structures but differed in their internal structures; an extreme example being a Mozart with a horse-shoe kidney. Other Mozarts might have identical internal and external morphologies, but differ by a single memory or irrelevant piece of information more or less. Although identical at the level of their genetic specifications, when information at other levels is taken into account, the population of Mozarts is clearly heterogeneous.

If we were now to explore some of the other sub-regions of *Mozart Space*, we should not be surprised to find further seas of morphologically identical Mozarts. In one of these we might find a sea of infant Mozarts, a sea of fifteen-year-old Mozarts, twenty-year-old Mozarts, or Mozarts enduring their last few moments of existence. We might also be surprised when we stumble upon sub-spaces which contain grotesquely deformed Mozarts, or Mozarts which look more like

Marilyn Monroe or Billy the Kid than the Mozart with which we are familiar. But why should these kits have been included in *Mozart Space*? The answer is that when cataloguing the information stored within a given kit, the *Mozart Space* librarians do not give bias to any particular type of information. It might turn out that purely in terms of lines of mathematical specifications, Mozart's genetic information is trivial compared with his extragenetic information. It is consequently necessary to clarify our *Mozart Space* inclusion criteria and to decide exactly what makes a Mozart, *Mozart*. We might also decide that it is not reasonable to weight each piece of information equally. A Mozart with a mutation in a region of junk DNA would clearly be quite different to one with a mutation that resulted in diabetes. A Mozart that had memorized the entire contents of the Encyclopaedia Britannica would, similarly, be different to one that had not, or was unversed in the rudiments of musical theory.

From the imaginary perspective of the *Information Zoo*, living creatures are symbols which stand as ambassadors for their underlying mathematical edifices. Given that the information of life can be expressed more explicitly in terms of substrate-free information, one might begin to wonder whether it may be possible to discern a 'life pattern' in the informational structures which specify living things: a set of mathematical specifications that are an essential prerequisite for any process of life and constitute a mathematical signature of an underlying potential life process. One may also wonder whether it might be possible to formulate a new system of biological classification based upon 'species' of information clusters. A taxonomy based upon information structures would certainly be constructed on sounder logical foundations than the Linnaean system, which is based exclusively on morphology. If it were indeed possible to discern a life pattern hidden within abstract informational structures, then it should be possible to peruse the *Information Sea* and identify those kits that belong to the *Information Zoo* and those which do not. If the pictures on the information kits were systematically erased, we should, simply by examining the broad nature of the mathematical description in the box, be able to distinguish a kit which encodes a potential angle-poise lamp from one encoding a potential rhinoceros, porpoise, or ant. We should also be able to distinguish the mathematical specifications that describe the types of rhinoceros with which we are familiar from those describing impossible rhinoceros-like creatures, which would not result in the formation of a viable organism.

One might imagine a 'potential-life detection machine' which is able to generate and define the entire contents of the *Information Zoo* by systematically examining all of the information clusters arrayed within the *Information Sea* and determining which of these could generate life. We might choose to wheel our potential-life detection machine down the aisles of the *Information Kit Hypermarket* and to feed the information contained within each information kit box into the machine. Having examined the information, the machine would tell us whether the kit in question should be placed within the *Information Zoo*. That is to say, whether the information in the kit describes a process for the construction of a living thing. Of course it is one thing to generate a living thing and quite another to generate a viable living thing.

If we imagine that the *Information Kit Hypermarket* is organized so that each kit is located at a unique point on a three-dimensional grid, then each kit may be assigned a unique address according to the information it contains. We might decide to employ a cartographer to draw a map of the *Information Kit Hypermarket* and to colour-code it, so that all the kits which describe a potentially living thing are coloured black, while those which do not are coloured white. A map of this sort would enable us to explore all the possibilities that life has to offer, without having to construct huge numbers of irrelevant kits. We are also likely to find that the black kits cluster predominantly to one region of the *Information Kit Hypermarket*, though the sea of white kits might occasionally be punctuated by a fine seam of black. We might also expect the kits to be organized according to their complexity. One measure of an organism's complexity is the amount of information needed to describe it. If we had two Julius Caesars identical in every respect other than in their knowledge of the texts of Socrates, then the one that had memorized the works of Socrates would require several more lines of mathematical specifications to complete his description than the one who had not. The relatively ignorant Julius Caesar is a more compressed version of the more educated version. Lines of information may in fact be carved off each level of Caesar's complete description in order to achieve a maximally compressed version of this historical figure.

There is a sense in which the creatures in the *Information Zoo* have an existence which is independent of their realization. The information cluster which describes a potential dodo, for example, has a timeless mathematical existence irrespective of the fact that there are no living dodos remaining today. Potential dodos thus existed in a very real sense, long before the first dodo was discovered by history and realized in flesh and blood. Furthermore, if it were possible to retrieve a dodo information kit from the *Information Zoo*, we could in principle reconstruct an exact historical dodo. Similarly, although leopards continue for the time being to roam the Earth, when one day they become extinct, as it appears they inevitably will, their information kits will persist unchanged in the archives of the *Information Zoo*. There are likewise huge numbers of creatures whose kits have never and most likely never will be opened and 'blooded'. There simply isn't enough time, space and matter in the universe to construct them all. But despite the fact that they have never been realized by history, these creatures are in a sense no less 'real' than a dodo.

Before we finally leave the *Information Kit Hypermarket* and the *Information Zoo*, there are two final points we should address. The first relates to the question of whether it is meaningful to discuss the informational structure of a potential organism without referring to the environment in which it will be constructed. The second relates to the question of whether it is possible, even in principle, to turn all of the grid spaces within the *Information Kit Hypermarket* a definitive black or white. Might it be the case that some of the spaces on the map must be coloured grey? That is to say that our machine is in some cases unable to decide whether to colour a given box black or white and thus colours it grey, in order to indicate this intrinsic undecidability.

A leopard would, for example, instantly vaporize if it were constructed on the surface of the Sun, as would all the other creatures that have existed on Earth. It would similarly freeze if constructed in Antarctica. It might, on the other hand, be constructed on the pavement in Oxford Street, or on the runway of a major international airport, in which case it may be shot, in the Sahara Desert in which case it would dehydrate, or a thousand metres under the Pacific Ocean, in which case it would drown. It might be constructed in a world with no other leopards, in which case it would find no mate. It may even be constructed in a universe with different physical laws, in which case it might implode, or have its flesh ripped apart into a million pieces. It should be clear then that in attempting to draw limits to the extent of the *Information Zoo*, we make several broad assumptions about the type of world in which the potential creatures might be realized. It should also be clear that the nature of the world in which the kits might be constructed will introduce constraints that will influence the nature of the mathematical creatures which populate the *Information Zoo*.

The question of whether it is possible to design a machine which is able to turn all of the boxes in the *Information Zoo* an irrefutable black or white is not only of theoretical interest, but of considerable practical importance. In the future we will almost certainly enlist the help of machines to help us search for new forms of unrealized potential life. But without a good method for detecting such kits, these mathematical journeys will be tedious and incomplete. What is needed, is a general method for scanning the informational structures of information kits and determining whether they describe a potentially living thing or not. Each kit thus constitutes a hypothesis which our method must either falsify, or prove correct. If correct, then the description encodes a potentially living creature, whereas if false, it does not. Our requirements of this method are that: (1) it should be internally consistent and should never try to colour the same information kit simultaneously black and white, (2) it should be complete and thus be able to turn every kit within the hypermarket either black or white, with no kit escaping categorization and (3) it should be finitely describable, that is concise and compact, and able to make the decision as to whether the kit should be coloured black or white in a finite amount of time. Unfortunately Gödel's incompleteness theorem indicates that no such method exists. If we apply our method of choice to the *Information Kit Hypermarket*, there will always be some kits which will never be turned black or white and must consequently be coloured grey.

It appears that no machine can be programmed to print out the complete list of information kits which describe true creatures that belong inside the *Information Zoo*. We are thus in practice forbidden from delineating its boundaries and cataloguing its contents. To make matters even worse, the work of Alonzo Church indicates that given any particular method for discriminating between information kits, there is no easy way to predict which ones the method will be able to turn black or white, and which will be coloured grey. We find ourselves in a position where the best thing we can do is devise a method, release it within the *Information Kit Hypermarket* and sit back and observe whether the target kit that we are interested in is turned black or white. If the method is unable to decide whether the target kit describes a 'true' or a 'false' creature, we might have to wait indefinitely, never knowing whether the method is just about to make a definitive decision. Indeed, Alan Turing has demonstrated that it is impossible to predict how long a method of this sort will take to complete its task.

It would appear, then, that the only fully exhaustive method for locating the complete set of information kits that are able to encode living things, is to actually construct each of the kits and then sit back and see what happens. If we rely on machines to reveal potential life kits, we are likely to leave boxes which contain seams of glittering jewels untouched and unopened and in so doing run the risk of consigning them to obscurity. We should, however, be aware that although we appear to be precluded even in principle from attaining a complete rationalization of the *Information Kit Hypermarket*, we should in practice be able to design procedures which will greatly facilitate our hunt for curious, exotic and occasionally useful forms of potential life. Information kits which, from the very first moment of the creation of the *Information Kit Hypermarket* some fifteen billion years ago, have remained silent; progressively gathering dust as the machinations of a capricious history touch some boxes, whilst leaving the greater majority undisturbed.

5

The Origins of Life

We live in the world of form, surrounded by a profusion of complexity. Trees, leaves, giraffes, antelopes, lichens, fish and algae. The spirals of a pine cone, the petals of a daisy, the geometry of a cell, the architecture of the brain. All these structures demand an historical explanation. Some of this complexity is created artificially. Buildings, vases, cave paintings, Christmas carols, stone axes, bridges, steamships, telephones and street lamps. All symbols, fixations of logic and facts. How easy to invoke a designer. An omnipotent, omnipresent, omniscient creator of form. How wonderful this form. But how much more wonderful the potential for form. And yet without the possibility for representation, the ghosts of potential existence would never have their day. The *Bismarck* would not have been constructed and sunk, and the pyramids would remain forever unknown; obscure corners of formless information embedded in a dusty and irrelevant logic. For it is symbols which breathe existence into the information clusters of the *Information Sea* and life into the potential creatures of the *Information Zoo*. They are the machines which transform possibility into history and allow mathematical spaces to be navigated and explored. They are both the generators and turbines of actuality.

The search for the origins of symbolism takes us back to the explosion which occurred simultaneously everywhere around fifteen billion years ago, when the space and matter of the universe was created from nothingness and time began. For it is in the origin of space and matter that both symbolism and the potential for symbolism have their origin. Following the creation of space, energy, matter and the infinite archive of mathematical descriptions which we have called the *Information Sea*, the universe began to expand and evolve. Although the first one-hundredth of a second following the 'big bang' remains something of an enigma, observations made with powerful telescopes that enable radio astronomers and astrophysicists to look back into the past, have helped construct an account of the early universe. And it is here, at the very beginning of the universe, that we will begin our quest for the origins of that particular type of complex symbolism which we call life.

It is possible that we will never fully understand the earliest part of the first one hundredth of a second that followed the origin of the universe. Amongst other things, our ability to do so may ultimately be limited by our notions of causality; it does after all seem counter-intuitive that something should have emerged spontaneously from nothing. The models of causality with which we are familiar use time as a dimension in which events are placed in a certain relationship to one another. This leads us to believe that when we observe an effect, it must have been preceded by a cause. If, however, time began with the origin of the universe, no causal events could have preceded its origin, as they would, by definition, have to exist outside time, which is clearly impossible. One of the most significant impediments to our understanding arises from our ignorance of microscopic physics, particularly at the extraordinarily high temperatures and densities which characterized the very early universe. This is because beyond the relatively cool temperature of one hundred billion degrees centigrade (about seven thousand times hotter than the sun) that prevailed at the end of the first one hundredth of a second, a class of elementary particles known as hadrons and anti-hadrons would have been present in large numbers. These particles are greatly affected by the extremely strong short-range interactions which hold the nuclei of atoms together. But there is no generally accepted way to model their behaviour. Modelling the weak electromagnetic and gravitational interactions at extremely high temperatures and densities turns out to be equally problematic.

About one hundredth of a second following the origin of the universe, the primordial cosmic soup was scorched and baked at around one hundred billion degrees centigrade. This unimaginably high temperature prevented the formation of atoms, and the elementary particles from which the cosmic broth was composed continued to rush apart independently of one another. The amount of matter in the universe at that time was not fixed, but in a continuous state of creation and destruction. These changes in the material content of the universe may be imagined as having been mirrored within the *Information Sea*, which underwent strange multidimensional twists, contortions, expansions and contractions as the extent and limits of mathematical possibility were continuously delineated and revised. About three minutes following its origin, the universe cooled to a temperature of around one billion degrees centigrade and the heavier particles that form the nuclei of atoms began to accumulate in significant quantities. These eventually joined to form the atomic nuclei which were the precursors of atoms. It was not, however, until hundreds of thousands of years later that the universe cooled sufficiently to allow the formation of the atoms in the periodic table. These constitute the atomic tool kit from which the macroscopic contents of the universe are constructed. This new atomic soup was the precursor of both stars and galaxies. As the universe continued to expand and cool, it eventually gave rise to a more complex soup which consisted of combinations of atoms joined together to form molecules. It is somewhere within this complex chemical soup that we must search for the precursors of life, the unfossilized chemical antecedents of a class of symbols known as genes.

As the universe continued to expand and cool, nucleosynthesis on the cosmic scale became almost imperceptible. But the production of heavy nuclei from the fusion of lighter precursors did not cease completely and indeed continues at the interior of stars today. As a consequence of both this and the fact that the space of the universe increases as it expands and matter continues to be created and destroyed, the *Information Sea* can be imagined as remaining in a state of continual revision; its mathematical boundaries and the nature of possibility itself changing from moment to moment. If the universe was infinite from its inception, then the *Information Sea* must have been as well. All subsequent cosmic events are thus accommodated by the *Information Sea*, as it is clearly not possible to expand or remodel something which is itself infinite.

The first stars began to form as a result of the condensation of clumps of simple atoms such as hydrogen and helium under the influence of gravity. As the number of stars increased, clusters of stars known as galaxies began to form. One of these, known as the Milky Way, was created just under ten and a half billion years following the origin of the universe. This contained amongst its one hundred billion or so other stars, one which we call Earth. The Milky Way is itself just one of millions of other galaxies, each created as a result of condensation events scattered liberally throughout space. A galaxy consists of an independent system of stars. But the number of stars in each galaxy varies greatly. The panoply of stars visible in the sky at night represent only a fraction of the stars in the Milky Way. And the Milky Way is itself a rather unremarkable galaxy, containing only a fraction of the stars distributed across the countless other galaxies in the universe. It is distinguished from these solely by the fact that it is our current home and the only known repository of life in the universe.

If one were to cycle at the speed of light, which is about one billion kilometres per hour, it would take approximately a hundred thousand years to cross the Milky Way. Proceeding at this rate, it would take around two million years to reach Andromeda, which is the nearest neighbouring galaxy. In universal terms a two-million-light-year jaunt is equivalent to nothing more than a short trip down to the corner shop at the end of the road. If feeling energetic, you might decide to cycle into town, which may be compared to pedalling at the speed of light for one billion years. During this time you would navigate a volume of space which contains about a hundred million different galaxies. If you set off at the speed of light, you might be able to cycle in a straight line indefinitely. If, however, space is finite and curved, you might eventually find that you arrived back where you started.

About 0.7 billion years following its creation, the Earth had cooled sufficiently for the first rocks to form. Its various components such as the core, mantle, crust and eventually the oceans then began to differentiate and the homogeneous parent Earth begat its increasingly heterogeneous daughter. During this time the Earth was likely to have been bombarded by meteorite showers. Unfortunately the geological record of the first five hundred million years of the Earth's history has been erased, leaving us largely ignorant of the events associated with its early maturation. Around 3.85 billion years ago (0.85 billion years following the origin of the Earth), the first organisms emerged. But it was not until about seven million years ago that the prototypes of modern man made their debut. The generation of all past and present biological macro- and micro-complexity has thus occurred in a period which extends from some time before the origin of the first cellular micro-organism, at least 3.6 billion years ago, to the present day. The period of biological creativity during which the patterns of ancient life were generated from their cellular precursors may be known as the Archaean.

The question then, is whether we can construct a plausible account of the chemical events that occurred between the formation of the first rocks 4.6 billion years ago and the origin of cellular life around 3.6 billion years ago. For it is at some time during this apparently cell-free age, which we will call the Hadean, that the pre-cellular 'proto-life' scaffolding for the cellular life of the Archaean was assembled. It is important to emphasize that this is not the same as asking whether it is possible to trace the exact historical pathways that were taken by proto-life. It is indeed unlikely that we will ever be able to infer the actual historical meanderings through the space of chemical possibility. It might nevertheless be possible to establish the broad co-ordinates of these pathways. But this level of historical detail need not concern us at present. What is important is that we are able to construct *any* reasonable account of how acellular life might have originated at all.

The hypothetical acellular proto-life of the Hadean provided the informational foundations on which the 3.6-billion-year history of conventional cellular life was constructed. Without the invisible skeletons of these organisms, which dissolved long ago into the anonymity of ancient history and left no discernible trace, it is unlikely that cellular life would have originated on Earth. Although the chemical details of the flora and fauna of the Hadean may elude us, the processes that shaped them do not. One of the principal organizing principles was the process of evolution by natural selection as formulated by Alfred Russel Wallace and Charles Darwin. The possibilities available to proto-organisms were not however open-ended, as evolution operates within the context of several constraints. These include the laws of physics and chemistry, the available micro-environments, historical contingency, the need to maintain a capacity for evolvability, and as yet poorly defined laws of complexity which are likely to underpin the self-assembly and self-organizational dynamics of complex systems.

When we search for the precursors of cellular life, we are searching for sets of candidate symbols that are able not only to represent information, but to transmit it in a flexible manner across time. For it is only by perpetuating the core information of a structure that a symbol can become fixed in the material world for a sustained duration. But even this is not enough, for in order to function as a substrate for natural selection, the symbol must incorporate a capacity for change. The ability to generate variants on the theme of the information and logic that they represent enables populations of evolving structures to explore alternative versions of themselves. A potentially successful substrate for an evolutionary process must thus be able to generate a faithful replica of itself, whilst simultaneously allowing modifications to be introduced into the informational structures of its offspring. The replication process must thus be accurate, but not too accurate, otherwise the symbols would be inflexible and unable to evolve. But if the fidelity of replication is too low, the information is likely to melt away and the daughter symbols will bear little resemblance to their parents. It seems reasonable to suggest that proto-life on Earth came of age with the formation of the first informational structures that were able to balance these requirements. Mechanisms which regulate the fidelity of information storage and replication are consequently likely to have played an important role in the generation of lineages of biological machines.

Christopher Langton and Norman Packard have used the term 'evolution at the edge of chaos' to convey the notion that adaptive systems must incorporate the correct balance of rigidity and flexibility in order to maximize their ability to generate, process and store information. Indeed, computer simulations suggest that evolving systems are drawn naturally and inexorably to the ordered region at the edge of chaos. By sailing close to the wind of chaos, evolving systems are able to maintain a delicate balance between the ordered inflexibility necessary to preserve the information of the past and the plasticity necessary to respond to the future. But if evolving systems become too flexible, their information is at risk of dissolving away much like a sugar-lump in hot tea.

Each of us could, in principle, construct a family tree which starts with our parents, grandparents, great-grandparents and great-great-grandparents and stretches back to the Stone Age and beyond to the earliest moments of human history. From there the unbroken chain of life continues back even further, initially to the first hominids and then to our vertebrate precursors, the unicellular organisms of the Archaean, and finally to the inception of proto-life in the Hadean. We would need a very large dining hall to hang the portraits of all these ancestors and there would doubtless be some we might choose not to display. But ancestors nevertheless they all are, each one a necessary and essential link in our personal evolutionary histories. If the portraits were arranged chronologically so as to form a genealogical 'rogues gallery', we should not be surprised if our own portraits led to those of single cells, and then to creatures made from RNA, peptides, or crystals of clay.

We all walk tall on the shoulders of an unbroken chain of ancestors and a string of historical events which date back fifteen billion years or so to the beginning of the universe. But if we were to rewind, erase and then replay the formless chemical tape of the Hadean, the history of life on Earth would almost certainly be different. One meteorite or other, or one random event more or less, and life may not have originated at all, or might have generated a bestiary quite different to the one with which we are familiar. We might indeed have found ourselves patiently munching popcorn in the cinema of life for tens of billions of years, only to find that we were still looking at a blank screen and that the main feature *Life* had not yet begun. There would not furthermore be any alternative movies to view, or indeed any prospect of getting our money back. As if we examined the small print on the back of the ticket, we would find that satisfaction had never been guaranteed. And if by chance the movie did start, there would be no way of predicting what course it would take and no certainty that any individual life-line would lead to a system that could eventually be captured by DNA. Even if the generation and perpetuation of self-replicating systems is inevitable, it is not necessarily the case that they would proceed along life-lines able to generate the patterns of biochemical organization that define modern cells. Indeed, many precursors of life would have explored blind alleys that eventually led to their extinction.

It is reasonable to assert that life as we now know it, originated with the first replication event. The fidelity of the earliest replication events was however likely to have been very low. The information of the most ancient symbols would have decayed into meaningless noise almost instantaneously. The degree of fidelity that a replication process must attain in order to qualify as a process of life is to some extent arbitrarily defined. A rock which splits into two has in one sense reproduced itself, but has not in any sense replicated its information. It is neither able to initiate the process, nor to produce an exact copy of itself. Indeed, the informational content of the parent rock is only poorly represented in the structure of its siblings, and successive fissioning of daughter rocks would degrade even this residual information. It might be more accurate to say that life began with the first self-replicating event that was able to generate a copy of the parent structure, which preserved its core informational features. Life is in one sense an autonomous process for replicating informational structures with a defined degree of fidelity. In order to implement a replication process and to ensure that a threshold level of copying fidelity is attained, a copying machine is needed. The key to understanding the origin of life, is thus to define the minimal architecture sufficient to implement a process of this sort. In order to do this, we must trawl the archives of possible chemistry for historically plausible candidates. But we should not begin with any preconceived notions about the nature the complexity of the very first replicating machines.

An interesting consequence of defining life as a process, is that the process itself can be considered independently from the hardware in which its essential logic is represented. The primitive biological machines that were the precursors of cellular life are unlikely to have been made from the DNA and protein building-blocks from which contemporary life is constructed. It was, nevertheless, these simple hypothetical precursors of the Hadean, that used life technologies now thousands of millions of years past their sell-by date, which carried the torch of life for the 1.1 billion or so years that it took to produce living things embodied in the familiar software and hardware technologies of DNA and proteins.

Our search for the first self-replicating entities, those pioneering symbols that had the power to sustain a prolonged trajectory across mathematical space, will take us deep into the imaginary space of all possible chemical reactions, which we have called *Chemistry Space* and of which *Gene Space* occupies a vanishingly small but nevertheless highly significant corner. A space so immense that it swallows *Gene Space* with ease, like a drop of paint dripping back into the tin from which it was taken. *Chemistry Space* itself owes its existence to the possibility for the existence of atoms. For chemistry is only possible as a result of the unique properties of atoms, which are able to associate with one another in a combinatorial manner to form constellations of connected atoms, known as molecules.

It is worth emphasizing that our aim is to search *Chemistry Space* for plausible mechanisms by which self-replicating processes might have originated and handed the torch of life onto DNA, proteins and cellular technology. It is unlikely that we will ever be able to reconstruct the actual historical pathways through *Chemistry Space* that resulted in the generation of the first unicellular organisms. These will almost certainly remain lost forever in the *Space of All Possible Routes Through Chemistry Space*. All that remains of these conduits are disjointed intellectual fossils that include biochemical pathways, the genetic code and the DNA and RNA sequences of contemporary and a limited selection of extinct organisms. But our inability to reconstruct the exact proto-life lineages of the Hadean is largely immaterial. As, if we are able to find evidence for the existence of even a single plausible life-line through *Chemistry Space*, then we have demonstrated the validity of the hypothesis that cellular life evolved from simple chemical precursors which had little or no resemblance to contemporary living things.

At this point, I decide to enter *Chemistry Space* by sliding down a piece of rope. I descend for what seems like ages, until the rope stops, short of the floor. I jump the last few feet, landing firmly on sticky ground. Steam and strange vapours, some purple and others an iridescent green, permeate the dank and humid atmosphere. I half expect to see witches stirring bubbling copper cauldrons. But as I look around, all I can see is boxes. Rows and rows of them, stretching in every direction. Is it possible that I have arrived in yet another kind of hypermarket? I wander aimlessly around the apparently endless boxes, until eventually I notice a sign above my head which reads *Welcome to Molecule Space* and realize that I have entered the first dimension of *Chemistry Space*, which is the *Space of All Possible Combinations of All Possible Atoms*. At last things seem to make more sense, as it is combinations of atoms bonded together to form molecules that form the chemical building-blocks of life. The shelves of *Molecule Space* are packed tightly with molecule construction kits. Each contains the atoms needed to assemble the molecule depicted on the box, as well as details of how these should be connected and a supply of energy sufficient for its construction.

The number of molecule construction kits that might in principle be utilized to build living things is extraordinarily large. This is because there are over a hundred different types of stable atoms in the atomic 'tool kit' of the periodic table and molecules may be composed of huge numbers of atoms. It is in this context interesting to note that the complexity of living things is achieved using only a fraction of the combinatorial power of the periodic table. In fact, the great majority of the molecules used in the construction of living things are assembled from combinations of just six types of atoms: carbon, hydrogen, nitrogen, oxygen, phosphorus and sulphur.

The key player in the historical game of life creation is the promiscuous carbon atom. Carbon forms not only a myriad of chemical associations with a wide variety of other atoms, but also readily bonds with itself. It has a predilection for forming long chains, and may also twist and contort to form ring structures, each of which has unique chemical properties. Carbon is in fact a virtuoso within *Chemistry Space*, having the greatest repertoire of chemical performance. There are, for example, over sixty trillion different molecules which could in principle be constructed from a starting kit consisting of only forty carbon atoms and eighty-two hydrogen atoms. Each of these results in the formation of a unique molecule with a characteristic profile of properties. It is both the composition and sequence of the component atoms that confers a molecule with its unique properties. The set of all possible combinations of carbon and hydrogen atoms alone is staggeringly large. If you happen to venture into the *Carbon Chemistry* subregion of *Chemistry Space*, you would be advised to take a good map.

In order to enter the higher dimensions of *Chemistry Space* which contain all of the possible combinations of interactions between different molecules, one need only open a molecule construction kit and step inside. The higher dimensions accessible from any given box represent the complete set of chemical reactions in which the molecule in question might participate. The number of chemical reactions that have been historically realized by any given class of molecule is, however, likely to represent only a fraction of those which might have been explored. Located somewhere within the higher dimensions of *Chemistry Space* are the chemical skeletons of the proto-life antecedents of modern life. Indeed, somewhere within this vast space lie the secrets which underlie the origins of life on Earth, chemical life-lines that stretch for unimaginable distances in unbroken, ramifying chains.

We are now ready to peruse some of the candidate chemical structures that might have constituted the first link in this unbroken chain, which originated somewhere between the beginning of the Hadean and the inception of the Archaean and extends to the present day. As we have said, there were doubtless numerous false starts. The multiple decimations of partially or fully plausible lineages that are likely to have occurred prior to the establishment of a successful lineage, have however left no easily discernible imprint. Some of these invisible life-lines may have stretched far into the late Hadean before being obliterated, whereas others may have stretched deep into the Archaean before vanishing without trace. In our attempts at assessing the plausibility of any putative life-line through *Chemistry Space*, we will apply two criteria. First, is the life-line plausible? That is to say, does it constitute a reasonable proposition for how proto-life might in principle originate, even if it is not obvious how such a pathway might have been historically realized. And second, is the life-line historically feasible? That is to say, given the chemical conditions that we think prevailed on the early Earth, might such a putative life-line have formed the *actual* route through *Chemistry Space* that led to the contemporary repertoire of living things?

Although modern genes represent their essential information using DNA, this might be nothing more than a quirk of history. If we were to return to the inception of proto-life in the Hadean and let life unfold once again, it is possible that we might never again witness the emergence of living things that utilize DNA technology. Indeed it seems unreasonable to suppose that there are laws which prohibit the expression of life's information in other technologies. Nevertheless, one cannot rule out the possibility that the range of chemical technologies able to support processes of life is very limited. The notion that all the roads of possible life lead inevitably and inexorably to DNA cannot yet be confidently dispelled.

Much in the same way that mankind has passed through ages defined by the technology of the day, might not the information of life also have utilized several different technologies? Iron Age man would, for example, have found it hard to imagine a time when everything was made of stone. Bronze Age man might similarly have puzzled over the primitive technologies of the Iron Age. There is a sense in which modernity has its roots in the Industrial Revolution, which was built on the foundations of several new materials and technologies. The iron, coal, cotton mills and steam engines that were so essential to the nineteenth century, have been marginalised in the post modern age by technologies such as silicon chips and nuclear power. Each precursor technology nevertheless contributes to the scaffolding upon which all subsequent technologies are constructed.

There is no doubt that, like iron, steel, plastic or aluminium, DNA is a remarkable construction material; a fantastic technology not easily outdone when it comes to storing, replicating and modifying the information of life. Indeed its unique properties have played a central role in generating and sustaining the complex pageant of form which has defined the history of life on Earth. But like all technologies, it has its own idiosyncrasies and constraints. So might DNA one day be supplanted by a more powerful and versatile information-storing hardware? Could it be that DNA technology, that seemingly indispensable edifice of modern life, is just a stepping stone to some other design solution? Might the 'DNA Age' turn out to be as finite and ephemeral as the Iron Age?

Perhaps in years to come, organisms that utilize a different and more complex hereditary material will look back at the dark ages of DNA technology with the same curiosity that we experience when we look back at the precursor proto-life technologies that are likely to have underpinned the Hadean. How difficult it will be for them to acknowledge that their biochemical roots date back to a time when DNA formed the essential informational backbone of all living things.

Should a new life technology come to supplant DNA, we should not be surprised if, in the early stages, DNA-based organisms coexist with the post-DNA creatures that they spawn. This would be similar to the way in which record players coexisted with compact disc players, until the last record player was finally carried up to the attic to become yet another technological curiosity of history. Furthermore, if at some hypothetical point in the future when DNA-based life has been extinct for thousands of millions of years, a molecular anthropologist from another planet happened to visit Earth on a collecting expedition and stumble upon a DNA artefact, they might be struck by the simplicity and elegance of its design. Yet however primitive DNA might appear next to its own more advanced information-storing substrate, it would be clear that DNA technology is, in an engineering sense, fairly complex. This results from the fact that DNA information-generating, -storing, -perpetuating and -modifying machinery is made up of several components. Each of these plays an essential role in the processes by which DNA self-replicates and furnishes information for the assembly of proteins. These components include: (1) proteins that initiate and implement DNA replication; (2) 'checkpoint' proteins that recognize DNA replication errors; (3) proteins that repair damaged and misreplicated DNA sequences; (4) a genetic code for translating DNA sequences into amino acid sequences; (5) proteins which copy the DNA sequence of the gene into an mRNA sequence; and (6) machines that are able to assemble amino acid sequences according to the mRNA program fed into it, by systematically applying the rules of the genetic code.

The DNA replication and repair proteins regulate the spontaneous error rate associated with replication, and in so doing prevent core aspects of the information from melting away. They must, however, allow some changes to be fixed into the genetic material so as to generate the raw material for evolutionary change. A high fidelity

DNA replication process should produce identical or near-identical daughter DNA sequences. Non-identical sequences arise only very rarely, with a low but nevertheless finite probability. Modern DNA technology is based on the principle of creating a functional division of labour between the DNA sequence (genotype), which is the information-storing software or 'program' of an organism, and proteins, which are the hardware or physical embodiment (phenotype) of the information encoded and stored within genes. Ancient DNA technology, however, did not necessarily employ this principle.

Checkpoint 'quality control' proteins ensure that each step of the DNA replication process proceeds correctly. If the DNA is unable to produce a 'visa' to demonstrate that it has completed a replication event successfully, it is sent back from the checkpoint until such a time as these discrepancies can be attended to. The information from checkpoints is communicated to 'cell cycle' proteins, which delay progression through the cell cycle until the necessary corrections are made. A large number of separate checkpoints are passed during a single cycle of DNA replication. If a mis-replication event cannot be corrected, a self-destruct 'apoptosis' program is activated within the aberrant cell.

These requirements define a minimal complexity essential for the construction of a genetic system that utilizes DNA technology. Given that all these elements are necessary for such a system to function, one is compelled to ask how they might have been historically assembled. A close examination of this issue appears to lead to a chicken and egg paradox. Without an appropriate genetic code and the proteins necessary for its synthesis, repair and faithful replication, DNA would be unable to preserve its information content for any sustained period of time. There are, however, good reasons for believing that proteins which are sufficiently complex to effect these functions cannot themselves be generated and regulated without the agency of DNA. This indicates that DNA is too complicated a technology to have encoded the essential informational nuts and bolts of the proto-life of the Hadean. We are obliged to look for precursor technologies in which the complexity inherent in the division of labour between genotype and phenotype is not present. Our search for the very first living creatures will thus necessitate exploring chemically and historically plausible precursor technologies that are likely to have been able to store and utilize information simultaneously, without the need to codify and translate it into different chemical languages.

It is possible that life has used several different technologies during its three to four billion year sojourn on Earth. Graham Cairns-Smith has suggested the term 'genetic takeover' for the process by which one life technology might supplant another. How humbled this must make us feel. For not only is man an historically contingent and apparently improbable player in the game of life, but even his last refuge and bastion of dignity, the very DNA and protein building-blocks from which he is constructed, may be dispensable and constitute just one of several alternative life technologies. Indeed DNA may have no special significance at all, other than the fact that history happened to have stumbled upon it and cobbled together a DNA-based life technology to which all subsequent processes of life have tenaciously stuck. It might indeed be possible to store the information of living things in materials other than DNA, whilst at the same time retaining the protein building-block materials into which the information is translated. Genes constructed from DNA might, on the other hand, dispense with proteins and utilize some other building material. However, although the end products might be broadly similar, the choice of construction material will have a profound influence on the nature of the final product. An axe head made from iron is not, for example, the same as one fashioned from bronze. There may well be no DNA or protein mimics able to generate the repertoire of living creatures with which we are familiar and, if this is the case, DNA and proteins may be more fundamental to life than an argument based exclusively on the conservation of information might suggest.

If we imagine that the current world of DNA-based organisms represents a sort of genetic Bronze Age, what might the precursor technologies of the genetic Iron and Stone Ages have looked like? And how, via a series of successive genetic takeover events, might these technologies have generated contemporary DNA-based technology? Although there are many possible explanations for how the modern *DNA World* came into existence, the hypothetical precursor worlds we will explore in our search for the chemical antecedents of DNA-based life are: (1) the *Extra-Terrestrial Genetic Invasion World*; (2) the

RNA World; and (3) the *Geneless World*. This is not to say that there are not other hypothetical worlds to explore such as *Clay World*, which constitute equally plausible candidates. The aim, however, is not to give an exhaustive account of all the possible alternative life technologies, but rather by focusing on just a few examples, to demonstrate the general plausibility of primitive ancestral creatures that did not use DNA to store their essential information.

Most of the candidate technologies are thought to have been directly, or in the case of the *Extra-Terrestrial Genetic Invasion World* hypothesis, indirectly, rooted in a prebiotic chemical soup which formed the crucible for proto-life. This primordial chemical library was generated by the rocks, oceans, deep-sea vents and atmosphere of the neonatal Earth, with the assistance of geochemical and meteorological events such as volcanic activity and lightning storms. Meteorites, composed of both inorganic and organic material, contributed to the diversity of the soup by delivering additional matter from space. The impact of the meteorites may have provided an additional mechanism for the synthesis of organic molecules. Exogenously delivered and impact-generated material may thus have contributed to the chemical complexity of the prebiotic broth and introduced a random extra-terrestrial component to the history of life on Earth. This sterile prebiotic broth is thought to have given rise to proto-organisms by means of a sustained and incremental process of chemical evolution. It has, however, been suggested that life may have originated at the surface of iron sulfide molecules within deep-sea vents and consequently that life did not originate directly from a prebiotic soup at all.

The bombardment of the early Earth by meteorites forms the basis of the first and most controversial explanation of how life may have originated on Earth. According to what may be known as the *Extra-Terrestrial Genetic Invasion World* or 'panspermia' hypothesis proposed by Fred Hoyle and Chandra Wickramasinghe, life did not originate on Earth but arrived ready-made on meteorites. But even if this inter-planetary hitch-hiking hypothesis is one day verified, it will remain intellectually unsatisfying. For if life did originate on another planet or within another galaxy, the problem of life's origins is not solved, but simply deferred. The hypothesis is also uneconomical, as evidence suggests that there was both sufficient time and chemical opportunity for life to evolve *de novo* on Earth. Of course that is not to say that life did not originate simultaneously or at different times in other regions of the universe. It is even possible that life on Earth has a dual ancestry, with one strand originating from an Earth-bound precursor and another from an extra-terrestrial source. But given evidence which suggests that all contemporary living things can be traced back to a common ancestor which lived between 3.2 and 3.85 billion or so years ago, the two hypothetically different strands of life would have to have converged at some point before this date.

Although evidence to support the panspermia hypothesis is at best rudimentary and tangential, there are some points of interest. Meteorites are, for example, known to be rich in organic material, which includes extra-terrestrial amino acids and nucleic acid bases. Furthermore, the discovery of the amino acid glycine floating deep in space at the centre of the Milky Way within a dense cloud of gas known as Sagittarius B2, provides further evidence to suggest that amino acids can be made elsewhere in the universe. In the words of Christian de Duve, such findings demonstrate that 'organic chemistry' far from being unique to Earth, 'is the most banal chemistry in the Universe'. More recently, Geoff Marcy and Paul Butler have studied distant planets which, according to Marcy, 'almost smell like planets in our own Solar System'. These new planets in the Big Dipper and the Virago constellation are so similar to those in our Solar System that their discovery has been described as representing the culmination of 500 years of intellectual history. The findings so excited astronomers, that even the waiters at a coffee shop near the site in San Antonio where the discovery was first announced were reported to have been able to recite details of the planets. Could some type of life have originated independently on these planets, much in the same way as it is thought to have originated on Earth?

The credibility of exobiology (which is the study of extra-terrestrial life), received a tentative boost from the Martian meteorite ALH84001. Having been prised off Mars by the impact of an asteroid or comet, it travelled across space for around fifteen million years and then fell to Earth, where it remained encased in sheets of ice for about thirteen thousand years before being discovered in the Allan Hills region of Antarctica in 1984. It was not however until August 1996 that a group of researchers led by David McKay at NASA's Johnson Space Center in Houston announced that their investigations of this grapefruit-sized meteorite had led them to conclude that tiny bacteria-like creatures may have lived on the early Mars billions of years ago, at a time when conditions on this now freeze-dried planet were more hospitable. But the researchers were careful to emphasize that their evidence did not represent definitive proof that primitive life had existed on Mars, but rather 'pointers in that direction'. They also acknowledged that further independent evidence would be needed to verify their interpretation of the studies of ALH84001 fracture surfaces.

Their evidence consisted of three principal components. First, the rock contained large, complex organic residues called polycyclic aromatic hydrocarbons (PAHs) which are known (amongst other non-biological mechanisms of synthesis) to be produced by the degradation of living things. Second, a characteristic combination of carbonate, iron sulfide and magnetite was found, suggesting underlying biological activity. Third, minute ovoid and elongated 'blobs' were found within the rock which, the researchers claimed, were fossilized examples of the ancient Martian creatures themselves. As they would have been around a hundred times smaller than modern bacteria, the putative fossilized remains of these creatures were referred to as 'nanofossils'. Although each of these observations taken alone may be explained by non-biological processes, McKay and his colleagues argued that when considered together in the context of their spatial associations, the chemical, mineralogical and morphological evidence strongly suggests that primitive nanolife may once have flourished on Mars.

The jury is still out on this interpretation of the results, and critics were not slow to criticise their conclusion that the blobs represent fossilized forms of ancient Martian life. One major problem stems from the fact that the so-called nanofossils lack the types of organized structures usually seen in bacterial microfossils. Furthermore, the putative Martian nanobacteria would have been so small that they are unlikely to have had sufficient space to accommodate the biochemical machinery necessary for their survival. Studies have also suggested that the nanofossils formed at temperatures of up to 700°C, which is several hundred degrees higher than the maximal temperature (120°C) tolerated by the most robust types of terrestrial bacteria. One group of geologists argued that the blobs are inorganic crystal growth steps of non-biological origin. Others have suggested that the structures are artefacts and that the PAHs arise from terrestrial contamination. Even the basic premise that the meteorite is of Martian origin has been contested. But despite this scepticism, no obvious flaws have as yet been found in the work. It is clear that more studies will have to be conducted and extensive international investigations of ALH84001, which include a search for amino acids and a more detailed examination of the putative nanofossils which McKay has called 'little critters', are underway.

The studies of ALH84001 tacitly assumed that the chemical signature of putative Martian life will be similar to that of terrestrial creatures. But it is possible that the biomarkers for life which may have evolved independently on other planets, will be quite different to those that are appropriate for the detection of terrestrial life. Some of the evidence for life in extra-terrestrial rocks may consequently have been overlooked. The implications of finding the imprint of life on Mars, if indeed this interpretation of the evidence is corroborated, are very considerable. If the Martian-life hypothesis is vindicated, it will have a profound influence on both our conception of the unity of life on Earth and of its historical connectivity with life elsewhere in the universe. We may have to re-evaluate our conceptions of how common a phenomenon life actually was in the early Solar System and whether in some cases these early experiments with life persisted until the present day. Amongst other things, studies of putative extra-terrestrial life may furnish information about the range of biochemical structures able to implement life processes. However, given the poor evidence for panspermia, we will adopt the more conventional assumption that the chemically fertile but biologically barren primordial soup of the early Earth was not seeded by an exogenous source of life ensconced upon a meteorite. Instead, terrestrial and any putative extra-terrestrial life will be considered as representing discreet and historically non-mingling lineages.

Given the vastness of the *Information Zoo* and the influence of contingency and adaptation to unique micro-environments, it seems unlikely that any two strands of life originating independently on different planets would have utilized the same genetic material, structural building-blocks, or biochemical logic. The chemical details of any particular primordial soup will furthermore constrain the set of possible pathways which lead into the space of life. The historically available repertoire of life-lines associated with life on Earth is thus likely to represent only a fraction of all possible such routes. If any putative form of extra-terrestrial life was found to use DNA as its genetic material and proteins interconnected into a biochemical logic reminiscent of terrestrial biology as its executive machinery, this might suggest that living material had at some point been exchanged between Earth and the source of that life. This would remain the case even if the composition of the extra-terrestrial and terrestrial primordial soups were nearly identical. But it might conversely imply that there is just one chemically privileged convergent route from the diverse collection of abiological chemical soups that might have been found in our region of the early universe, leading to simple DNA-based forms of life. If this were the case, then the possibilities for rudimentary existence are significantly more constrained than might be imagined.

This hypothesis might be tested by searching for life or evidence of past biological activity in other regions of our galaxy, such as on the Jovian moon Europa or Titan, Saturn's largest moon. The search for habitable worlds is at present biased towards planets and moons that have a supply of liquid water, as all terrestrial life is utterly dependent on this. Although the surface of Europa is frozen, there is evidence of a liquid ocean beneath its surface. But it is unreasonable to assume that all possible forms of life are dependent on liquid water. Carl Sagan and his colleagues have demonstrated how unmanned fly-by spacecraft exploration of other planets might be used to detect the presence of surface life, without making any assumptions about the nature of its chemistry. The presence of gases profoundly shifted from thermodynamic equilibrium, pigments that cannot be explained by natural geochemical processes and modulated radio wave emission can be taken as a signature for an underlying process of life. This type of fly-by strategy would be unsuited though for the detection of extinct life, sparsely populated life, subsurface life, or life which was only weakly coupled to the surface environment.

If other sources of DNA and protein-based life were discovered within our galaxy, putative kinship relationships between terrestrial DNA and protein sequences and those from extra-terrestrial sources, could be examined. If sequence similarities were found, it might be possible to infer historical connectivity between different lineages. This would greatly strengthen the panspermia hypothesis. Alternatively, the patterns generated by life processes might be so heavily constrained that any observed similarities between the DNA and protein sequences might be explained on the basis of analogy rather than homology. For example if African lions were found on Titan, one could not conclude that there was any historical connection between Titan lions and their terrestrial counterparts. It may merely reflect the fact that life processes based upon DNA and protein technologies will eventually, in an inexorable and invariant manner, deliver up creatures which resemble African lions. This would, however, appear extremely unlikely.

As we shall see later, everything that we know about the unfolding of processes of life across vast expanses of time contradicts this notion. Although there are likely to be several factors which constrain the nature of any life process whose information is expressed in the languages of DNA, RNA and proteins, there is still enough freedom to generate an extraordinary degree of variation. The meanderings of history across mathematical space and time does not appear to be programmed, or controlled by the invisible strings of a divine artificer. But neither is it necessarily completely unpredictable. Hypothetical laws of complexity which are as yet only poorly understood, may, at some coarse level, constrain the overall nature and fluidity of biological patterns. The existence of such laws may grate away a small portion of the intrinsic improbability associated with any particular embrace of information and logic. But given the statistical foundations upon which life is based, it is likely that even if every planet in the universe had received the same seeds of life as Earth, you would have to search huge expanses of the universe before you found a naturally occurring creature which had even a passing resemblance to an African lion. Such is the immense improbability of any particular beast being delivered up from the sleeves of the *Information Zoo* by the capricious and indifferent artificers of Darwinian evolution by natural selection and chance.

Before leaving the subject of ALH84001, it is worth mentioning a French meteorite named Orgueil, which fell to Earth in 1864 and was at the time examined by several experts, including the famous microbiologist Louis Pasteur. Experimental investigations conducted almost a hundred years later at Fordham University in New York led to Bartholomew Nagy's announcement that he had identified organic material within the meteorite. He also claimed that it housed microscopic particles which resembled algae and tested positive for DNA. However, after a heated fourteen-year debate, it was concluded that the DNA tests were inadequate and unreliable and that the so-called extra-terrestrial exotica were in fact microscopic particles of terrestrial ragweed pollen contaminants. In a final blow to Nagy's claims, it was decided that the organic matter was more likely to have been created by inorganic processes than by a process of life. By 1975 even Nagy himself was obliged to concede that the particles were unlikely to represent extra-terrestrial microfossils.

More conventional theories about the origin of life on Earth are unified in that they place the first steps towards life firmly within the chemical complexity of the prebiotic soup incorporating both the primordial oceans and chemical films that covered the mineral surfaces of the neonatal Earth. These theories are rooted in the ideas of Alexander Oparin and J. B. S. Haldane, who came independently to the same conclusion in 1924 and 1929 respectively. According to Oparin and Haldane, the primordial soup of organic molecules constituted a construction kit which contained all the chemicals needed to get life going. The soup itself was envisaged to have been formed as the result of an interplay between geochemical processes and energy sources such as lightning and ultraviolet light. The origin of life was, furthermore, thought to have occurred in an atmosphere that was principally Jovian in nature; that is to say one in which unoxidized gases such as methane, ammonia and hydrogen predominated. This was in keeping with the conclusion of the chemist Harold Urey, that the atmosphere of the early Earth was similar to that of Jupiter, which lacks oxygen and is thus reducing.

In 1953 Stanley Miller, who was at the time a young graduate student working in Harold Urey's laboratory, published the results of an experiment which marked the beginning of the experimental study of the origin of life. Miller attempted to mimic the conditions on early Earth by filling a glass flask with a reducing mixture of methane, ammonia, hydrogen and water. He then simulated the action of lightning, with sparks generated by two electrodes. When the tarry material that formed in the flask was analysed, it was found to contain a variety of organic compounds. These included formaldehyde (which is a precursor of the sugar ribose that is a constituent of RNA), cyanide (which is a precursor of one of the nucleic acids) and four of the twenty amino acids from which proteins are constructed. Miller's experiment provided the first demonstration of how the amino acid constituents of proteins might have been produced by inorganic processes. It also showed how the complex molecules which are of fundamental importance to living things, may have been generated as a result of chemical reactions between simpler molecules. Naturally occurring synthesis of this sort may have generated the biochemical foundations for the first processes of life on Earth. By 1987, using a variety of different conditions, Miller and others had managed to produce a total of ten out of the twenty different amino acids which are found in proteins. They also managed to synthesize twelve amino acids that are not constituents of modern proteins.

However, despite the initial promise of these experiments and the ease with which certain biologically important molecules can be produced by inorganic processes, the synthesis under prebiotic conditions of DNA, RNA and the fatty molecules thought to have been essential for the formation of the walls of the first proto-cells has proved difficult, and often impossible. Furthermore, recent suggestions that the atmosphere of the early Earth was neutral and not reducing contradict Urey's hypothesis, and if correct, would undermine the results of experiments conducted under reducing conditions. Unfortunately, there is no good evidence to resolve this issue. But as Leslie Orgel has said, 'it is hard to believe that the ease with which sugars, amino acids, purines and pyrimidines are formed under reducing atmospheric conditions is either a coincidence or a false clue planted by a malicious creator'.

The general failure of attempts to synthesize the essential chemical components necessary for life under prebiotic conditions appears to support the notion that DNA, RNA and proteins represent too complex a set of technologies to have been present from life's inception. Although chemical evolution within primordial sub-soups in the depths of oceans, at the surfaces of rocks and perhaps in deep-sea vents was essential for the generation of life, the earliest organisms are unlikely to have utilized DNA, RNA or proteins in their contemporary manner. Indeed the assumption that life sprang from the prebiotic broth wildly brandishing its new and complex DNA, RNA and protein technologies like a confused child playing with a sophisticated toy, seems about as likely as an archaeologist finding a BMW car amongst a collection of hominid artefacts. It would appear then that at least some of the assumptions that underpin prebiotic experiments are invalid. A more biologically plausible approach to the question of the origin of life appears to be required. Given the inherent complexity of DNA and DNA-encoded protein technology, might the first organisms have used simpler chemical technologies to establish the informational, metabolic and energetic principles on which all subsequent life was based?

6

Hunting for Intellectual Fossils

When confronted with the multitude of complex biological shapes and patterns that imprint themselves upon the space of the Earth, it is something of a curiosity to imagine what the very first organisms might have looked like. The signature of modern life is written in the language of form, so it is not unreasonable to conjecture that the earliest organisms also possessed this apparently fundamental characteristic. How strange then to realize that it might be possible to trace our family trees back to shapeless ancestors which even the most gifted painter would have difficulty capturing on canvas. But it is to the efforts of these formless and long forgotten cousins that our glasses should forever be raised. Indeed these impassive veterans of the 'battle for first life' that have been cast into a historical wilderness and whose chairs at the banquet of life have remained empty for so long, deserve the seats of honour. Their presence at the feast is indeed sorely missed. For had it not been for the efforts of these bold and intrepid pioneers, exploring the frontiers of biological possibility and trekking across fearful and unknown mathematical landscapes of terrible impossibility, the surface of the Earth would still be sterile.

We should, however, be a little cautious before we humble our pre-cellular ancestors. Creatures that are foreign and distant, but nevertheless part of the family. Our blood of blood, to whom we owe so much. But what exactly do we mean when we say that our most ancient predecessors may have lacked form? Do bacteria lack form merely because we are unable to see them with our naked eye? The magnification of a bacterium with a microscope reveals the very opposite to be true, for although lacking the complexity of an animal cell, a bacterium is highly sophisticated. If we were to build one as large as an elephant, we would have no problems acknowledging its structural complexity. And if an elephant-sized bacterium were to cross our path, we would doubtless pay it the appropriate respect and think twice before crossing its way.

Earlier we discussed the remarkable complexity of a single haemoglobin molecule, which is so small that you could stuff vast numbers of them into a space far smaller than that occupied by even the tiniest bacterium and still have plenty of legroom. If we were to build a giant haemoglobin molecule, its structure would be impressive enough to warrant its inclusion in any museum or gallery of contemporary art. Indeed, a scale model of a whale myoglobin molecule housed within a perspex case in the Cambridge University Medical School Library is sufficiently aesthetically compelling to have remained on prominent display for years.

It is clear then that when we talk about our formless, pre-cellular chemical ancestors as if they were poor cousins, we are doing them an injustice. Furthermore, it takes but a little consideration to realize that our ancestors were far wealthier than we might have imagined. For what they lacked was not form, but rather the types of rigid cellular structures that are visible to the human eye. Our ancestors might in fact have been acellular, 'liquid organisms'; unclothed and non-compartmentalized creatures that may have been constructed from technologies far simpler than DNA, RNA and proteins, or which utilized these components in a novel and rudimentary manner. Because they may have lacked the rigid, well-defined, membrane-bounded structures which we associate with modern unicellular organisms and are unlikely to have utilized DNA, RNA and proteins in the contemporary manner, one might be tempted to call these pre-cellular creatures 'quasi-organisms'. But in using such belittling terms, we may not be doing our ancestors sufficient justice. I am reminded of a Thomas Hardy poem pondering a sleepy fly rubbing its legs, in which he remarks: '*God's humblest, they! I muse. Yet why? They know Earth-secrets that know not I*'. Instead of regarding these hypothetical acellular organisms as impoverished precursors of modern life, might we instead then marvel at what life might have been like devoid of skin, blood, teeth, bones, cells and membranes?

Simple and purposeless perhaps, but shapeless and formless most definitely not.

In using the term liquid organism, I am implying that the very first organisms may have lacked any type of compartmentalization at all. That is to say, they lacked not only the type of complex membranes and cell walls that characterize modern cellular life, but any other type of membrane which may have partitioned them off from the surrounding chemical world. This hypothetical phase of acellular or liquid life is likely to have been extremely short-lived and confined to a restricted set of privileged micro-environments; perhaps small indentations within the surfaces of rocks or in minute rock pools able to concentrate the chemical skeletons of these liquid creatures. Although lacking cell walls and membranes, these pre-cellular organisms would have been defined by the formal logical structure of their chemical circuits.

Before considering some of the ways in which these apparently formless, acellular chemical creatures might have been constructed, it is worth addressing some general issues raised by using phrases such as 'bold and intrepid pioneers' and 'battle for first life'. It is appropriate to compare the first organisms to the pioneers of the American Wild West, who trekked in their horse-drawn wagons across vast expanses of unknown and often hostile terrain. The phrase should not be taken literally, however, for two reasons. First, the phrase 'bold and intrepid pioneers' appears to attribute agency to the earliest organisms, whereas clearly they had none; it was to take thousands of millions of years for even the most rudimentary type of awareness to evolve on Earth. Second, although the terrain of the American Mid-West was unknown to the pioneers, it was already known to the native Americans. This implies that the *plains* on which the first living things evolved were not entirely devoid of life, as they must indeed have been.

It is important then that we mentally replace any suggestion of purpose, free-will or agency with 'as if' statements. The first organisms may thus be imagined to have behaved *as if* they were involved in a fictitious primeval conflagration which we may call the battle for first life. One might similarly imagine that they explored the uncharted terrains of chemical possibility *as if* they were bold and intrepid pioneers. The fact of the matter, however, is that they were neither intrepid nor bold in any literal sense, but rather almost certainly blissfully unaware. These non-compartmentalized creatures had no understanding of the important chapters of prehistory in which they were to be active and essential participants. Indeed they lacked even the most basic awareness. The principal causal agents in the game of organism creation were purposeless chance, coupled with Darwinian evolution by natural selection, historical contingency, the laws of physics and chemistry, and laws of complexity which may influence the dynamic behaviour of complex structures.

If it were possible to manufacture exact facsimiles of our non-compartmentalized ancestors and the partitioned progeny that they eventually spawned and place them in front of us, who would believe that they harboured such formidable potential? It would be almost inconceivable that amongst the multiplicity of seeds of possibility inherent in their informational structures lay the potential to generate not only modern man, but also all of his artefacts including: stone flints, paintings, language, poetry, philosophy, citadels, mathematics, music, amphitheatres, suspension bridges, telephones, helicopters and spacecraft. One cannot help wondering what other unbroken lineages these early precursors might have produced had things been slightly different. There is only one thing of which we can be certain; the historical pattern of form which has issued from our precursors represents only a fraction of the set of all the possible patterns and pageants that might have issued from such unlikely beginnings. An acknowledgement of this humbles us in much the same way that our hypothetical formless pre-cellular ancestors are themselves humbled in the presence of their compartmentalized descendants.

If it were possible to run time backwards, to erase the imprint of thousands of millions of years of history and then let the clock run its course once again in the way that Simon Conway Morris has suggested, it is most improbable that anything even remotely resembling mankind would ever walk the surface of the Earth again. Our ancestors would be far more likely to proceed along one of the huge number of alternative trajectories that historical circumstances *have* prevented them from exploring. How different life might have been. The realization of any particular pathway from the *Space of All*

Possible Historical Events is thus vanishingly unlikely. Furthermore every possible potential lineage of interconnected historical events has as much intrinsic validity as any other. There are apparently no privileged pathways through the *Space of All Possible Historical Events*. No route has more imprimatur than another. This is not, however, to say that every position within this potential space is equiprobable with respect to its realization. But there does not appear to be any pre-determined pattern or program to history, although broader patterns might be discernible. These types of 'deep' influences might constrain the nature and extent of the possible alternatives in some very general sense. Such considerations lead us to conclude that nature may in one sense be defined as that with which we are familiar. For given the huge extent of the *Space of All Possible Living Things*, how can we ever be intimately familiar with anything other than the smallest sample of possibility?

In the next chapter we will examine some of the candidate technologies from which the ancestral creatures that occupied the earliest rungs of our family trees might have been constructed. For the moment, however, we will focus on two issues that will complete the intellectual foundations necessary to help us search for our most ancient ancestors. The first of these is the hypothetical process of genetic takeover, by which aspects of the logical form or informational structure of living things might in principle be transferred from one type of technology to another. For it would be pointless to speculate that DNA technology was preceded by a simpler precursor technology if there were no mechanism for making this possible. We will now examine the general nature of the type of mechanism that allows the design logic of a biological or non-biological machine to be conservatively or semi-conservatively transformed from one type of material representation into another. We will also discuss why it is unlikely that the essential structural elements of the earliest forms of life were built from a technology as sophisticated as DNA.

Despite the fact that it is being realized in a different technology, in each hypothetical genetic takeover event the informational structure of the design is broadly conserved. Of course, the differences between alternative technologies may in some cases be so great that the substitution of one type of hardware for another necessarily embodies a change in the logical structure of the machine. Imagine a raincoat made from metal sheets. If you could support its weight it would doubtless keep you dry, and in this sense incorporates one of the key aspects of the logic underlying the concept of a raincoat. But one would be hard pushed to say that a metal raincoat would be particularly light, cheap, warm, or comfortable to wear. It might even eventually begin to rust.

In order to understand the process of genetic takeover in a more general sense, let us consider the evolution of a non-biological machine. One might argue that the precursor of the modern motor car was not the Model T Ford, rickshaw, palanquin, chariot, horse and cart, or any other such contraption. It was instead the horse, mule, donkey, ox, elephant, camel or indeed any of the other quadrupeds that have at some time been incorporated into a transportation device. The invention of the harness and saddle was, in this sense, the first technological embellishment of the principle of using the locomotory capacity of an animal for transportation. The transition, however, from sitting directly on the back of one of these creatures to sitting in a chariot drawn by them represents a conceptual change of a more fundamental nature which we will call paradigmatic. A paradigmatic transition can be defined as one which results in a change to the essential logic of the process in question. Once you have invented a chariot, the transition to similar, though qualitatively different contraptions such as a cart, coach or wagon, amounts to a change in the design, but not to a change in the logic of the design. In a non-paradigmatic transition the essential logic of the design is thus conserved. Build something with wheels, attach it to a powerful and docile animal, feed it, water it, crack the whip, and there you have it: a locomotory machine. The transition from a horse-drawn coach to a motor car, however, constitutes change of a far more fundamental nature and represents another example of a paradigmatic transition. For, unlike all previous locomotory machines, motor cars dispensed with biological motors and replaced them with non-biological devices.

It is interesting to note that the essential blueprint for modern motor cars was inherent in Ford's prototype. Despite being constructed from different materials and equipped with new gadgets and engines that in some cases consume different fuels, most modern motor cars do not deviate from the basic design established by Ford in his prototype. The logic is intact. Four wheels, two or four doors, an engine, steering wheel, boot, fuel tank, seats, windscreen wipers, mirrors, horn, driver and passenger seats. Ford's prototype will doubtless at some point suffer the same fate as the palanquin and be relegated to the status of historical curiosity. The logic of the lineage nevertheless continues to flow freely and uninterrupted, directly from its source in Henry Ford's garage notebooks to the latest showroom example. Meanwhile, deeper down within the essential logic of Ford's prototypical design, lie more subtle conceptual structures; barely discernible intellectual fossils that date back to the moment when man first jumped onto an ox, bridled and saddled the first horse, or attached the first animal to a cart.

Prior to the mass production of motor cars and the adoption of Ford's basic design as the industry norm, it is likely that large numbers of other designs were explored by inventors. These would have been formulated on paper and then constructed so that their performance could be tested. In the early days, the logical distance between each design is likely to have been considerable. There were presumably good reasons why the majority vanished without trace and were consigned to the scrapyard of automobile history. Chance doubtless also played a role. Not every engineer would, for example, have been able to secure funds for their projects and some potentially excellent designs may have fallen victim to poor marketing strategies. Individual elements of the logical structure of these early designs may nevertheless have persisted. Once the basic logic and topology of the modern motor car had been established, the rate at which significant innovations could be introduced into the design was much curtailed. Despite having the potential to redesign each generation of motor cars from scratch, it seems as if engineers are able to deviate from the basic logic of the design to only a limited extent.

The major motor car designers and manufacturers nevertheless continue to produce new models with remarkable virtuosity. Despite the constraints imposed by the basic design, it appears that there is an almost limitless library of potential motor car-like structures to explore. Indeed, the flexibility of Ford's prototype is likely to be one of the main reasons for its endurance. Although new models differ from one another in minor ways, such as their fuel consumption, gadgetry, or chassis design, it has proved difficult to design a successful car which deviates from the basic ground plan in any really significant way. The prototype is both a constraint that prevents the exploration of qualitatively different designs and the source of innovation. If the size of the motor car increases too much, it becomes a van and then a truck, coach, lorry or articulated lorry. If it acquires ladders and blue lights it becomes a fire engine, and if it acquires armour, a tank. If the engine become too powerful it becomes a racing car and if it acquires a barrel for mixing concrete, it becomes a concrete mixer. The *Space of All Possible Automobiles* is not a rigid space, and its boundaries and inclusion criteria are determined only by our somewhat arbitrary definitions; it is hard to imagine that people will ever drive to work in a coach, but if family sizes increase in the future this might become routine. If we were to take the time, we could unpack the set of assumptions that define the notion of what a motor car actually is.

In the world of non-biological machine construction, trying and testing wildly new designs is a matter of making sketches and calculations and then realizing them. If they do not work and you are fortunate enough to obtain further funding, you can simply start again. Nature does not permit such luxuries. Redesign from first principles is precluded and only a limited number of the components of biological machines can be significantly modified at any time. Each contemporary design is rooted in the history of its precursors and cannot easily transcend this legacy. The initial design fixes and constrains all future designs. This contrasts with motor cars and other artificially constructed machines, as although rooted in the history of their precursors, they are not in this sense rigidly constrained by them. Artificially constructed machines of a biological or non-biological nature can be reinvented at the whim of the designer and there is no intrinsic reason why the motor cars or artificially constructed organisms of tomorrow, should be similar to those of today. Furthermore, each element of a biological structure exists in a cooperative relationship with its accompanying parts. Because of this interdependency, a significant change to one component may compromise the function of several others.

The introduction of significant changes into biological machines by natural engineering processes is clearly no trivial affair, and the potential for change is much more limited than that of their artificially engineered counterparts. Nevertheless, logical structure appears to be conserved regardless of the mechanism by which the machine is engineered. So although each new type of motor car is designed from scratch, it will usually bear a resemblance to its precursors. And, if provided with the specifications for a number of modern motor cars, it should be possible to infer a set of candidate architectures for a Model T Ford, even if there were no remaining examples and all existing plans had been destroyed. It is not immediately obvious, however, whether it would be possible to infer a saddle or horse and cart from a motor car. These types of paradigmatically distant structures are much harder to infer and it might take some time to realize that the seat of a car is a heavily disguised saddle, the chassis a modified cart, and the engine an artificial horse, ox, or donkey.

In the case of natural biological machines, there is also a more general sense in which particular types of organisms are merely modifications of the essential constructional logic that underwrites all living things. All known cells are, for example, constructed from DNA, RNA and protein components. Furthermore, within the broad kingdoms of plant and animal, organisms share large numbers of genes and metabolic networks. This reflects both a commonality of descent, with a design logic being passed across multiple branching lineages, and convergence towards similar design solutions. Once a successful innovation has been discovered, it can be subverted for use in several different guises. The machinery of any particular metabolism may, for example, be used to power a wide range of different cells. One might similarly reflect on the range of creatures that can be constructed around the basic design of four legs, a body and a tail.

It appears then that much like artificially constructed machines, in the biological world redesign to effect what we have called a paradigmatic change is so difficult to achieve that it occurs only very rarely in the history of any design; and only then after the passage of a considerable period of time. The rate of paradigmatic change in artificial systems greatly exceeds that of natural systems. This is because the natural refashioning of biological machines can only occur as a result of the addition, loss, modification or substitution of individual components or groups of components across large expanses of time. It is, however, a trivial thing to understand how a horse and cart, or chariot was changed into a motor car. Ford did not have to engineer a horse to look like an internal combustion engine, he simply dispensed with horses altogether. Biological engines were replaced by non-biological components which were assembled into artificial, non-biological machines. Things are not so easy for nature. The implementation of a paradigmatic change in a biological machine is a monumental task, and it is by no means inevitable that a function realized in one technology can be transferred to another without complete redesign. The rate at which paradigmatic changes are able to occur within a particular design is itself likely to decrease as the number of constraints accumulate.

But why all this talk about motor cars, design logic and paradigmatic transitions? The point is that the history of locomotory devices addresses the important question of whether DNA technology, and indeed the general molecular logic of DNA-based information systems was, like Ford's prototypical design for a motor car, present right from the very beginning. If not, then what were the key paradigmatic changes in the history of life? What types of logical structures and chemical technologies might have underpinned the putative molecular saddle or horse and cart of early life, and how might one type of technology have replaced another? Before we address these questions, it will be necessary to examine some of the idiosyncrasies of DNA technology.

DNA and the proteins they encode form the informational and structural nuts and bolts of contemporary living things. DNA is a highly specialized molecule which encodes and maintains an archive of genetic information. Every nucleated cell contains a complete archive, but in organisms which contain more than one cell type, the information is differentially accessed in order to furnish the information profile needed to generate a specialized cellular function. If the impact of the encoded instructions allows the organism to survive long enough to reproduce, then aspects of the archive will be transmitted to future generations. But DNA must do much more than this if it is to be anything more than a frozen archive incapable of assimilating new information and avoid becoming outdated, uncompetitive and obsolete. In a clockwork universe where everything was predictable, a genetic archive might in principle be transmitted across an infinite number of generations without acquiring modifications, but the universe is continuously changing. Perhaps the only thing of which we can be certain is that our chronologically distant tomorrows are unlikely to bear any more than a passing resemblance to our yesterdays and todays. It is thus essential that the information stored within DNA remains contemporary, whilst at the same time retaining a logical picture of the causal texture of the past. The simplest way of achieving this is to incorporate mechanisms that update the DNA archive. Information updating need not, however, be restricted to the genetic archive itself. Development, learning and culture are able to supplement genetically encoded information and to furnish formidable capacities for extra-genetic information storage, replication and generation. These additional tiers of information processing are furthermore sensitive to very high frequencies of change which cannot be tracked by DNA.

All information is subject to corruption and the DNA archive is no exception. This susceptibility arises from its continuous exposure to potentially damaging chemical agents such as the oxidizing chemicals produced by metabolism and to physical processes such as irradiation by UV light, which introduces harmful cross-links into DNA. Even the water molecules that are so essential for maintaining its structural integrity, attack DNA and cleave its chemical bonds. Damage to the informational content of DNA also occurs during its replication as a consequence of copying errors. The degree of fallibility of a copying process may in one sense be taken as a measure of its complexity. If we were to ask a class of four-year-olds to copy the complete works of Milton into their exercise books, we would be surprised if even a small percentage of the words resembled the original texts. Our expectations of a professor of English literature would, however, be quite different. Even the most competent scholar is likely to make the odd mistake here and there; perhaps as a result of boredom, or because they were distracted at a crucial moment. A photocopier on the other hand should be faultless. So surely what we need is a miniature photocopier to replicate the information stored in DNA?

In fact this is absolutely not the case. And if DNA did replicate with the fidelity of a photocopier, it is unlikely that there would be any DNA-based life at all. What DNA actually requires is replication machinery with an accuracy somewhere between that of a sleepy English don and a state of the art Rank Xerox photocopier machine. The balance between fidelity and error is both critical and finely balanced. Too much error and the information would slowly melt away into nonsense; our precious texts would become a meaningless word soup. Too little error, however, and life becomes rigid, frozen and a thoroughly untenable proposition.

DNA is thus a complex machine for both the perpetuation and generation of information. It is a generator of information, not because it learns directly from experience, but because a population of related DNA archives may acquire and accumulate random changes to its informational content. Different versions of the same information may then be selected on the basis of their relative abilities to represent critical elements of the causal structure of their environment. The tension arising from the conflicting need for both stability and change, lies at the core of molecular evolution.

In an earlier chapter we discussed a simple model for understanding the informational structure of DNA in which the four nucleotide bases of DNA were represented by four coloured beads threaded onto a long piece of wire to assemble a DNA necklace. Although suitable for our initial purposes, we must now modify this model slightly in order to elucidate the mechanism by which DNA replicates. Our new model will help us to understand what is known as template-based replication. This is generally considered to have been the basic design standard for all historical genetic systems, including those of the very earliest organisms. We should not forget though, that other possible paradigms for the replication of genetic information could have preceded template-based replication. The concept of genetic takeover should not thus be restricted to meaning only a change in material technology, as it may also entail the introduction of changes into the *logic* in which processes of life are embedded.

The key to understanding the information-storing and template-based replicative capacities of DNA lies in its double-helix structure. This was determined in Cambridge by James Watson and Francis

Crick in 1953, using X-ray diffraction photographs taken by Maurice Wilkins and Rosalind Franklin at King's College, London. Watson and Crick demonstrated that modern DNA molecules are not single-stranded, but double-stranded. One strand functions as a master or 'coding' strand, which stores genetic information, while the other 'non-coding' strand forms a negative copy of the master. The master strand and its corresponding negative strand lie side by side and are joined by weak chemical bonds between the corresponding bases in each strand. This is achieved by the implementation of a simple chemical rule that enables a **C** to bond with a **G** in the neighbouring strand, and each **T** to pair with corresponding **A**'s. The converse is also true; a **G** may pair with a **C** and an **A** with a **T**. Every **C**, **T**, **G**, or **A** in the master strand is thus able to form a weak hydrogen bond with the complementary **G**, **A**, **C**, or **T** in the negative strand. The result is that the two strands lie side by side, held together by the weak inter-strand bonds that link the partners of each base pair. The two strands then twist around one another to form a double-helical structure, which resembles a spiral staircase. This then twists around itself like the supertwists of a rope, to form a higher order and more compact superhelical structure.

The sequence of the master strand can be inferred from the negative strand simply by reading the sequence of bases in the negative strand and then implementing the base chemical conversion rule. A sequence **CCTGAC** in the negative strand will, for example, correspond to the complementary **GGACTG** sequence in the master. The sequence of the negative strand may similarly be inferred from the master strand. Each strand thus functions as a template for the synthesis of its complementary partner.

Let us now go back to our bead necklace model and imagine that we have constructed a long necklace from the four differently coloured beads. When we have finished making the 'positive' coding necklace, we begin to make a second non-coding or 'negative' necklace. The sequence of the beads in the second necklace is determined by the sequence of the beads in the first, so that its sequence is complementary but not identical to the master necklace. This is achieved by the implementation of a colour conversion rule, analogous to the chemical conversion rule. It states that if there is a green bead in the master necklace, we should insert a red bead in the corresponding position of the negative necklace. Similarly, if there is a red bead in the master necklace, we should insert a green, if there is an orange we should insert a blue and if there is a blue we should insert an orange. If the information of the master necklace is then used to template the construction of a second necklace, the newly synthesized necklace will be a negative copy of the master. The complementary necklaces may now be aligned and the corresponding beads in the positive and negative necklaces joined using loosely fitting press-studs which represent the weak hydrogen bonds between each of the bases within a given pair. If the bonds between complementary bases in the two strands were visible, the structure would look something like a rope ladder. If we held onto one end of the two strands as if they were a double skipping rope, asked a friend to hold the other two ends and then both twisted in opposite directions, the two strands would wrap around themselves to form the characteristic double-helical structure of DNA.

We might now decide to construct a third necklace using the negative necklace as a template and implementing the colour conversion rule, so that we end up with a new necklace with a sequence identical to that of the master. This third necklace could function as a messenger or ambassador for the information in the master strand. In summary then, if part of the master necklace contained the sequence **red**, **green**, **orange**, **blue**, then the corresponding region of the complementary necklace would read **green**, **red**, **blue**, **orange**. If we then synthesized a new necklace using the negative necklace as a template, we would end up with the original **red**, **green**, **orange**, **blue** sequence of the master. It is worth noticing that in order to gain access to the information stored in the negative strand, we would have to unclick the press-stud bridges which linked the two complementary necklaces.

We are now in a position to appreciate the complexity of DNA-based information storage technology. This can itself be divided into two principal components; first, that associated with replicating the two strands of DNA to produce an identical copy of the genetic material and second, the complexity associated with the process of copying the fields of information stored within the master template, which are known as genes. This incorporates the complexity associated with translating the abstract information of genes into a concrete material representation. All these elements are however inextricably interconnected. The work of John von Neumann, who in 1948 presented a general theory of self-replicating machines which he called automata, helps clarify this.

According to von Neumann, self-replicating machines, whether naturally occurring or artificially constructed, may be divided into two logically separable sub-components which in modern computer terminology are known as software and hardware. Whereas software encodes the information of the program, the hardware is responsible for its implementation. For a machine or automaton to have self-replicatory competence, both components are essential. In the language of modern biology, software equates with a DNA genetic principle, whereas hardware equates with the networks of protein workhorses which run the errands dictated by the genetic principle within the automaton. The protein networks are connected into a metabolism and constitute the executive machinery which ensures that the functions necessary for the survival of an organism are implemented successfully. It is important to emphasize that the software and hardware elements which form the prerequisites for the construction of self-replicating automatons are neutral with respect to the technology in which they are realized. The requirements of these two components may thus be met by any number of different technologies. DNA and proteins just happen to be the particular software and hardware technologies with which contemporary life is familiar. This does not mean, as we have noted, that history has not experimented with other technologies. The logical structure of an automaton is more fundamental than any particular representation of its logic.

The double-helical structure of DNA indicated that it might self-replicate by unwinding its two strands and then initiating a replication process in which each strand functioned as a template for the synthesis of its complementary partner. This would result in the production of a new copy of DNA. Experimental evidence soon confirmed that this was the case, but it also established that despite its immense ability for storing information, DNA is inept at replicating information. It is consequently obliged to enlist the help of a host of highly trained protein 'whipping boys' to accomplish the task. The behaviour and structural complexity of these proteins portrays a finesse indicative of a considerable degree of historical processing in the form of adaptive specialization. Each protein has in fact been programmed over thousands of millions of years by a process of Darwinian evolution. The replication of DNA requires the coordinated action of a menagerie of highly specialized protein enzymes, each of which form an integral part of a metabolic network.

Enzymes function as biochemical facilitators which accelerate the rate of chemical reactions by a factor of between one million and one septillion (10^{24}). By determining which chemical reactions are allowed to proceed, enzymes help organize and define the chemical architecture of living things. Most of the molecules that are essential to living things are produced by enzyme-catalyzed reactions. Enzymes function as tiny machines which perform chemical reactions on the raw materials or 'substrates' that they bind and transform them into a new product, or set of products. These are associated with unique functions which play a central role in generating the organization of living systems. The repertoire of proteins involved in DNA replication includes enzymes that unwind the DNA helices and superhelices and enzymes that melt the inter-strand bonds between corresponding bases. They also include high fidelity template-based enzymes, which synthesize a complementary copy of each strand according to the information in the template they are reading. These enzymes incorporate proof-reading and editing properties which enable them to check the accuracy of synthesis as they proceed. Mismatched bases are for example excised and replaced with the correct base needed to meet the requirements of Watson-Crick pairing. A host of other enzymes are also involved in checking and maintaining the informational integrity of DNA sequences, both at rest and during DNA replication and transcription.

These DNA repairmen, much like their human railwaymen counterparts who labour day and night to repair damage to the tracks, ensure that the information of genes is not degraded. Damage, as we have already said, is caused by the general wear and tear that results from exposure to the chemical products of metabolism, foreign chemicals, ultraviolet radiation and misreplication events. In the absence of these essential repair enzymes and the accompanying checkpoint and apoptosis proteins, the informational structure of DNA would rapidly be corrupted and melt away into nonsense. Manfred Eigen has called the point at which the error rate exceeds the threshold value compatible with information conservation the 'error threshold'. It would be impossible to keep DNA with any significant information content below the error threshold without the help of repair enzymes and proof-reading capacities. In the absence of a highly specialized and complex metabolism, DNA is not only inept at self-replication and unable to realize its encoded information, but is also unable to conserve its informational structure. Without enzymes to run its synthetic and reparative errands, DNA would resemble the empty shop floor of a factory in which the entire work force was on strike.

This returns us to the paradox of how DNA in its present state of dependency could have self-replicated and conserved its informational integrity without a host of specialized pre-enzymes. The fact that DNA cannot replicate or repair damage to its informational structure without the assistance of enzymes indicates that DNA technology has always depended on enzyme activity. But if the protein technology from which modern enzymes are constructed is assembled according to information stored within DNA, then logic would dictate that enzymes, which have been processed and programmed by a sustained process of evolution, must always precede DNA. This is because in the absence of such enzymes there can be no DNA replication, repair transcription, or translation of the codified information into protein sequences. And yet, paradoxically, the highly programmed protein enzymes that subserve these essential functions are entirely dependent on DNA for their evolutionary education and existence. In other words, without DNA, it would appear that there can be no educated enzymes at all.

This conundrum takes us to the second component of DNA complexity. That is the mechanism by which the information encoded within genes is copied and decoded in a process known as transcription, and then used to synthesize protein molecules in a process known as translation. Our discussion of DNA structure will give us a more detailed understanding of the mechanism by which the codified information stored within genes and their associated non-coding regulatory sequences determines the informational structure of a metabolism.

Without a mechanism for deciphering and implementing the information encoded within the sequences of its bases, an archive of DNA would be like an instruction manual written in an unknown language. Raw instructions alone have no agency. The executive machinery which enables genetic instructions to be realized is metabolic hardware and it is thus proteins that empower genes. The mechanical process which links genetic information to the metabolic machinery it specifies is, however, underwritten by immense complexity. This applies both to the processes by which the information in genes is made available, copied and transported to the appropriate target destination and also to the process by which information written in DNAese is translated into the amino acidese language of proteins.

In most genes, the information that specifies amino acid sequences is located in the coding strand. When the gene for a given protein is activated, mRNA steals a copy of the gene and carries its information to a ribosome. The DNA double helix first unzips in the region of the gene which is being copied. Then with the assistance of a template-based DNA polymerase enzyme and an appropriate supply of energy and raw materials, the corresponding region of the complementary non-coding strand is copied, to produce an mRNA molecule. Because mRNA is synthesized using the non-coding strand as a template, it has the same sequence information as the coding strand. The transcription process in which information encoded in DNAese is copied into mRNAese requires the participation of multiple enzymes. The second component of this process involves the extraction of the information written in mRNAese, and its translation into the amino acidese language of proteins.

This task, as we have discussed earlier, is accomplished by translation machines known as ribosomes. These are constructed from a specialized type of RNA known as rRNA (ribosomal RNA). On attaching to a ribosome, the mRNA is read sequentially from beginning to end. This process is tightly coupled to decoding and simultaneous translation of the encoded genetic message. Decoding of the genetic cipher is achieved by the application of the genetic code, which stipulates how each mRNA base triplet 'word' translates into a corresponding amino acid word, or into a punctuation START or STOP signal. Synthesis occurs with the assistance of another class of

RNA molecule known as tRNA (transfer RNA). These molecules function as molecular sherpas which carry a cargo of a single amino acid of a given type, and deliver it to the correct position on the ribosome. Each tRNA molecule then releases its amino acid, which is immediately joined to the others in the newly synthesized chain to produce an amino acid string. The newly synthesized protein is then released and transported to the appropriate cellular location. In many cases, the nascent one-dimensional amino acid chain will fold into its two- and three-dimensional structures unassisted. In some cases, however, chaperone proteins and other specialist 'foldase' enzymes are needed to assist the folding process and to prevent protein aggregation and misfolding.

The logical and informational structure of the genetic code is so subtle and complex that although only indirectly visible through the agency of its consequences, it may be compared with any of the other structural edifices of the macro world, including the stripes of a tiger, the neck of a giraffe, or the spirals on a pine cone. It presents yet another challenge to the hypothesis that DNA was the first genetic material. For without a decoding principle, the genetic ciphers would remain as foreign and unintelligible as the most obscure and ancient human language.

The successful utilization of DNA as a genetic technology for the construction of living things thus requires the following elements to be in place: (1) a set of enzymes able to replicate and proofread DNA sequences; (2) a set of enzymes for transcribing DNA sequences; (3) an mRNA interlocutor to carry the information of genes to the site of translation; (4) a genetic code to translate the digital language of mRNAese into the digital amino acidese language of proteins; (5) machines to implement translation; (6) an adequate supply of raw materials and available energy; (7) a set of DNA checkpoint and repair enzymes, able to ensure that the informational structure of genes is conserved across time; (8) enzymes that are able to unwind and rewind DNA double helices and superhelices; and (9) in the case of multicellular organisms, apoptosis programs to initiate self-destruction of DNA that is so damaged it cannot be repaired.

The considerable interdependence of DNA and proteins in the processes of replication, transcription, translation, repair and in some cases folding, would seem to preclude DNA from being a reasonable candidate for the technology in which the genetic principle of the first organisms was clothed. If DNA had any significant scope for enzymic activity that could be implicated in these processes, then this might change matters somewhat and it may be possible to imagine ancestral organisms constructed exclusively from DNA. Similarly, if proteins could be synthesized according to information stored within other proteins, it might be possible to imagine precursors constructed exclusively from protein. But given that double-stranded DNA is thought to have only the most basic catalytic competence, and that the evidence for self-replicating protein sequences is rudimentary, are there any alternative chemical technologies which might have provided the logical requirements of software and hardware necessary for the production of self-replicating automatons?

The answer is yes, and it is lying on our doorstep. Being a highly specialized piece of software, DNA appears to lack the ability to function as a hardware technology; it remains dependent on the presence of an extraneous metabolism to represent its information. RNA, on the other hand, is capable of generating enzymatic activity and, like DNA, is also able to encode genetic information in a combinatorial manner. It would appear then that RNA should, at least in principle, be able to represent its own information without the need for proteins and thus to perform the functions of hardware and software simultaneously. Given that at least some RNA molecules appear to have the potential for this remarkable all-in-one property, is it possible that the genetic and metabolic principles of the first biological machines were constructed exclusively from RNA? And once the logical skeletons of living creatures had been cast in the form of RNA, might these have eventually been recast in DNA as the result of a process of genetic takeover?

Unfortunately, unlike bones, shells and the walls of some unicellular creatures, genes and metabolic principles do not fossilize. Although it is possible to extract DNA from the remains of creatures hundreds of thousands of years old, it is unreasonable to expect such molecules to have remained intact for the billions of years that have elapsed since the emergence of life on Earth. The only reasonable way forward is to try and infer the chemical structures of our most ancient ancestors by searching for intellectual fossils of the type we have discussed. Our intellectual fossil hunt will thus continue with an examination of the evidence in support of a hypothetical *RNA World*, which we will imagine was populated by creatures quite unlike modern DNA-based forms of life, and which were fashioned exclusively from RNA.

7

A World Without DNA

We are now entering a new world. A world in which DNA has no place. A world populated by creatures so foreign and strange that we will need special glasses to see them. But this apparently absurd and unnatural world is not a new world. It is very ancient. The forms of life which surround us here are not constructed from DNA and proteins, but from RNA. This is *RNA World*, a hypothetical world replete with molecular RNA fossils and boneless RNA creatures, whose existence, if indeed it ever did exist, has left no discernible trace.

We are now deep within the *RNA World* and, with our special glasses in place, are able to begin our exploration. In front of us is a cast-iron gate and above it an arch. On examination of the iron letters that span the gate, we are just able to discern the words 'Welcome to the *RNA Zoo*'. We enter cautiously and as if expecting to see some strange creature throw open its wings and run across our path, we clutch our cameras to the ready. But unfortunately, or perhaps fortunately, none arrives.

I have to say that the *RNA Zoo* is an odd place and I feel uneasy here. It is, I think, that feeling of not belonging, or perhaps of having once belonged. But I am compelled to continue my journey. Yes it is true, I do not belong here. Furthermore there is even a sense in which I cannot belong here, in this wretched *RNA Zoo*. For in addition to RNA, I am also made from proteins and DNA. Take away the proteins and DNA and I would be much less than a pale shadow of my former self. In fact the rules of the *Zoo* state quite explicitly that I *cannot* exist here. For the *RNA Zoo* is just another name for the *Space of All Possible RNA Organisms*. A pamphlet which I pick up at the gate informs me that the RNA organisms housed within the *Zoo* are called riborgs. This is an abbreviation of their full name, which is ribonucleic acid organisms.

At this moment a disturbing thought enters my head and I slap my legs and howl quite terribly. Have I been cheated of my DNA and become a riborg? A disembodied ghoul whose veins pulsate with the chemistry of RNA? I am relieved when the *Zoo* keeper arrives and assures me that I have not. He has made an exception to the rules of the *Zoo* and has allowed DNA-based organisms which he calls deoriborgs (or deoxyribonucleic acid organisms) into the *Zoo* on a limited twenty-four-hour pass. This is on the condition that we pin a green circle onto our jackets, so that we are easily recognizable and distinguishable from the riborgs. We comply with this request and pass through the gates. But then I become preoccupied with yet another thought. Might there be an exact RNA copy of myself lurking somewhere within the *RNA Zoo*? And if there is, would it look like me? Would it think like me, might it even be like me? Or is the *Zoo* devoid of creatures which have anything like my degree of complexity? This thought is more chilling, as if this is the case then even my logical structure cannot exist here.

Before disappearing, the keeper informs us that the *Zoo* is divided into two main sections. The *Possibly Historical RNA Zoo* and the *Potential RNA Zoo*. The latter is much larger than the former, as it contains the complete collection of all possible riborgs. But it is the *Possibly Historical RNA Zoo* in which we are interested, as this is the area which contains the candidate RNA creatures that may have inhabited the Earth at a time when DNA and proteins were not associated with life in their contemporary manner. The map that the keeper has given me indicates that the *Potentially Historical RNA Zoo* is itself divided into two sections. One of these contains hypothetical riborg proto-organisms, which are non-compartmentalized and lack form. The other contains hypothetical riborg organisms, which are discreet, compartmentalized structures.

Are such ideas purely fanciful? Can organisms really be made exclusively from RNA? Did the *RNA World* ever really exist and have any of the potential RNA creatures housed within the *RNA Zoo* ever been realized? Does the *RNA World* even have a *Zoo*? We have already encountered von Neumann's notion that the logical pre-requisites for life include both an informational and a metabolic principle. In the *DNA World* with which we are familiar, these roles are carried out by DNA and proteins respectively. Two separate technologies for two discrete functions. Is it possible, however, that a single RNA-building material might have performed both of these functions at once? If this were the case, then we might have chanced upon the solution to the chicken and egg paradox of how the informational and metabolic principles essential for modern life could have originated independently, but nevertheless simultaneously. It is clear that the *RNA World* hypothesis will to some extent rest or fall upon this point.

We may never know what the earliest forms of life were actually like. The greater part of our molecular history has been erased from the geological record, and the repertoire of creatures which present themselves as plausible ancestors is vast. But if we are able to demonstrate that it is, in principle, possible to fashion organisms exclusively from RNA, then until such a time as a more compelling candidate presents itself, the notion of an *RNA World* must be considered a serious historical proposition. At the very least, such creatures might constitute an approximation of our hypothetical DNA-lacking ancestors. This obliges us to come to terms with the somewhat uncomfortable notion that there may indeed have been a time, several billion years ago, when the Earth was inhabited by riborgs. The torch of life may indeed have been carried by these awkward creatures for thousands of millions of years. It was not, furthermore, inevitable that our RNA-based ancestors should ever have renounced their exclusively RNA-based lifestyle in favour of one based on a combination of DNA, RNA and proteins. Had events been different, the history of life may have been written and executed exclusively in RNA. The *DNA Zoo* may have remained an untouched and irrelevant subregion of the *Information Sea* with no historical significance whatsoever. Tigers, pelicans, daffodils and man would have remained no more than untested gene kit specifications for hypothetical DNA-based organisms.

To accept the historical plausibility of an early Earth inhabited by RNA creatures entirely unlike ourselves or other contemporary living things, has profound consequences. For once we accept that life might be cut from the cloth of RNA, we also accept that the logical processes which underpin life can be abstracted from any particular material representation. It would appear then that we will need to redefine our terms. When we use the word gene, we will need to specify the *type* of gene to which we are referring. Is it an RNA gene, or one made from DNA? We must learn to transcend our DNA-centrism and to see a gene for what it is, as opposed to what it is made from. When we talk about *Gene Space*, we will need to specify the type of *Gene Space* to which we are referring. Is it an *RNA Gene Space*, a *DNA Gene Space*, or some other sort of *Gene Space*? Different types of information storage technologies will have their own associated *Gene Space*. In some cases, these may be connected by life-lines delineating plausible routes that genetic takeover processes could follow. In others, there may be no connection between the different types of *Gene Spaces*. The description of each type of potential *Gene Space* encompasses both their interconnectivity, and relationship with some type of *Genetic Code Space*. We may imagine that the potential linkages within and between every possible type of *Gene Space* under every set of possible conditions are detailed by an infinite set of maps.

An awareness that living creatures might be constructed from different types of building materials highlights the unique advantages and constraints that different technologies have on offer. Given a situation in which organisms could be constructed artificially from first principles, a designer might choose to incorporate a subtle blend of different technologies. The properties that we demand from a construction material depend on the uses to which it will be put. An iron saucepan, for example, is unlikely to be as efficient as one made from copper, and a concrete suspension bridge would collapse in the first gale. It is hard to imagine a brain made exclusively from RNA, but that is not to say that a computational device of equivalent or greater power could not be assembled from this material. We must learn to abstract the logical principles of life from any particular manifestations of those principles. It follows, then, that although the absence of protein and DNA technology from the *RNA Zoo* prevents me from existing there, it is possible that I might be able to locate my logical self in one of its infinite cages. I shudder to think of the grotesque shapes that my RNA facsimile might adopt. It is also terrifying to think that the logic of the mind I hold sacrosanct, could exist intact but disguised in some fearsome RNA manifestation located in a far and distasteful corner of the *RNA Zoo*.

We will need to abandon any narrow definition of biology that does not allow for the possibility of life clothed in technologies other than DNA and proteins. But what exactly is RNA, and how is it related to DNA? What, furthermore, are the differences between DNA and RNA that limit one technology to function exclusively as an information carrier, whilst allowing the other to combine information-storing and metabolic functions? We might also be curious to know whether RNA is as proficient at information storage as DNA, or as adept at generating biomolecular structures and machines and performing the intricacies of metabolism as proteins. These considerations lead to the question of how living things came to be built from a combination of DNA, RNA and proteins. To answer some of these questions we must return to the structure of DNA.

A bead necklace may be divided into two main components. The invariable component is the wire along which the beads are threaded, and the beads themselves, which form the variable component. The variable component encodes information by combining beads in different ways to generate a repertoire of sequences. The longer the necklace and the more beads threaded onto it, the greater its information content. The wire plays an exclusively structural role and functions as the scaffolding upon which the sequence information is arrayed. The physical properties of the necklace are influenced by this backbone. A necklace threaded onto a piece of string would, for example, break if given a tug or placed in water: a gold wire, on the other hand, would be more durable. Similiarly if the beads were threaded onto a piece of cast-iron wire, we would have difficulties getting the necklace to sit in its usual manner. The introduction of modifications into a necklace's information-free skeleton, may thus alter its physical properties without affecting its one-dimensional information content. It may, however, introduce higher-dimensional changes. Imagine, for example, that you were a tortoise trying to decode the information in a necklace strung onto cast iron. As a consequence of its rigidity, it would take you several hours just to move from one end to the other. The necklace may also twist across itself and make some aspects of its information inaccessible.

Like coloured bead necklaces, DNA and RNA necklaces may also be separated into variable and invariable components. The variable bases of DNA and RNA require invariant skeletal structures on which they can be displayed. The bases which encode the information of genes are not the only structural components of DNA and RNA. Each base is in fact chemically attached to a sugar molecule and each sugar-base complex is linked to neighbouring complexes by chemical bridges known as phosphodiester bonds. The unit consisting of the information-free sugar-phosphate complex attached to an information-encoding base is known as a nucleotide. The string of either deoxyribose or ribose sugars linked by phosphodiester bonds forms the invariant structural backbone of DNA and RNA respectively. In DNA this is made from a sugar called deoxyribose, whereas in RNA it is made from ribose. Ribose sugars contain a chemical hydroxyl group (known as the 2′-OH) which is not present in deoxyribose sugars. Nucleotides made from deoxyribose sugars are known as deoxyribonucleotides, whereas those containing ribose sugars are known as ribonucleotides.

The differences between the sugar skeletons of DNA and RNA are responsible for many of the distinct characteristics of these otherwise similar nucleic acids. But DNA and RNA differ structurally in two other ways. First, RNA uses **U** (uracil) bases in place of the **T** (thymine) bases found in DNA. The chemical structure of **U** is almost identical to that of **T**, except that it lacks the methyl group found in **T**. Like **T**, **U** is also able to form Watson-Crick base pair bonds with **A** bases. Second, unlike DNA, in RNA short double-stranded helical segments of ten or less base pairs are interspersed between single-stranded regions that do not form helices. This lies at the heart of RNA's ability to function as a molecular double agent that is able both to encode information and to participate in metabolism.

Although the double-stranded structure of DNA is suited for high-fidelity information storage, it restricts the molecule to adopting only three similar structures. These are known as: A-DNA, B-DNA and Z-DNA, of which the B structure is by far the most common. RNA is much more flexible and adopts a host of complex geometries, which are assembled from a limited set of local structural elements. These include hairpin loops and bends, bulges, triple-stranded interactions and pseudoknots. Jennifer Doudna and Jamie Cate have shown that some RNA molecules form globular folds reminiscent of those found in proteins. So although limited to only four different nucleotide building-blocks, it appears that RNA can, at least to some extent, emulate the diverse and complex repertoire of three-dimensional structures generated by protein sequences

The suggestion that the earliest organisms were constructed from RNA was made independently by Carl Woese in 1967 and Leslie Orgel and Francis Crick in 1968. The phrase *RNA World* which refers to this hypothetical era of life's history was coined by Walter Gilbert in 1986. In the *RNA World* hypothesis, the first RNA organisms are imagined to have self-assembled from an unstructured, primordial alphabet soup of **A**, **C**, **T**, and **U** nucleotides. These RNA organisms are thought to have eventually acquired mechanisms for partitioning themselves from the surrounding chemical world; encasing themselves, for example, within spontaneously forming fat droplets. The acquisition of a compartmentalized reaction space was the first step towards the formation of a modern cell. Partitioned RNA organisms may have been able to generate and sustain a repertoire of metabolic functions and to evolve, as under appropriate conditions and length restrictions, their information store would be stable, despite being subject to continuous corruption by synthetic errors and damage.

The catalytic complexity of such hypothetical RNA organisms is likely to have increased over time, both through changes to their RNA sequences and by assimilating chemical species that enhanced their catalytic scope. RNA organisms are thought eventually to have acquired the ability to encode and synthesize proteins. No clear mechanism has been proposed for this however. For reasons which we will discuss later, proteins are much better catalysts than their RNA counterparts and would, over time, have taken over most of RNA's catalytic functions. In the final stages, the information archiving function of RNA is thought to have been subverted by DNA, as life moved from the darkness of the *RNA World* into its *DNA World* renaissance.

The processes that underlie the putative transitions which enabled life to progressively reinvent itself remain elusive. Exactly how, for example, did the *Pre-RNA World* give rise to the *RNA World*? How, furthermore, did organisms that relied exclusively on non-encoded RNA-based metabolic machinery come to update and replace these with digitally encoded protein technology? Is it even reasonable to call an RNA sequence which lacks a mechanism for the symbolic *codification* of information a gene at all? What degree of metabolic complexity were RNA organisms able to attain before their metabolic functions were usurped by proteins? How was the genetic code able to evolve and how was the archive of life's information transported from an RNA storage facility into one made from DNA? These questions cannot be brushed aside as if they did not exist. But although detailed models for these processes are for the larger part lacking, we can only assume that it will in time be possible to subject these problems to experimental analysis. Given the lack of direct evidence, we will focus instead on the general plausibility of the *RNA World* hypothesis. For if the hypothesis does not stand the test of cross-examination, then details become irrelevant.

Although the *RNA World* hypothesis predicts that RNA may once have had catalytic functions reminiscent of those found in modern proteins, it was assumed that these had been lost many hundreds of millions of years ago. Like a family of former tribesman long ago transported to urban life, RNA had apparently forgotten the ancient catalytic craft on which it once relied. Its modern incarnation was envisaged as being a watered down version of a past when RNA was truly a lion in the kingdom of molecules. How surprising then when the first molecular fossil evidence for RNA's primordial metabolic glory was discovered not underground, or in a cliff face, but in a test-tube. In 1982 Thomas Cech's laboratory discovered an RNA molecule that was able to perform catalysis. It is hard to convey what a surprise this was at the time. Until then, the words protein, enzyme and biological catalyst had been synonymous. Although this was only an isolated example of an RNA-mediated catalytic activity, itself restricted to RNA processing, it nevertheless represented a challenge to the notion that proteins held a monopoly over catalytic function. The term 'ribozyme' was introduced for catalytic RNA molecules, to emphasize that catalysis was no longer exclusively confined to the domain of proteins.

Since Cech's discovery of ribozymes, several other naturally occurring catalytic RNA molecules have been discovered. However, the known repertoire of naturally occurring ribozymes catalyzes only a very restricted set of chemical reactions. With a few exceptions, for example the ribozyme activity in ribosomes which catalyzes the formation of bonds between amino acids and hints at the possibility of a potentially wider catalytic repertoire, almost all of the naturally occurring ribozymes appear to be involved in RNA processing. There are two explanations for the restricted catalytic performance of contemporary ribozymes. Either there is an inherent limitation on the types of chemical reactions that they are able to catalyze, or naturally occurring modern ribozymes represent only a tantalizing glimpse of RNA's ancient, but now considerably attenuated catalytic virtuosity. If the former were the case, then the *RNA World* hypothesis would be implausible and untenable, as a wide range of catalytic activities are needed to construct even the most basic biological machines. If, however, the latter is the case, then the fossilized catalytic activity written into the chemistry of contemporary RNA molecules, despite being an impoverished record of its ancient past, might be used to reconstruct some of the features of the enigmatic ribose-based denizens of the *RNA World*.

How might this be achieved? Might it be possible to artificially recreate some small corner of the *RNA World*; an artificial frosted-glass window through which we could peer down into the expanses of frozen evolutionary time and ancient history? Or have the hypothetical RNA catalytic networks which may have defined and motored our ribose-based ancestors as they explored the *RNA World* been irretrievably lost, never to be recovered by the chroniclers of the Earth's biomolecular history?

One of the recurrent themes of our journey has been the exploration of mathematical spaces containing all possible examples of defined types of things. We began by considering an incomplete set of plastic Indian chiefs' heads amassed by my friend Conrad, and went on to discuss other spaces which included: *Gene Space*, *Crocodile Space*, *Protein Sequence Space*, *Chemistry Space* and *Apple Space*. All of these are infinite libraries which contain the complete collections of different versions of the item in question. The information needed to construct every item in a given library is described mathematically. To pluck a crocodile from *Crocodile Space*, we need only retrieve the appropriate information cluster. With this mathematical description in our hands, we could walk into an imaginary shop not dissimilar to the printing shops in any high street, which contains a machine able to translate the mathematical description into the flesh and bones of a real crocodile. Every kind of space of all possible things, be it *Piranha Space*, *Pterodactyl Space*, *Potted Plant Space*, *Portuguese Galleon Space*, *Stickleback Space* or any other sort of space, is an information space. The currency of information is universal and allows both living and non-living items to be defined, melted down and manipulated in a single abstract language. It also gives us the power of a fictional Dr Dolittle, who is able to talk not only to animals, but to fish, insects, bacteria, postboxes, snowflakes and street lamps.

The King of Corinth was condemned to pushing a stone up a hill for eternity. Unfortunately the stone had a habit of falling down to the bottom of the hill just before he reached the top. You might be forgiven for thinking that this was a cruel fate. But he should have been grateful, as things could have been much worse. He could, for example, have been forced to start work on the construction of *Camel Space*. Even if alternative mythology had been kind and limited his task to the construction of *Camels With Only One Hump Space*, I dare say that the universe would be stuffed full of camels long before he had constructed even a small region of this potential space. His popularity in the underworld would doubtless suffer as Hades filled with camels and the other occupants were forced to procure food for them. If, however, he had been handed the information cluster for a camel the size of Jupiter, he might spend billions of years without completing even a single camel.

The economy of storing potential things as information clusters should now be evident. The strategy is so compact that the whole universe of possibility occupies no space whatsoever. I suggested earlier that the natural home for the collection of all of these infinite information spaces is the *Information Sea*, which is the *Space of All Possible Information Spaces*. I have also tried to communicate the somewhat eerie fact that there is a very real mathematical sense in which these spaces actually do exist, even though most of the items they house will never be realized. Indeed, constraints of time, space, matter, and the nature of the processes by which items are realized in the world of form permit only the most cursory dip into this sea of possibility.

This is all very well, I hear you say, it is easy to talk about alternative worlds, mathematical spaces and collections of all possible things. But none of these are at the end of the day tangible, or have any bearing on the real world. I will now consequently endeavour to demonstrate the physical realizability of one mathematical space from the infinite set of all possible information spaces. If I succeed, it should not take too much of an imaginative leap to appreciate that the same might be true of other mathematical spaces.

The space I would like to consider is the hypothetical *Space of All Possible RNA Sequences*, which we will call *RNA Space*. This is much the same as the *Space of All Possible DNA Sequences*, which we called *Gene Space* or *DNA Space*. As *RNA Space* contains the complete set of all possible RNA sequences, it must by definition contain the skeletons of all the organisms that inhabited the *RNA World*. It is a huge museum, filled principally with meaningless RNA junk, but if we look close enough, we will find that it is replete with rich seams that contain the fossilized remains of RNA organism components. Were we able to search the space efficiently, we should be able to find the hypothetical RNA organisms themselves. Might we not then use this space to lure the creatures of the *RNA World* from the darkness of their millions of years' slumber and cajole them to dance before our eyes?

Before we attempt to do this, we must first locate the space. However, where does one start looking for a ghostlike space which lacks a material incarnation, and how is it possible to find something, that doesn't exist? But what if it were possible to prove that *RNA Space* exists? What if we could defy history and artificially synthesize a small portion of this space which until the moment of its realization had only a disembodied and abstract mathematical existence? Would that not constitute evidence for both the general reality of these theoretical spaces and the fact that the information and logic necessary to realize them precedes their realization? If we could conjure up *RNA Space*, even in a fractured or incomplete form, for but a few hours or days, might we not then take the opportunity of asking our

RNA Space genie some questions? Would our artificially constructed *RNA Space* not constitute a sort of biological time engine, enabling us to travel back through the depths of possible and actual history and to observe at close hand aspects of an ancient world which, like Gloucester's eyes, were extinguished so very long ago?

Let us for a moment imagine that, in a perversion of the logic of the House of Hades, we have been assigned by Pluto and his wife Persephone the fearsome punishment of constructing *RNA Space*. How might we go about doing this? How might we scavenge from a treasury far more extensive than that containing the incarcerated spectres of Hades the resources necessary to breath life into the *RNA Space* genie? Any attempt at creating the space in one go would clearly be impossible. *RNA Space*, is after all, infinite and gods that think nothing of having you torn apart by wild horses are unlikely to condemn you to a punishment that could be completed with alacrity.

Rather than trying to illuminate the whole of *RNA Space* with a single lamp, let us instead try shining a spotlight onto a small, well-defined region. But how should *RNA Space* be partitioned and where should we focus our searchlight? To amuse ourselves, we pass the hours searching for the molecular skeletons of extinct and never-before-tested *RNA World* creatures. The relics we find do not represent complete skeletons of RNA organisms, as these may only be found in the *Space of All Possible RNA Organisms*, which we called the *RNA Zoo*. They are instead the individual ribozyme building-block 'bones' of RNA organisms, the metabolic and informational cogs of potentially living RNA machines.

The plausibility of the *RNA World* hypothesis rests on the assumption that a broad repertoire of ribozymes can be found within a combinatorial library of RNA sequences. These should be capable of catalyzing a wide range of chemical reactions, some of which may form components of rudimentary RNA metabolisms. It was mentioned earlier that a small number of modern RNA sequences are known to have ribozyme activity. It seems reasonable, then, to speculate that others might as well. As *RNA Space* contains the complete collection of all possible combinations of RNA sequences of every possible length, it should be possible to fish ribozymes out from *RNA Space*. As contemporary ribozymes are around 40 to 400 ribonucleotide bases long, it seems prudent to use RNA sequences of this size to rationalize *RNA Space*. Furthermore, rather than focusing on a broad size range, it might for reasons which will become apparent, make more sense to focus our spotlight onto RNA sequences of only a single defined length.

RNA Space can be constructed in exactly the same way as *DNA Space*. Each position in an RNA sequence may be occupied by either a **U**, **C**, **G** or an **A** ribonucleotide. There are thus 4 (4^1) sequences of length one ribonucleotide (**U**, **C**, **G**, and **A**) and 16 (4^2) necklaces of length two ribonucleotides (**UU**, **UC**, **UG**, **UA**, **CU**, **CC**, **CG**, **CA**, **GU**, **GC**, **GG**, **GA**, **AU**, **AC**, **AG** and **AA**). Fortunately, as was the case with *DNA Space*, we can use simple mathematics to calculate the number of possible combinations for sequences of a greater length. In general for a sequence of length n, where n is a sequence of a specified length, there are 4^n different possible combinations of ribonucleotides.

If we were to place our spotlight on the region of *RNA Space* containing sequences of length 25 nucleotides, we might just be able to illuminate the entire array of 4^{25} (10^{15}) different RNA sequences. If we were to focus the light onto the region housing sequences of length 300 ribonucleotides, our beam would become so diffuse that it would illuminate only a minute fraction of the astronomical 4^{300} (10^{180}) possible sequences. Of course in comparison to sequences that are 3,000 ribonucleotides long, 300,000 long or even more, the astronomically large sub-space of sequences 300 ribonucleotides long becomes trivially small. Given that potential ribozymes might lie in RNA sub-spaces containing ribonucleotide sequences of any length, it is perhaps sensible to start the search for historical and ahistorical ribozymes in an RNA sub-space which houses sequences of a length comparable with those of modern ribozymes.

New technology has allowed the dream of building entire subregions of *RNA Space* and restricted samples of longer sequences to be realized. In fact, it is now possible to artificially synthesize combinatorial RNA libraries of a predetermined small size in test-tubes. These can then be screened for ribozyme activities. The intellectual roots of this type of experiment may be traced back to the work of Sol Spiegelman who, in the early 1970s, generated diverse populations of RNA molecules in test-tubes. He then showed that it was possible to derive RNA molecules with specific desired properties from artificial pools of RNA molecules, by an iterative process of diversity generation coupled with rigorous selection. The principle of searching for ribozyme activities outside living organisms in RNA populations confined to test-tubes and maturing them by a process of artificial evolution, was expounded more explicitly by Leslie Orgel in 1979. These ideas were not, however, realized until 1990, with the independent work of Gerald Joyce and Jack Szostak. Since then, large numbers of different ribozymes have been discovered in artificially synthesized RNA libraries.

Although the first set of ribozymes isolated in this way were restricted to the catalysis of reactions involved in RNA splicing, in 1995 Wilson and Szostak demonstrated that ribozymes can catalyze a far wider range of reactions. In fact it now appears that ribozymes able to catalyze almost any chemical reaction can be isolated from synthetic RNA libraries of an appropriate size and length. These include: ribozymes which catalyze carbon to carbon bond formation, ones that ligate RNA molecules together, and others that ligate amino acids and form the basis of a protein synthesizing-system. The fact that ribozyme hunting excursions in imperfectly sampled combinatorial sequence spaces of a restricted size can yield ribozymes with complex structures and functions, indicates that the density and variety of such species in the entire *RNA Sequence Space* is likely to be high. The general plausibility of the *RNA World* hypothesis has thus been vindicated. But the holy grail of such studies, namely the discovery a ribozyme that can replicate RNA molecules according to the information in an RNA template, remains elusive. The demonstration, in 1996, by Eric Ekland and David Bartel of a ribozyme with a rudimentary activity of this sort, however, bodes well for the eventual attainment of this goal.

I will now enter the *RNA* sub-space which contains the complete collection of RNA molecules of length 300 ribonucleotides and which I will call *RNA 300 Land*. Because of the enormity of this space, I abandon the idea of trying to navigate it by foot and borrow the space capsule that we used to explore *DNA Space*. I am now flying high above this region of *RNA Space*, and from the window of the capsule am able to look down at the ground. Below me lies an endless expanse of RNA sequence boxes, each of which contains a unique RNA molecule. I fly for what seems like hours and still the view does not change. I am reminded of railway journeys across the steppes of Russia and of the monotonous view from a carriage window which is indifferent to the passage of time. I continue patiently in this way, systematically surveying the terrain, until something very strange happens. The boxes below me begin to change into keys and a lock appears in my hand. Now that I have a lock that lacks a key, I resolve to find a key that fits it. If all of the boxes have turned into keys, then it would appear that I now have 4^{300} (10^{180}) keys to choose from. Surely one of these must fit my lock, even though the library of keys was created blindly without any prior knowledge of its existence?

Lest the keys should undergo another transmutation, I set off on my quest without further delay. Having no obvious place to start, I land randomly in a field of keys and begin to test them one by one. After testing a mere ten billion (10^{10}) I am delighted to find one that fits. But it doesn't fit well and opens the lock with great difficulty. Since I am able to travel great distances quite effortlessly in my space capsule, I resolve to continue the search in the hope that I may locate a key which fits better and opens the lock more efficiently. I continue to walk in a straight line, and for a distance of about another one billion keys, each successive key fits fractionally better than the one before. After this, they do not fit at all and I look elsewhere. I fly off to a different region of *RNA 300 Land* and after testing another one hundred billion (10^{11}) keys, I find one that opens the lock about as well as those in the previous field of keys. What is now clear, is that if I wanted to locate the best of all possible keys, I would have to test the entire collection of 10^{180} keys one by one, as the omission of a single key might undermine the whole project.

Before setting off on this new mission, I design a machine that is able to measure the extent to which any given key fits my lock. The result is translated into a scale of 0 for a key that does not fit at all, to one billion (10^9) for the best of all possible fits. As I test each key I build a skyscraper on the site of the key, such that its height indicates the efficiency of the fit. If the key does not fit at all I build nothing, and if the key fits with a value of one hundred, I build a skyscraper one hundred storeys high. The best of all possible keys corresponds to a skyscraper which is one billion storeys high. After testing all 10^{180} keys and building vast numbers of skyscrapers, I climb back into my space capsule and fly high over the metropolis I have created. The landscape resembles New York City and there is a downtown region in which the majority of the skyscrapers are clustered. There are also smaller downtown regions scattered across the landscape which correspond to the high-street buildings of a medium-sized town. Towering above the rest of the skyscrapers is the Empire State Building of the landscape, which represents the site of the best of all possible keys. Although the possibility of a World-Trade-Center-like situation in which there are two identical highest peaks is not precluded, in this particular landscape there appears to be only one optimal solution.

On closer inspection, the transition from one skyscraper to another is generally smooth, so that it is possible to set off from a building one million stories high and jump along the roofs of successive skyscrapers until you end up at the Empire State Building. But some regions of the landscape resemble the urban sprawl of Los Angeles. Individual one-hundred-thousand-storey skyscrapers may, for example, be surrounded by clusters of skyscrapers only one thousand storeys high. A ninety-nine-thousand-storey leap from one to the other would challenge even the most agile athlete. There are also other downtown-like regions located at huge distances from the Manhattan region. Using conventional means of transport, these skyscraper clusters can only be reached by driving great distances across flat, featureless *plains*. These irrelevant regions contain keys which bear no resemblance to my lock. It would appear then that some of the sub-regions of the landscape are more user-friendly than others. It is easy to traverse a landscape in which you can find a path that leads from low to high ground. If, however, you were caught on a minor skyscraper situated on an otherwise featureless landscape, it would be difficult to move to higher ground.

If we were now to fly to a height which enables us to take a satellite photograph encompassing the whole of *RNA 300 Land*, we could make a map. This could be used to locate the Manhattan of the landscape and from there the Empire State Building. It would also enable us to trace tours through the metropolis which lead from lower ground, gently across the roofs of the skyscrapers to the highest penthouse of the Empire State Building. In the natural world, tours of this type constitute the evolutionary routes by which systems may change with time. In the artificial world of synthetic libraries, however, no such constraints apply. If the entire space can be created, then the entire space may in principle be searched simultaneously without the need for passing sequentially through intermediate steps.

The most remarkable thing is that any useful keys can be found at all in randomly generated libraries. In a strange way, the keys pre-empt the existence of their corresponding locks. In our *RNA 300 Land* analogy, the keys located at the top of the highest skyscrapers represent ribozymes able to catalyze the chemical reaction represented by the lock that they open. The landscape in our example corresponds to the distribution of ribozyme keys that are able to catalyze the chemical reaction represented by the lock. If we were to search *RNA 300 Land* for the key to a different lock, that is to say a ribozyme which specifically catalyzes a different chemical reaction, the structure of the landscape generated by our search would be quite different. The old Manhattan would most likely become a flat plain and a new one arise, phoenix-like, from some unexpected corner of the landscape. The general nature of this terrain might also be quite different. Rather than having all the major skyscrapers clustered together in one metropolis, they might, for example, be randomly distributed across the whole landscape. This is because each type of chemical reaction is catalyzed by a different type of three-dimensional structure. The shape of the ribozyme that catalyzes one type of reaction is unlikely to be the same as one that catalyzes another, and they are consequently likely to be housed in very different regions of the space.

Our satellite photographs would be invaluable in helping us determine how user-friendly any given landscape might be. It would also be indispensable should we wish to navigate a given landscape by artificially mimicking a natural evolutionary process. By examining the contour lines, we would be able to trace out routes with gentle ascents to the peak. But a technical constraint presents a major obstacle to the efficiency with which artificially generated landscapes can be searched, and precludes the construction of complete sub-regions of

R*NA Space* which exceed a sequence length of around 25 ribonucleotides.

The constraint arises from the fact that it is currently extremely difficult to synthesize artificial RNA libraries that exceed a size of around 10^{15} different RNA species. In the case of RNA sequences of length 25 nucleotides or less, this is not a problem. The number of different RNA sequences of length 25 nucleotides is 4^{25} (10^{15}), so it is technically possible to realize the entire contents of *RNA 1 Land* through to *RNA 25 Land*. Artificial libraries of this size consequently have the potential to be complete. If a lock is thrown into *RNA 25 Land* and the synthesis has been representative and efficient, it is theoretically possible to locate Manhattan, and the best of all possible sequences within that Manhattan, in one go. The likelihood of this happening increases as the size of the *RNA* sub-space that you are searching decreases. It should be much easier to locate the Empire State Building in *RNA 10 Land* than in *RNA 25 Land*. This is because in *RNA 25 Land* each key will at best be represented by a single copy, which gives you roughly a 1 in 10^{15} chance of locating the Empire State Building when a single molecule is removed from the pool. But in an unbiased *RNA 10 Land* of size 10^{15} items, multiple copies of each key should be present. This means that the landscape should be dotted with several identical Empire State Buildings, which will consequently be easier to locate.

The reverse, however, is true when we search RNA sub-worlds which contain sequences longer than 25 ribonucleotides. *RNA 300 Land*, for example, contains 4^{300} (10^{180}) different items. In order to have even a fair chance of testing each possible key, we would have to create and screen a minimum of 10^{165} successive libraries of a size equivalent to 10^{15} items. This is a tall order, which is far from realizable. Given the fact that we are technically unable to efficiently and exhaustively search RNA sub-spaces containing sequences of any significant length, what other strategies might we adopt?

One strategy might be to artificially synthesize a micro-*RNA Space* of size 10^{15} items, which represents a random sample of the space in question. Within this random sample of the landscape's variability you might by chance net one element of Manhattan, or perhaps a little Manhattan located at some distance from the downtown area.

The majority of the items, though, are likely to represent samples from the featureless plains that separate regions containing skyscrapers. This random sampling strategy is a bit like a side-stall at a fair, where you have to throw a plastic loop over a wooden block in order to win a prize – only in this instance you would be throwing the hoop with your eyes shut and the prizes might be separated by thousands of miles. The 10^{15} throws which you are allowed in our game might seem like a lot, but given the enormity of the space that you wish to explore, they are in fact trivial. In our *RNA 300 Land* analogy, it would be like flying over the landscape and occasionally throwing the hoop out in the hope that it might encircle a Manhattan skyscraper, or one located elsewhere. Another strategy might be to bias the construction of the library, so that despite being random in most positions, the ribonucleotides at some positions are constrained. By reducing the variability of the library in this way, the concentration of each item can be increased. The intelligent design of non-random libraries, would however, depend on our having some theoretical notion of where the Manhattan regions of a landscape are likely to be located.

Given that this random sampling strategy is unlikely to bag the Empire State of a landscape or even a Fifth Avenue skyscraper, and that iterating the process is both tiresome and unlikely to improve matters, might it not be possible to upgrade some of the minor skyscrapers that we discover? Is there not a process by which a minor forty-storey skyscraper could be changed into one with four thousand, or even forty million stories? If one of the hoops happened to land on a skyscraper located within a metropolis region, even if this was not located within the Big Apple itself, one might be able to improve the fit of your key simply by jumping across the rooftops of the neighbouring skyscrapers until you reach the highest skyscraper in that region. But once you had arrived at the highest local skyscraper, you might find that the next highest skyscraper was located a good aeroplane's journey away and was thus inaccessible. You would have located the local maximum, but the best of all possible keys would remain beyond your reach. Of course, without the benefit of a satellite map, you might never know that the Manhattan you had stumbled upon was not *the* Manhattan. If on the other hand the hoop had landed on an isolated skyscraper surrounded by one-storey houses and four-storey office blocks, there would be no natural route leading to higher ground.

This type of random sampling strategy has been successfully used to generate artificially synthesized mini-*RNA Space* test-tube worlds. In practise, a random collection of 10^{15} RNA molecules of a given length will almost always contain the ribozyme activity that one is looking for. But the quality of the ribozyme is invariably poor, equivalent say to an eighty-storey skyscraper in a world where a billion stories are attainable. Might it not then be possible to optimize this rudimentary activity in the same way that minor skyscrapers can be upgraded by exploring their immediate surroundings? Can we artificially mimic the natural process by which populations of things change with time? The answer is yes. Darwin and Wallace's principle of evolution by natural selection is not narrowly confined to the evolution of whole organisms, but applies equally well to their components. Populations of RNA molecules may evolve by a process of natural or artificial selection in the same way that populations of butterflies, buffaloes or armadillos change with time by flowing across their own information spaces. Whereas in naturally evolving systems the pattern of change is determined by imperatives arising from natural circumstances, artificially evolving systems may be coaxed into preferred directions by extraneously applied selection pressures.

Test-tube strategies for selecting and artificially evolving synthetic populations of RNA molecules may be understood in the following way. First, a random collection of 10^{15} RNA sequence keys is synthesized. Rather than arraying the newly synthesized keys on the ground to generate a flat landscape, they are mixed into a complex key soup. Multiple copies of the target lock are attached to fishing lines and dipped into the soup. Keys that fit the lock are then fished out by pulling back on the rod. The population of keys selected in this way represents a privileged set which, unlike the unfortunates left to languish in the soup of historical failure, have a measurable affinity for the lock. The stringency of selection is easily controlled. A gentle knock on each lock and key complex as it is pulled from the soup will encourage poorly fitting keys to fall out. The harder each set of lock and key complexes is knocked, the greater the number of keys that will be dislodged. Other refinements may be introduced to increase the stringency of selection and thus the quality of the keys derived from the first round of selection. The number of rods might for example be increased, so as to give high-quality keys more of a chance to collide with the locks, or the duration of the fishing expedition extended. If we were to wash each rod carefully before harvesting the collection of lock and key pairs, most of the low-stringency binders would be displaced. Functionally inept keys which hitch a ride by sticking non-specifically to the line or locks could then be discarded. In such ways we would be able to minimize the number of undesirables in our selected set, which undermine the quality of a catch.

The keys which successfully jump these selection hurdles represent only a tiny fraction of the starting population. As in the real world we do not have the luxury of satellite maps, it is not possible to determine their landscape address and thus the local structure of the terrain in which they are situated. The only assumption we can safely make is that because they fit the lock, the keys must represent regions of the target-specific landscape which contain skyscrapers, rather than underdeveloped wasteland. Although skyscrapers are rare within these landscapes, they are not as rare as high skyscrapers. If the library of keys is complete, this is not a problem as it will always contain the best of all possible keys. But as their size is limited, artificial libraries are usually incomplete.

To synthesize and maintain a complete collection of keys would be both costly and time-consuming. The collection would also occupy an immense amount of space, most of which would be irrelevant to any specific lock. So rather than generating the complete library of all possible keys, it is more economical to make do with a random sample of its contents. There is a good chance that it will not contain the best of all possible keys, but it is likely to contain at least one that opens the target lock. The quality of the keys in any random sample will depend on its relative size and the nature of the library from which it was taken.

For these reasons, keys which pass the first round of selection are more likely to have originated from regions of the landscape corresponding to skyscrapers of a low or intermediate height than from the downtown monstrosities whose sheets of mirrored glass glisten in the summer sun. But as the population of selected keys is derived from a random sample of the whole landscape, it might be expected to contain ambassadors from disparate major and minor regions of skyscrapers. Some might represent isolated skyscrapers surrounded by featureless plains which extend in all directions. Others may be located in a peripheral metropolis and surrounded by globally minor but nevertheless higher and more readily accessible skyscrapers. Still others might be located at the edge of *the* Manhattan and have access to a walkway that extends high up above the taxi-cabs, delicatessens and street musicians, and terminates at the Empire State Building. Because we lack information about the structure of the local and global landscape in which each key is located, it is difficult to know how high any given skyscraper is, both in relation to its immediate neighbours and to the global maximum. It is possible that the loosest fitting and most inadequate member of the initial set of selected keys might be the one with the greatest potential for improvement and thus the brightest potential evolutionary future. Conversely, the key with the best initial fit may have only a rudimentary capacity for improvement. It is therefore prudent to think carefully before discarding any of the keys that have been selected.

Now that we know that superior keys are, at least in some cases, located close to the poor-fitting neighbours we have fished out, how might the first generation of survivors be refined to produce second-generation structures of a subtler and better adapted nature? How indeed might each of the good but nevertheless sub-optimal keys be empowered and given the opportunity to realize the potential inherent within the region of the landscape in which they are situated? A number of different strategies might be employed to construct the type of artificial evolution machine that could achieve these goals, indeed several have already been realized. The simplest strategy for implementing artificial evolutionary change uses nature's own procedure of generating a theme, introducing variations on that theme and then selecting a set of mutants from the newly generated repertoire.

Any Manhattan-like cluster of skyscrapers in a given RNA sub-space corresponds to a set of related RNA sequences. In combinatorial landscapes of this sort, similar items will always be found together.

Therefore if one particular RNA sequence has ribozymic activity, it is likely that many of its immediate neighbours will as well. Once you have located a Manhattan-like region of RNA sequences, movement from one skyscraper to another may be realized by changing individual ribonucleotides at discreet positions within the target sequence. For example in *RNA 6 Land*, running across the tops of a series of skyscrapers in the direction of the Empire State Building might involve movement across the *RNA Space* sequence landscape in the following way. The wild type **AAUAAU** sequence might first be changed to the mutant sequence **AAAAAU**. This may then be changed to **AAAGAU**, which itself may then be changed to **AAAGCU**. The process by which one nucleotide is replaced by another is known as mutation. In the example just given, the change from the wild type sequence to the mutant involved the sequential substitution of three ribonucleotides, or what we might call a three-step mutational walk. The information distance between the wild type parent sequence and the mutant has a numerical value of three. If we were fortunate enough to have a satellite map of the landscape, then we would be able to construct better fitting keys by tracing out mutational walks that lead from low skyscrapers to higher ground. The realization of such routes through *RNA Space* would generate sets of RNA sequence lineages, or family trees. We might decide to employ guides to help locate the most economical mutational walks through *RNA Space*. But we had better choose them carefully, as a poorly planned walk might drop us off the side of a skyscraper and leave us hurtling towards the ground.

Because nature has denied us access to maps of her theoretical spaces, it is not at the moment easy to design mutational walks through sequence landscapes that guarantee rich pickings. Although in some cases computer modelling and structural prediction enable us to design mutants with desired properties, the exploration of the unknown landscape that surrounds any individual item is best carried out by abandoning the principle of design and replacing it with an evolutionary strategy of exploration. This involves generation, selection, amplification with random modifications and then re-selection of a population of new variants. By taking a poor ribozyme and artificially introducing one or several mutations into its sequence, we are able to explore local alternatives in the surrounding sequence landscape.

If the repertoire of mutants is large enough, one of the neighbouring sequences may, by chance, lie on a mutational pathway that leads to higher ground. This upgraded ribozyme can then be amplified to produce a population which clusters around the new RNA sequence. By coupling the process of amplification to a process of mutagenesis, random changes can be introduced into the sequences of the progeny. The new population of mutants may then be fished in the same way as before. This time, however, each key and lock complex can be knocked a little harder, as survivors should fit with a greater affinity than those of their parent's generation.

This process of repertoire generation, selection, amplification and reselection coupled with mutation can be iterated any number of times. Mutational tours of defined regions of local landscapes enable the potential paths leading from low to high ground to be traced to their logical conclusion. The quality of the best possible obtainable key is limited only by the structure of the starting landscape and the efficiency of each evolutionary step. The structure of the landscape itself determines the ease with which it may be traversed by an evolutionary process. Whereas some landscapes are evolution-friendly, others are not so easily navigated.

If a landscape is uncorrelated and the skyscrapers are scattered randomly across the sequence landscape, movement from a poorly fitting key to an excellent key might involve crossing plains of diabolical keys. Given the constraints of evolutionary processes, a plain represents a forbidden area across which we cannot travel. The mutational path leading from the highest local skyscraper to a higher one in a neighbouring metropolis would involve the creation of a series of diabolical keys that would have no chance of fitting the locks at the end of the fishing lines. As each successive key must survive the trial of a rigorous test by the lock, these keys would rapidly be discarded and the lineage would become extinct. If, on the other hand, the landscape was correlated and the highest skyscrapers clustered together in a single downtown region, then mutational journeys in which every intermediate key was at least as good as its predecessor could be mapped out with ease. It is clear, then, that the success of both artificial evolutionary experiments and the historical explorations of nature is critically dependent on the structure of the landscapes across which the system evolves. We must presume that there are some landscapes which as a result of their lack of correlated structure, are totally unnavigable, despite their containing many optimal design solutions.

The extent to which landscapes may be efficiently searched by natural or artificial processes, is limited by the power of the mechanisms available for exploring their combinatorial complexity. The success of living things has depended on the evolution of such search engines. One important search mechanism is the reshuffling of genetic material that occurs as a result of a sexual process called recombination. This enables the search process to reach out to regions of the sequence landscape that are located at some distance from the wild type sequence. Exploration by point mutations, on the other hand, narrowly restricts the search process to the local neighbourhood. The success of artificial evolutionary strategies will depend on the development of techniques for searching information spaces more efficiently. Sexual-type recombination processes might, for example, be realized artificially. Indeed DNA recombination has been successfully mimicked in a test-tube.

One additional point relating to the structure of landscapes is a historical one. If landscapes had an 'all or none' structure, that is to say they were made up from either plains or one-billion-storey-high skyscrapers but nothing in between, then the historical possibility within the landscape would be restricted. For regardless of your initial position, you would have a choice between only two futures. With a landscape containing an array of skyscrapers of different heights, skyscrapers of equal heights might be located in widely separated regions. Although two separate skyscrapers of the same height might appear identical, the possible futures derivable from these two starting points may be quite different. The richness of the potential inherent within a given landscape is thus inextricably bound up with the nature and extent of its structural complexity.

If the RNA foundations of contemporary DNA-based life had originated from one skyscraper as opposed to a twin located in another region of the landscape, the history of life on Earth would have been quite different. Life as we know it has been influenced by a host of frozen historical accidents that resulted in lineages originating from one set of skyscrapers rather than near-identical alternatives located elsewhere in the degenerate landscape. The impact of these differences depend on the extent to which the landscape is correlated. In a perfectly correlated landscape, all the skyscrapers will be located in a downtown Manhattan region of the landscape. In such a landscape, different solutions to a problem are clustered together and the potential for connectivity between different alternative futures is high. In an uncorrelated landscape, however, the skyscrapers are scattered randomly across the landscape and the connectivity between possible futures is negligible.

It will soon be time to leave *RNA 300 Land*, *RNA Space*, *RNA World* and the *RNA Zoo* where our journey began. There are many further important points which we will not have time to address. How, for example, did the hypothetical *RNA World* meet its eventual demise and how did life based on explorations in *RNA Space* come to be replaced by a new type of life which flowed through the informational reservoirs of *DNA Space*? We may never know the answers to these questions. But we will doubtless eventually be able to construct plausible models of these processes, based upon theoretical and experimental considerations. How furthermore were RNA molecules able to replicate themselves and to prevent their information from decaying? And if these hypothetical RNA organisms were composed of a network of interacting ribozymes, how were individual self-replicating RNA sequences able to cooperate with one another? Some of these questions will be addressed in a later chapter. But, before we leave the *RNA Zoo* there are a few loose ends that need to be tied. Why did the RNA molecules in *RNA 300 Land* transmute into keys? How are keys and ribozymes related, and are there situations in which DNA molecules might function as catalytic keys? I mentioned that ribozymes are metaphorical keys which are able to unlock chemical reactions. Ribozymes, like enzymes, catalyze chemical reactions by increasing the rate at which they occur. But might catalytic RNA molecules and keys be related in a more literal sense?

The metaphor of a lock and key was introduced by Emil Fischer in 1890 to describe the mechanism of enzyme function. Fischer suggested that the shape of an enzyme is complementary to the chemical entity it associates with, much in the same way that the shape of a key is complementary shape to the lock it opens. The specificity of an enzyme for a chemical reaction results from the unique shape of its binding site. A space containing the shapes generated by all possible types of protein enzymes might thus be called *Protein Binding Site Space*. We might similarly imagine the *Space of All Possible Active Sites* generated by proteins which we could call *Protein Active Site Space*. In contrast to the binding site of an enzyme which is the region responsible for substrate recognition, the active site of an enzyme is the region where catalysis actually occurs. As it is the three-dimensional structure of a protein which is critical for its enzymic function, might not other technologies, such as RNA, mimic the three-dimensional shape of binding sites and active sites and in this way map directly onto *Protein Binding Site Space* and *Protein Active Site Space*? Indeed might not the repertoire of potential molecular shapes able to function as binding sites and active sites be more crucial than the technology in which the shape is realized?

Let us now imagine a binding site shape space called *Catalytic Shape Space*. This is the theoretical *Space of All Possible Three-Dimensional Molecular Shapes* able to catalyze the chemical reactions that comprise the complete catalytic repertoire housed within the *Space of All Possible Metabolisms*. *Catalytic Shape Space* documents both the abstract three-dimensional shape that a molecule forms in three-dimensional space, and exists independently of any particular technology in which these shapes and associated catalytic activities might be realized. Each three-dimensional molecular shape has a unique associated electrostatic surface potential. We might thus equally well call the space of all possible catalysts *Catalytic Electrostatic Surface Potential Space*. It is clear that the protein technology from which enzymes are constructed is able to map onto some of the shapes or electrostatic surface potentials that are contained within these spaces. There will doubtless be holes in the shape repertoire of proteins which represent potential molecular shapes and catalytic activities that no theoretically possible protein can realize. Some of these holes may, however, be filled by paying a visit to the *Protein Sequence Coupled with Molecular Chaperone Space*. Within this framework of reference, it seems reasonable to suggest that the catalytic abilities of ribozymes arise because RNA sequences are able to generate some of the biologically relevant architectures contained within the *Catalytic Shape Space* repertoire. Experimental evidence bears this out and suggests that enzymes and ribozymes catalyze chemical reactions in essentially the same way. Enzymes, however, appear to be better at any given catalytic task than their ribozyme counterparts.

The ability of RNA sequences to fold into complex three-dimensional binding sites has been demonstrated by several research groups. Dinshaw Patel has demonstrated that the small ATP molecules that are the universal currency of cellular energy can be accommodated within a ribozyme binding site. It appears then, that like the primary amino acid sequences of proteins, the linear sequences of RNA encode higher-dimensional information that dictates how the sequence will self-assemble into a three-dimensional shape. The structural and functional mimicry between RNA and proteins provides a physical basis for the metabolic competence of RNA molecules. But it is not shape alone that enables ribozymes to mimic the functions of proteins. It is also their ability to bind the metal ions that are essential components of active sites. By forming a scaffold able to display metal ions in a defined spatial orientation, RNA molecules can mimic this other important feature of enzymes. The differences in catalytic competence between enzymes and ribozymes may consequently, to some extent result from their differential abilities to bind and orientate metal ions in particular ways. Our *Catalytic Shape Space* thus accommodates a *Metal Ion Three-Dimensional Spatial Array Space*, which is itself just a subregion of *Active Site Space*.

Whereas proteins are constructed from combinations of twenty different building-block units, RNA sequences are assembled from combinations of only four building-blocks. How then, given their relatively limited chemical complexity, are RNA sequences able to generate repertoires of compact three-dimensional geometries reminiscent of the structural virtuosity seen in the binding sites of proteins? Evidence suggests that the folding of one-dimensional RNA sequences into higher order structures is a strictly hierarchical process. As with proteins, this is likely to occur according to a fundamental set of as yet poorly understood folding rules. The RNA folding problem may be broken down into the sub-problems of generating secondary structure and then collapsing a collection of secondary structure elements down into a stable three-dimensional structure. As the principal secondary structural elements of RNA sequences are short, independent, double-helical segments, the problem of RNA tertiary structure reduces to a question of RNA helix packing. But how do RNA sequences pack their short, multiple, disconnected and modular helical regions and organize them in space so that they can form active site architectures that map onto *Catalytic Shape Space*?

Unlike the double helices of DNA, that display their chemical side groups along the outside of the helix where they are optimally poised for chemical interactions, the unique chemical groups of the ribonucleotides in RNA are buried within each helix. RNA must overcome this structural handicap in order to generate complex geometries. This is achieved in a number of ways. The bread and butter units of RNA structure are the large stacks of helices which, as the result of contortions in space, are often positioned on top of one another to form a longer, but disconnected helix. These contortions are possible because of the many conformational degrees of freedom that its single-chain sugar-phosphate backbone allows. The organization of the stacks of helices themselves is accomplished by several means.

The single hydroxyl group (2′-OH) that distinguishes the ribose sugars of RNA from the deoxyribose sugars of DNA plays a critical role in facilitating RNA helix-packing. Whereas the chemical groups of the RNA bases are hidden within helices, the 2′-OH groups which line the outer edge of the ribose sugar chain are optimally positioned for chemical interactions. These are able to form an extended network of hydrogen bonds between one another, collectively known as the ribose zipper. Unlike DNA in which Watson-Crick base-pairing rules only allow **A** to **T** and **G** to **C** bonding to occur, in RNA, non-classical base-pairing is common. It is often found within helical regions of RNA and produces perturbations within their structure. Particular types of helix perturbations appear to be conserved in RNAs from different species, suggesting that they are important elements of RNA structure. The importance of non-classical base pairing events such as **A-A**, **G-A**, **G-G**, **C-U** and others, is that they make normally inaccessible chemical groups available for chemical interactions. The generation of helical imperfections by non-classical base pairing is thus essential for the formation of RNA tertiary structure.

In addition to their role in catalysis, metal ions with two positive charges, such as magnesium, also help produce RNA structure. They function both to neutralize the negative charges of the sugar-phosphate backbone, which allows closer packing of adjacent helices, and to form anchor sites around which secondary structural elements may be organized. In addition to these mechanisms, long-range stabilizing interactions between helices are mediated by a host of higher order motifs. Helices may, for example, dock against one another using specific sequence motifs and their corresponding receptor regions. One example of such a motif is the **GNRA** loop (and its corresponding receptor), and the **GAAA** loop. In the **GNRA** loop, the **N** can be any base and the **R** either an **A** or **G**. These types of motifs staple themselves to their corresponding receptor sites and in so doing, introduce kinks and bends into the RNA sequence. Although these mechanisms provide ample opportunity for RNA sequence architectural complexity, evidence suggests that the chemical diversity of primordial RNA sequences may have been further increased by the incorporation of chemically modified bases which may have formed under prebiotic conditions.

I mentioned earlier that the principal reason why DNA is unable to generate a diverse repertoire of three-dimensional structures is because of its double-helical structure. The *RNA World* theory relies upon the fact that double-stranded DNA is unable to generate architectures that map efficiently onto *Catalytic Shape Space* and perform catalytic functions. RNA, on the other hand, is single-stranded and therefore a molecular jack of all trades that can both store information and effect its biochemical consequences. Unlike the ribozymic catalytic activities found in some naturally occurring RNA sequences, no biocatalytic DNA molecules are known to occur in nature. The apparent absence of a naturally occurring DNA enzymes is, as we have already said, likely to reflect the fact that nearly all the DNA in modern organisms has a double-stranded structure which prevents it from generating complex architectures. That this is the case in modern organisms, however, does not mean that ancient organisms did not utilize single-stranded DNA sequences for informational or structural purposes. Given the chemical similarities between DNA and RNA, it seems reasonable to wonder whether if liberated from the confines of a double-helical environment, single-stranded sequences of DNA might also be able to generate complex three-dimensional architectures which map onto *Catalytic Shape Space*.

As single-stranded DNA molecules lack the 2′-OH groups that form the ribose zipper interactions so critical for the generation of RNA structure, it might seem unlikely that single-stranded DNA could produce a repertoire of complex structures that even approached the extent of the potential ribozyme repertoire. But although single-stranded DNA sequences would be unable to form structures realized in an identical manner to their RNA counterparts, they might be able to use other types of interactions to achieve the same overall folds.

The catalytic potential of DNA has been investigated using artificially generated single-stranded DNA libraries that have been subjected to artificial evolution techniques. A range of artificial DNA enzymes, or what we can call DNAzymes or deoxyribozymes, have been found within combinatorial libraries, including sequences which ligate short single-stranded regions of DNA and sequences that cleave RNA sequences. The performance of DNA enzymes is furthermore every bit as good as that of their ribozyme counterparts. Preliminary structural studies indicate that single-stranded DNA is able to generate a surprising diversity of tertiary structures. These include several of the structural motifs found in folded RNA molecules. The full nature and extent of the *Space of All Possible DNA Enzymes* is, however, yet to be explored.

Are we not compelled to revise our notion of the historical plausibility of the *RNA World*? Might our early ancestors have had enzymes made from DNA and thus inhabited a *Single-Stranded DNA World*? Indeed, if DNA has the potential to function both as an informational molecule and as an enzyme, then we appear to have solved the chicken and egg paradox in a way that does not oblige us to invoke the agency of an ancient RNA technology. We cannot exclude this possibility, but the fact that RNA enzymes play a critical role in modern biochemistry whilst no DNA enzymes of natural origin have yet been identified and, moreover, that some viruses have RNA genomes, may be taken as evidence in favour of the *RNA World* hypothesis. If naturally occurring DNA biocatalysts are one day discovered, we might have to re-evaluate the plausibility of the *RNA World*.

The differences between DNA and RNA enzymes might be more

subtle than is initially apparent. The structures of protein enzymes are, for example, able to accommodate multiple point mutations without significantly altering their specificity and catalytic function, whilst single point mutations in ribozymes or deoxyribozymes are found typically to completely destroy their catalytic activity. This makes DNA and RNA enzymes appear more brittle, less robust, and consequently less evolvable than protein enzymes. The sensitivity of these technologies to small changes might, however, have been an early advantage, as it would have enabled sequences to take large leaps through *Catalytic Shape Space* as a consequence of only minimal sequence changes. But it would at the same time make such components extremely fragile and susceptible to corruption by random change.

So RNA is not unique in its ability to furnish informational and metabolic functions, but to date there is no direct evidence to suggest that artificial attempts at forcing DNA to behave like an enzyme has any direct relevance to the historical events responsible for shaping modern life. However, the fact that the specificity and affinity of deoxyribozymes appears every bit as good as their ribozyme counterparts, suggests that the apparent primacy of RNA molecules over DNA in the informational and metabolic structures of early creatures was more a consequence of the relative availability of RNA, than as a result of any intrinsic technological superiority of RNA over DNA. It is also possible, however, that single-stranded DNA technology was overlooked by history, as a consequence of its relative inability to furnish landscapes which could be easily and efficiently navigated by natural search processes.

The fact that the Watson-Crick base pair interactions of single-stranded DNA and RNA molecules are chemically programmed into their sequences, enables deoxyribozymes and ribozymes to form compact folded structures. These may be generated from very short sequences. Proteins on the other hand, generally appear to require far longer amino acid sequences to form complex folds. Thus whereas short prebiotic DNA or RNA sequences are likely to have been able to generate extensive repertoires of catalytic activities, short amino acid sequences are less likely to have been able to do so.

As we exit the *RNA Zoo* its iron gates slam shut. The air is sulphurous and the zoo keeper whose smile had previously appeared amiable, now seems quite wicked and perverse. Our pass has expired and we must disappear from this place. The iron letters that had read 'Welcome to the *RNA Zoo*' have been erased and replaced with featureless metal strips. As we leave the *RNA Zoo*, the potential RNA animals hiss and scream. Their desperate cries echo through the night-time air and follow us as we trace our way through the darkness. At one point I hear footsteps behind me, but resist the temptation to look back. Soon we are far away from the *RNA Zoo*, and the formless shadows of the wretched potential RNA creatures that it contains melt imperceptibly into the sound of rattling chains and the pall of dank smoke that begins to obscure our view.

8

Analog Creatures

On the 18th of December 1832, Charles Darwin and captain FitzRoy left the *Beagle* and rowed ashore. They had left Montevideo two days earlier and were now in Good Success Bay, near the southern tip of Tierra del Fuego in South America. The air had 'the bracing feel of an English winter,' but this did not deter them, for meeting 'savages' was to be one of the highlights of their trip. As Darwin observed these 'wretched looking beings', their wild behaviour and 'hideous grimaces' made them seem like 'troubled spirits of another world'. He also observed that, on seeing the bare arm of a member of his party, the Fuegian natives 'expressed the liveliest surprise, just in the same way in which I have seen the orang-utan do at the Zoological Gardens'. Just over two months later, on the 24th of February, as they lay anchored at Wollaston Island just above Cape Horn, Darwin encountered another group of native Fuegians. He found these 'unbroken savages' detestable and yet was at the same time intrigued by them. From whence had these people come? For 'although essentially the same creature', there was indeed a world of difference 'between the faculties of a Fuegian savage and a Sir Isaac Newton'.

Darwin was clearly shocked by the notion that modern Victorian gentlemen might have been derived from such humble stock. Where indeed was the dignity for humans who 'slept naked and scarcely protected from the wind and rain . . . on the wet ground coiled up like animals?' He also reflected that 'it is a common subject of conjecture what pleasure in life some of the lower animals can enjoy', but 'how much more reasonably the same question may be asked with respect to these barbarians!' Finally on leaving the cove at Woollya on the 5th of March, he noted, 'I believe, in this extreme part of

South America, man exists in a lower state of improvement than in any other part of the world.'

In the previous chapter, we considered the possibility that the earliest forms of life may have utilized RNA rather than DNA technology. Although the *RNA World* hypothesis remains speculative, a significant body of experimental evidence suggests that such a world could have existed. Furthermore, the *RNA World* resolves the chicken and egg paradox that overshadows the question of how life originated on Earth. But what existed before RNA technology came to monopolize the gene construction business? And what might the unsophisticated, miserable and degraded 'molecular savages' that preceded the *RNA World* have looked like?

Evidence suggests that any number of different molecules might have encoded information using the digital combinatorial strategy employed by modern genes. But to date, RNA is the only informational molecule known to have the potential to catalyze a wide range of chemical reactions in a manner similar to enzymes and which incorporates both genetic and metabolic dimensions. This obliges us to consider the possibility that life began with genes constructed from RNA, and that the family trees of all living things can be traced back to one or more ancestral RNA molecules.

But I would like to consider a different possibility, which is perhaps even harder to accept than the notion that the first genes were made from RNA. What if the very first organisms lacked genes altogether and in the words of Stuart Kauffman, '. . . large aperiodic crystals carrying a microcode' are '. . . neither necessary nor sufficient for the emergence and evolution of life'? What if rather than being present right from the very beginning of life, genes were a technological innovation that were only introduced at a later date? Genes would not then constitute a necessary feature of life, but an historical embellishment. That is not to say that geneless forms of life would have been able to achieve anything like the degree of complexity seen in their gene-utilizing descendants; there were indeed good reasons why primordial geneless organisms were eventually supplanted by genetic biomachines. Like Darwin on his first encounter with 'savages', we might revile the idea that creatures as lofty as ourselves could have had such humble beginnings. Indeed we might feel as if the very ground of biological certainty was collapsing beneath our feet. Nevertheless, address such issues we will and in doing so, we will, like Darwin, finally come to realize that our hypothetical geneless ancestors were probably far more sophisticated and refined than we might have imagined.

The place that we will shortly be entering is somewhere far more foreign than the *RNA World*. For at least RNA genes mimicked the essential template-based digital logic of their DNA-based counterparts. Furthermore unlike the *RNA World*, most of the molecular fossil evidence for this hypothetical period of life's history is likely to have been erased. The only way that we will initially be able to resurrect this enigmatic hypothetical world inhabited by geneless organisms is by enlisting the help of computers.

The existence of what I will call the *Geneless World* is purely hypothetical. Unlike the *RNA World* hypothesis which rests upon sound experimental foundations, the *Geneless World* hypothesis rests on theoretical and only very fragmentary and tangential experimental foundations. These are compelling enough though, for the possibility of a geneless era preceding the origin of the first genetic organisms to be taken seriously. Furthermore, although experimental evidence for its biological plausibility is rudimentary, this may simply reflect the fact that the appropriate experiments have not yet been attempted. It would indeed be far more worrying if the *Geneless World* hypothesis was not falsifiable.

Before examining the nature and structure of the *Geneless World* and the theoretical work with which it is underpinned, it will be necessary to clarify some of our terms. It seems reasonable to begin by examining what we mean by the word 'gene'. We have already seen that genes may be realized in more than one technology, so clearly neither DNA nor RNA need form part of the definition. What we require is a more general definition that captures the logical essence of genes. One might reasonably suggest that genes are: (1) discreet chemical polymers that (2) encode information using a combinatorial strategy that employs unique sequences of discreet digital symbols and which (3) can self-replicate with or without the assistance of other complex chemicals. (4) Their replication incorporates the principle of template-based complementary copying whereby (5) the sequence information should be conserved so that the products of replication share a core informational structure with that of their parents, but (6) should sometimes have slight, and occasionally large differences randomly or semi-randomly introduced into their informational structure by events such as substitution, deletion, insertion and recombination of the digital symbols from which they are made. (7) The rate at which these changes are introduced and fixed into the digital sequences should be closely regulated or fall naturally below a certain threshold value, so as to maintain a balance between maintaining the integrity of pre-existing informational structures and generating new information which (8) will occasionally affect their differential survival potential so that some daughter genes will compete more efficiently for representation in the next generation and the population of genes may evolve across time. (9) Although as a result of their chemical composition some digital sequences may have a direct survival advantage over alternative sequences, biological meaning is encoded in the one-dimensional pattern in which the symbols are arrayed. This abstract symbolic information is made explicit by the application of an external decoding and translation process. Genes thus possess a *semantic* dimension in addition to a purely *syntactic* dimension. In the case of DNA, this enables a one-dimensional structure written in DNAese to be translated into a three-dimensional structure written in amino acidese. Genes are thus (10) molecular ciphers in which biological information is systematically codified. They constitute a central information centre from which the programs that specify the dynamic structure of living things are operated. (11) They are also able to compress information.

A finite resource of a small set of discreet digital symbols may thus be combined in unique ways according to a simple code to generate a potentially infinite range of genetic meanings. This capacity to make infinite use of finite means by harnessing the power of combinatorial possibility is the hallmark of digital genetic strategies of information acquisition, storage and transmission. In the case of genes constructed from RNA, the information in the primary sequence is expressed in the same RNAese language in which it is encoded and does not need a translation mechanism. The linear RNA sequence itself embodies the folding rules that allow RNA molecules to form complex three-dimensional structures. Given the above definition, one could reasonably argue that self-replicating RNA sequences that did not encode proteins were not true genes, and that the RNA sequences which encoded the ribozymes of the first hypothetical RNA-based organisms that inhabited the *RNA World* only became true genes when they acquired the capacity for codification and an accompanying translation mechanism and machinery for protein synthesis. We might consequently wish to subdivide the *RNA World* into an early non-genetic era and a later transitional genetic era in which the metabolisms of these RNA creatures began to incorporate encoded protein elements. This might more accurately be referred to as the *RNA/Protein World*.

We might choose to make our definition of genes more general by, for example, dropping the requirement for template-based copying, or the requirement that genes need be chemical in nature. But given that DNA-based genes have provided the informational foundations for both contemporary and large chunks of ancient life, it seems reasonable to restrict our definition of genes in the DNA-biased way that I have suggested. This approach will also help highlight differences between the classic DNA and RNA genes that have been the historical life-construction industry standard for more than three billion years, and hypothetical non-classic informational replicating entities which although performing a function approximating that of prototypical DNA and RNA genes, are likely to have been structured in very different ways.

We have now considered a preliminary definition of a gene that is suitable for our current needs. In formulating this definition, we have established the boundaries of, and drawn a limit to a hypothetical space which we might call the *Space of All Possible Types of Genes*. Every item in this space must satisfy the inclusion criteria outlined above. According to this definition one might choose quite legitimately to exclude early RNA organisms that lacked symbolically encoded metabolisms and include only those which translated the information of their RNA sequences into protein sequences.

What I would now like to suggest is that the very first living things, the hypothetical molecular savages that established the informational foundations of modern life, may have been made from informational structures that were non-digital in nature and which, like the earliest

RNA organisms, did not encode their information symbolically. Violation of these two critically important inclusion criteria would exclude such entities from the privileged 'gene club' set. If this were the case, then it follows that the very first living things could not have been made from genes. It would after all be futile to search for a shoe in the *Space of All Possible Bicycles*. They must instead be hunted down in their own appropriate space or zoo, which may be called the *Geneless World* or *Geneless Zoo*. Given the importance we have attributed to the digital exclusion criterion in our definition of a gene, it is worth exploring the meaning of the terms digital and non-digital in more detail.

Perhaps the best way to understand the meaning of the term digital, is to consider its antithesis. That is a machine which acquires, stores and transmits information using what is known as an 'analog' strategy. A good example of such a machine is an old-fashioned horn gramophone and the records it plays.

The singer Enrico Caruso was born in Naples in 1873 and was the fifteenth of twenty-one children. He is thought to have been one of the greatest tenors that ever lived. When auditioning for his first major engagement as Rodolfo in the opera *La Bohème*, the composer Puccini, who was holding the audition, is said to have leapt from his piano and cried, 'Who sent you to me? God?' Caruso's performances set the operatic world alight. Fortunately for us, they coincided with the invention of the gramophone. Before his untimely death at the age of forty-nine, Caruso had recorded more than 260 titles.

Emil Berliner's invention of gramophone record technology made it possible for musical events to be captured for posterity. Indeed if one is lucky enough to own a horn gramophone, it is still possible to recreate Caruso's magnificent voice, immortalized forever in the complex grooves impressed onto circular pieces of black plastic. The horn gramophone and the Caruso record that it might be playing stand as ambassadors for the great tenor himself. In imitating Caruso, the machine brings certain aspects of his ghost into our sitting rooms. But let us now imagine that we have been fortunate enough to find an old horn gramophone machine. When we get home we pull a dusty Caruso record from its sleeve, wind up the machine and let it play. As we settle down and stretch back in our armchair, the air fills for a full four and a half minutes with the characteristic and unmistakably haunting sound of Caruso's voice.

The first 'talking machine' able both to record and reproduce sound was invented by Thomas Edison in 1877. In Edison's machine, the vibrations of air that underlie the sensation of sound are represented in the form of small indentations which are scratched by a stylus attached to a vibrating diaphragm onto a sheet of tinfoil wrapped around a rotating cylinder. The continuously changing information of the sound is encoded by corresponding variations in the depths of the indentations, which are traced in a helical pattern across the surface of the foil. Emil Berliner's gramophone technology which, towards the end of the nineteenth century began to replace Edison's talking machines, substituted the rotating tinfoil cylinder with a flat zinc disc covered with a layer of wax. The process by which the information of Caruso's historical performances was frozen in time and immortalized on a gramophone disc may be understood in the following way.

The performer, in this case Caruso, stood close to a conical recording horn which collected and concentrated the vibrations of the air created by the performer's vibrating vocal cords onto a small diaphragm located at its apex. The diaphragm itself vibrated in sympathy with the perturbations of the air, in a manner that reflected the frequency and amplitude of the incoming sound. A cutting stylus attached to the diaphragm translated these vibrations into an undulating spiral groove imprinted into the wax lining the surface of the zinc recording disc. The information of the sound was encoded either in the depth of the groove, or, by 1900, as lateral contours in a groove of constant depth. On completing the recording, the imprinted disc was dipped into an acid bath. The acid only cut into the zinc at points where the stylus had obliterated the protective wax layer. This acid-etched master disc constituted a permanent 'record' of the information of the sound generated at the time of the performance.

The zinc master 'positive' was used to make a copper template 'negative', known as a stamper. In the stamper, the grooves of the zinc positive were represented by an exactly corresponding and continuous multi-contoured ridge. The stamper was then used as a template in a plastic-moulding press for the mass production of near identical plastic daughter positives. If more than a hundred or so daughter copies were required, the stamper was used to make a copper positive, known as the mother copy. This was used to make several second-generation stampers for use in plastic record production. Thousands of copies of the original zinc master could be made in this way without much loss in the quality of the daughter's informational record, as any stamper that showed signs of deterioration was promptly replaced.

The diamond or sapphire reproduction stylus on a gramophone machine is able to mechanically access the sound memory imprinted into the spiral grooves on the surface of the record and recreate the pattern of air vibrations present at the time of recording. This is achieved by translating the information encoded within the contours of the grooves into mechanical vibrations of the stylus. These are then translated into vibrations of the gramophone's diaphragm, which establishes vibrations in the surrounding air. These in turn are amplified by the horn. In the case of electrical gramophone machines, the stylus is attached to a magnet surrounded by coils of wire. Movement of the stylus in the spiral groove produces a movement of the magnet which induces a corresponding current in the surrounding coils of wire.

It is interesting to note that in recreating certain aspects of the auditory information present at the time of recoding, hand-cranked gramophone machines and their electrical counterparts function as imperfect, but nevertheless acceptable engines for time travel within a hypothetical space which we might call the *Space of All Possible Sound Information*, or *Sound Space*. If it were possible to capture and exactly recreate the auditory information present at the time of a musical performance, then there is a very real sense in which the listener may be said to travelling in time within the dimension of sound. This would constitute an unconventional but nevertheless acceptable mode of time travel, as *Sound Space* exists outside time. Time, in this context, is just a yardstick with which to measure realizations of *Sound Space*, a peg on which the historical events that it allows may be hung; like a multidimensional row of hats in a cloakroom of infinite size.

Travel within this space is achieved by generating auditory information of sufficient complexity to mimic different regions of *Sound Space*. It might not be possible to do this to an infinite degree of precision, but this should not matter as, if the target information structure and its mimic are indistinguishable, they can for most purposes be considered as being identical. What is important, however, is that the resolution of the facsimile is higher than that of the device used to represent its information. The hair cells in a human ear are sensitive to sound of frequencies between 20 and 20,000 Hz. Although the vibrations of Caruso's vocal cords might produce sound of a frequency greater than 20,000 Hz, it would not be audible to a human listener. A passing cat or dog might, however, prick up its ears. So while a complete description of the auditory information would need to include both ultrasonic (above 20,000 Hz) and infrasonic (below 10 Hz) frequencies, these aspects of the sound do not form part of human auditory experience and might thus be omitted to produce a more compact description.

As *Sound Space* is an infinite hypothetical space and contains descriptions of all possible past, present and future auditory events, it is something of a treasure chest of curiosities. For example, in a subsection labelled *Historical Sound Space*, we might find the voice of the young Julius Caesar singing happily as he ran along the banks of the Tiber. We might also overhear secret conversations between the conspirators of the Gunpowder Plot, chance upon the cry of a dodo, or be an auditory fly on the wall in Hitler's bunker. We might similarly stumble upon the region which contains the *Complete Set of Sounds Which Might Have, But Never Did Happen*. Here we may find the conversations that we wish we had had, the unsung songs and the laughter that never rang out.

But these islands of auditory meaning are twinkling stars in a sea of cacophony. In order to find even a sliver of meaningful sound, we would have to traverse vast expanses of distasteful noise. Of course what constitutes noise is subjective, and today's noise may be tomorrow's music. We might choose to organize hunting excursions into *Sound Space* to retrieve the information of the complete set of historical Caruso performances. More refined adventurers might hunt for Caruso practice sessions, or search the *Space of All Possible Caruso Performances* for historically unrealized material. But as the search space is infinitely large and not necessarily correlated, it would be like searching for a specific molecule of water in an infinite ocean.

Such hunting excursions will thus remain plausible but unrealizable propositions until such a time as search algorithms of sufficient power can be devised.

The invention of the first vacuum tube amplifier shortly before the First World War, marked the beginning of a transition from an acoustic era of sound recording to one of electronic recording. Over time, the early acoustic gramophone players were replaced by electrical machines of increasingly higher fidelity. Much of the early impetus for technological innovations in gramophone machines arose from the competition generated by the introduction of public radio broadcasts in 1919. In modern recording systems, microphones are used to translate sound into voltage oscillations proportional to the frequency and amplitude of the incoming sound. These are then fed into an amplifying device.

The introduction of dynamic loudspeakers in 1925 did away with the need for clumsy and ineffective acoustic horns, while electric motors dispensed with the need for hand-cranking. In 1948, Columbia Records introduced the first long-playing records (LPs). These were made from vinyl and played at a speed of thirty-three and a third revolutions per minute (rpm) instead of the standard seventy-eight rpm. Fine microgrooves were cut into their surfaces, which allowed the storage of much more information per unit of disc space and enabled playing times to be extended from the standard four and a half minutes, to around thirty minutes per side. The first commercially available stereophonic systems were introduced in 1958. In the 1950s and 1960s a new breed of high-fidelity (hi-fi) enthusiasts emerged. These people went to extraordinary lengths to obtain perfect sound reproduction. Their methods included tinkering with electrical components and mixing and matching turntables, amplifiers and loudspeakers in order to produce the best possible sound.

Although the few remaining gramophones have undergone significant changes since the days of hand-cranked machines, the analog logic underlying their function remains the same. Analog processes of information storage, acquisition and transmission create a permanent physical record or representation of the target information which is directly 'analogous' to its informational structure. The information stored in this physical representation is retrieved through the implementation of a mechanical process. The closer the correspondence between the analog 'picture' of the target and the information itself, and the more efficient the machine, the higher the fidelity of the reproduced sound. In an ideal analog recording process, the physical representation of the information issuing from the sound source would be an exact facsimile of that information. A perfect analog machine would be sensitive to all the variation in the target sound wave that occurs at the infinite number of time intervals into which a performance can be divided. In a perfect electrical machine, these infinitesimally accurate values that change across infinitesimally small durations of time, would be translated into a set of infinitely detailed voltage values. The infinitesimally accurate value of the oscillating voltage in the output wire of the microphone at every infinitesimal instant would then be translated into a value of an infinitely variable physical quantity. In the case of a gramophone record, this would be the depth or width of the groove cut into the surface of the master. In an ideal material, the depth and width of the grooves would be infinitely variable so that infinitesimal differences in the numerical values of the voltage could be faithfully mirrored in a correspondingly infinitesimal change to the groove cut into the master. The stylus of an ideal gramophone machine would then need to be infinitely sensitive to the infinite set of values that the width or depth of the groove may adopt at any position.

Any recordings made by an idealized analog machine would provide a near-perfect channel of communication between the performer and listener. But even an idealized communication channel can create only a plausible illusion of the historical performance. There are several reasons for this, which include the fact that, unlike the human ear, microphones do not hear selectively and are sensitive both to the target sound and ambient noise. Furthermore, the loudspeakers used to reproduce recorded sound radiate into a considerably more restricted acoustic space than the original sources of sound.

Gramophone records have their own unique sound and charm. A newly-pressed record played on a state-of-the-art player may reproduce sound more faithfully than any other type of machine. Analog strategies of information storage, reproduction and replication are however characterized by several constraints and the ideal of a perfect analog communication channel is for all practical purposes unrealizable. These constraints arise as a consequence both of the mechanical nature of analog machinery and from the way in which such devices attempt to describe the target information in infinite detail. It is worth examining some of these constraints, so that we can later contrast hypothetical analog, geneless, biological systems with their modern digital genetic counterparts.

Analog systems use physical variables such as voltage or the shape of a groove to represent information. A mechanical process is needed to mediate between the information and its physical representation. Mechanical processes are invariably imperfect. Noise can be introduced into the communication channel between the source and the receiver and the information of the signal is also distorted. This undermines the fidelity of the processes by which information is acquired, stored and reproduced in these machines. The origins of this noise and distortion may be understood by examining the processes that generate the channel through which the information is communicated.

In gramophone technology, the information of sound is translated into a physical representation by a mechanical process. The accuracy of this representation is limited both by the translation efficiency and the storage medium. Although all of the voltage values can in principle be represented, in practice the recording stylus has only a finite degree of resolution and cannot discriminate between values which differ by very small amounts. The malleability of the master also constrains the fidelity of the process. Both the mechanics of translation and the storage medium thus fall short of the ideal of representing infinitely small differences between components of the target information.

Noise and further loss of fidelity is introduced into the system at many other points: in a perfect machine the drive mechanism which rotates the disc would be noiseless, vibration-free and would move at a constant speed; the arm which holds the stylus would also be perfectly counterbalanced so that movement of the stylus would be frictionless and influenced only by the shape of the groove; it would furthermore be unaffected by the vibrations of the stylus and would be adjusted to compensate for the tendency of the stylus to 'skate' and for the fact that it presses harder on the inner side of the groove than on the outer. Many other factors including the inertial forces that result from the up-and-down movement of the arm also need to be accounted for. The machines also have a tendency to deteriorate and become less efficient as a result of use. These types of factors also undermine the fidelity of the information which is acquired, stored, and reproduced in analog machines.

The noise and loss of fidelity in analog systems has profound consequences for the fidelity with which they are able to replicate information. Over time, the information that flows through them will decay until a point when the signal is indistinguishable from background noise. The imperfect nature of information replication in these systems sets a limit to the number of generations across which information can be faithfully transmitted. Information replication in analog systems is like taking a photocopy of a photograph and then making a photocopy of the photocopy and subsequently taking successive photocopies of this photocopy. After several generations the information of the parent structure would change beyond all recognition. With gramophone records, the imperfect and indirect nature of the process by which the information of the master is replicated means that even brand new and perfectly pressed daughter copies will never be exact facsimiles of the master. A perfect analog clone is not in fact realizable even in principle, as in order for it to be an exact copy of the master, a clone would have to be synthesized with an infinite degree of precision.

Plastic records cannot themselves be used as templates for the production of daughters. If a plastic positive was used as a reproduction template, the first generation of daughters would be quite different to the master stamper. The only way to make high-fidelity daughters is to press copies from the original stamper. But although the first generation of imprints from the master stamper may be faithful replicas, after second, third and more generations even the master will become worn and its information degraded. If the master positive is discarded or lost, there is no reference structure with which the degraded information of the master stamper can be compared. This might not be important if (for example) Caruso recordings, went out of fashion after only a few thousand records had been produced; but if Caruso remained in fashion for thousands of years, his recordings would eventually become distorted beyond all recognition. Once lost, this information would be irretrievable. The problem of reproduction across generations can to some extent be circumvented by producing large numbers of mother copies from the metal negative, but constraints of economy and storage space usually prevent this.

Because of the physical nature of its representation, the information stored within analog substrates is easily damaged or lost and thus unstable across time. Whether the information is copied or not, successive reproductions will be of a progressively poorer quality as the core information melts away. One has only to try and play a record which is severely scratched or warped to illustrate this point. Because of its continuous nature, damage to the information in analog systems is difficult to detect and repair. The fidelity of the information of analog systems is thus not easily maintained by the introduction of error-correction devices. Wear and tear introduces incremental changes into the representation which results in a cumulative corruption of its information. It is this potential for decay to a limitless range of infinitesimally different values which makes the detection and correction of errors so difficult. For this reason, information stored within analog systems is prone to irreparable damage.

The inherent instability of analog information makes analog machinery an unreliable storage medium. Information stored in this way has only a very limited capacity for evolution. Errors are not introduced onto a background of stability, as every part of the information may change at any time. The rate at which change may be introduced into the information cannot, furthermore, be regulated or tuned by error-correcting mechanisms. Analog systems are thus prone to an error catastrophe in which the rate at which change is introduced into their information exceeds the rate at which optimal mutants can be selected. The result is an informational meltdown and the core information of the system is destroyed. Analog modes of information storage are not, furthermore, compact and are thus uneconomical. Even relatively small amounts of information require large storage spaces. The limited storage capacity of analog systems results from their mechanical nature.

The information stored in analog systems has only rudimentary flexibility. The information stored in one type of analog representation is not easily accessed by a machine that stores its information in a different type of physical substrate. Information in analog machines is consequently a private currency, stored in a private analog language. Analog machines of one particular type are thus unable to communicate with other sorts of machines. The information and the substrate in which it is represented are inseparable. Furthermore, the nature of the process by which analog information is represented limits the extent to which it can be edited.

The point may once again be illustrated by gramophones. The information stored in a record cannot be extracted without a gramophone machine. The representation of the sound encoded within the grooves of a record thus constitutes a private language which is unintelligible to other types of machines. The potential for generating sound that is encoded in the furrows on a record's surface cannot consequently be realized by different machines. Indeed without a gramophone machine to translate the undulating language of plastic or vinyl grooves into sound, a record would be useless. The information stored within a record is thus locked into a symbiotic relationship with the gramophone machine on which it is played. The plastic groove language of a record is, for example, incompatible with the magnetic tape language of a tape recorder.

The master copy of a gramophone record may be replicated, but cannot be edited. The content of the information stored within the grooves is fixed and immutable, although it is inevitably corrupted by the wear and tear of playing. Individual records may vary in quality as a result of inconsistencies in the manufacturing process, but each copy produced in a given generation will be essentially identical. Not all analog representations are this inflexible. Magnetic tape, for example, can be cut and pasted. The precision, however, with which such procedures can be executed is constrained by the mechanical nature of the process by which such procedures are implemented.

Analog sound-recording and reproduction machines like horn gramophones, high-fidelity gramophone players, reel-to-reel tape recorders and cassette recorders have now been superseded by digital compact disc players, mini discs and digital audio tape (DAT) machines. These new digital machines store the information of sound in a very different way to their analog counterparts, and in so doing free themselves from the constraints of analog information processing. Unlike analog machines, which form a physical picture of the sound they are recording, digital machines do not. Instead, the continuously oscillating voltage in the microphone which corresponds to the information of the sound waves is sampled at discreet intervals. The value of the voltage at any given moment is then converted into a numerical value is encoded as a binary number. In binary language, any number can be represented by a string of 1s and 0s. Modern digital sound machines use binary words that consist of combinations of up to sixteen 1s and 0s, which are able to represent any numerical value between 0 and 65,536. The changing value of the voltage across time is thus represented as a series of instantaneous snapshot pulses of information. These are encoded as a pattern of binary words. Rather than being a copy of the target sound, these pulses of binary words are instead a set of abstract instructions for recreating the information of the sound. In selectively sampling the information of the target sound waves, digital machines capture only a fraction of the available information. Nevertheless, as long as the sampling rate is equivalent to or greater than the highest frequency in the target sound, the digital snapshots can be used to reconstruct the target sound. While gramophones represent the information of sound using a potentially infinite number of values, compact disc players and DAT machines do so using a relatively small set of finite values. Unlike the continuous grooves of a record, the sequence of 1 and 0 symbols which encode the information in digital representations is discreet. The symbols may be present or absent, but cannot adopt intermediate values.

Discreet chunks of digitalized information may be used to describe everything and anything. The English language constitutes a good example of a digital process of information acquisition, storage and transmission. The finite resource consisting of the twenty-six digital letters of the English alphabet may be sequenced in infinitely many ways, to produce a *Library of All Possible Books*, which the writer Jorge Luis Borges has called the *Library of Babel*. There is indeed a sense in which we need never write a book again, as every possible book already exists in a disembodied mathematical incarnation. If we were to program a computer to generate every possible combination of the twenty-six letters in the English alphabet, along with every possible interval between letter strings of different lengths and every possible combination of punctuation marks and then print every book out on a laser printer at home, we might generate a significant portion of the *Library of Babel*. There may not be enough space on Earth to house even a fraction of this library, or enough time and molecules to construct it, but that is beside the point.

We can view the *Library of Babel* as nothing more than a mathematical artefact. There is, however, a sense in which this library really does exist, even if most of its potential mathematical form will never be realized. This mathematical space is thus another invisible landmark which is 'out there' and as real as more familiar landmarks such as the Colosseum, or the Victoria Falls. If one espouses this type of mathematical Platonism, then the *Library of Babel* is a place that one can actually visit. If one day someone were to print out a large chunk of the *Library of Babel*, then somewhere within it we should be able to find the entire works of Shakespeare, an ancient bible, a copy of *The Times* newspaper dated 15th May 4004 and the complete works of Freud, Newton, Keats, Eliot and Hume, as well as every different version of these works. The greater majority of the texts, however, would be nonsense. The fine seams of sense would in fact be numerically irrelevant compared with the seas of incoherent texts, and, if the library was not ordered in some way, it is unlikely that you would ever find the book that you were looking for, or have any notion of the riches waiting to be discovered.

There is a very real sense in which Shakespeare's *Hamlet* existed billions of years before Shakespeare was born. In writing *Hamlet*, Shakespeare did nothing more than turn himself into a finely honed selection machine that was adept at sifting through the endless piles of nonsense within the *Library of Babel* and pulling out occasional pieces of prose. If you were to find yourself in the region of the *Library of Babel* labelled *Hamlet Space*, it might take you hundreds of thousands of years to pluck the original from the sea of alternative mutant versions. In order to locate a book containing the target sentence '*My hour is almost come, when I to sulph'rous and tormenting flames must render up myself*', we would have to traverse endless texts which contained versions such as '*My hour is almost sulph'rous, when I to come and tormenting flames must render up myself*', '*Mi hour is almooo come, whena I to sulp'hrous and tormanting flames -ust render uz myself*' or '*My rouh si tsomla come, when I ot sour'hplus and tormenting flames stum redner pu flesym*'. Of course as far as literature is concerned, one man's chaff is another man's grain and it is not difficult to imagine a cultural milieu in which anything written in the style of Shakespeare would be an abomination. The *Library of Babel* is, however, indifferent to meaning, value and utility, all of which are imposed by the user of the library. The discovery by history of the mathematical *Space of All Possible Human Brains* provided the key for unlocking the door to the mathematical space in which the *Library of Babel* is housed. Like Shakespeare, Dickens was also in one sense nothing more than an efficient machine for selecting pre-existing editions of *Oliver Twist*, *Great Expectations*, *A Christmas Carol* and other great works from the *Library of Babel*. These books would nevertheless continue to exist in a physically unrealized state, even if Dickens had not been born. Had he lived longer or worked a little harder, an even greater sample of Dickens-like works might have been selected from the *Library of Babel*.

The twenty-six letters of the English language can be substituted for one another, or recombined in different ways so that meaning is changed, but no individual letter may adopt an intermediate value. An 'a' can only be an 'a' and a 'b' only a 'b'. There is no intermediate state corresponding to an 'a/b', 'a/a/b' or 'b/a/b'. There are only unique combinations of discreet a's and b's. Furthermore every word has a precisely defined beginning and end. Thus the words 'cat' and 'sat' in the sentence 'the cat sat on the mat' represent two distinct units of meaning which can be recombined to generate qualitatively different composite meanings. Each word, however, always preserves its dictionary meaning; there are, for example, no intermediate states corresponding to a cat/mat, mat/cat, or cat/sat. Thus, despite the analog complications introduced into languages by semantics, one's tone of voice and the fact that when I use the word cat I might be thinking affectionately of a creature which is an amalgam of my current cats Dusty, Woollie and Bluebell and my old ginger Manx Mutchinka while you might have a cat phobia, the digital structure of language enables a certain amount of information to be communicated between individuals. Clearly the more fundamental the nature of the information being communicated, the easier it is to communicate it unambiguously.

Digital strategies of information representation have the considerable advantage that they are, for the greater part, free from the constraints inherent to analog representation. The information stored in digital systems thus has the potential to be immune to noise and distortion and may be replicated across many generations with high fidelity. Errors can be removed by specialized correction mechanisms that enable information to be reproduced many times without loss of fidelity. The digital nature of the stored information also ensures that it is highly flexible. It may be, for example, transferred from one type of digital machine to another and edited with impunity. Most importantly of all, digital mechanisms of information storage are amenable to updating by evolutionary processes, which differentiate between alternative versions of the same target information.

The lack of noise and distortion in digital systems results from the fact that, whereas in analog machines each number is represented by a physical quantity, in digital machines each number is represented by a binary word. This binary representation is itself represented by a series of physical markers. In a compact disc player, the 1 and 0 components that represent the sound wave at any given instant are stored as minute pits on the surface of an optical disc. The depth of a pit determines whether it represents a 1 or a 0. During a recording process, the binary pulses of information that issue from the sound source are translated by a laser into a spiralling pattern of pits. After the disc has been coded, it is dipped into an etching chemical which eats away at regions that have been exposed to the laser to produce the pits. In order to access these digital instructions for recreating the information of the sound, a laser in the playback unit scans the disc and converts the varying light intensities into digital signals. These coded numerical values are then used to recreate sound of the appropriate frequency and amplitude. Nothing touches the surface of the disc during playback and thus, unlike gramophone records, compact discs do not deteriorate as a result of the playback process. Furthermore, unless very severe, dust and scratches do not significantly affect the quality of the reproduction. The fact that digital machines represent information using only a very limited set of discreet symbols which are either present or absent but cannot assume intermediate values means that the information passing through them is high fidelity and essentially noise-free. This contrasts with the repertoire of graded physical values employed in analog representation. Whereas an analog value of 0.3 could slip unnoticed to a value of 0.3000001, a mutation which changes a 1 into a 0, or a 0 into a 1, would involve a large leap of inaccuracy that could be easily detected and corrected.

Digital information is thus more robust than its analog counterparts. This resistance to decay is one of the most important advantages of digital information-encoding strategies, as it enables information to be replicated across many generations without significant loss of fidelity. Information transmission across generations is like a game of Chinese whispers. If fidelity is low, the target information rapidly degenerates into incoherent noise. The degree of error in the replication of the information of living organisms must, similarly, be tightly controlled. Too much error and the information of the organism melts away into random noise. Too little error, however, and the organism is frozen, rigid and unable to field new genetic solutions to meet the challenges of the future. Digital technology provides the potential for a fidelity which approaches perfection. Once this state has been realized, the actual error rate can be tuned to a level compatible with evolution. This contrasts with analog systems that lack error-correcting mechanisms and are unable to prevent their informational structures from decaying across relatively short durations of time.

A bit of digital information is far more flexible than its analog counterpart. If, for example, compact disc players are one day replaced by a more advanced digital technology, the master information could be transferred to the new machines. Unlike the analog information encoded in the grooves of a record which is destined to remain private and accessible only to gramophone machines, digital information is in the public domain and accessible to any universal digital machine. Although the digital information contained within a DAT tape cannot be directly accessed by a compact disc player, the digital master can be used to manufacture a compact disc. The fact that digital sound information is independent of any particular physical representation allows it to be edited efficiently and extensively. We could, for example, paste sections of a Mahler symphony into a performance of a Mozart piano concerto by Annie Fischer. The end result might not be to our liking, but it should nevertheless be clear that information stored digitally is more flexible and versatile than the frozen archives of analog recordings. For this and all the other reasons we have discussed, we should not be surprised that modern biological machines have come to utilize a digital strategy for representing their essential information.

Having discussed the differences between digital and analog modes of information representation, what I would now like to suggest is that the first organisms were analog, not digital. Their information was not represented using the digital strategy of genes with its associated template-based replication and symbolism but was instead stored in a non-discreet, distributed, analog manner. The success of these hypothetical geneless creatures was, however, likely to have been limited by the constraints associated with analog technology. Most importantly, the putative geneless creatures are likely to have had only the most rudimentary capacity for heredity. Hypothetical analog lineages of primordial geneless organisms are unlikely to have been able to perpetuate their core information for more than a few generations before it melted away. The possibilities for generating variants by mutation coupled with selection are also likely to have been limited. This would have curtailed the extent to which such systems were able to modify their information by classic Darwinian processes.

In the absence of an adequate mechanism for perpetuating their information, the hypothetical analog creatures would have been continuously created and destroyed. Any particular lineage may have been extinguished after only a few generations. It is tempting to speculate that these successive, ephemeral waves of analog lineages were eventually ousted by a new class of digital creatures that were able to jettison their outdated and inefficient analog technology and replace it with digital technology. By utilizing self-replicating digital polymers to store their information, these primordial digital creatures acquired the potential for unlimited inheritance, which is the signature of modern life; that is to say the potential to transmit their information in a stable manner across many generations and thus to explore efficiently and sustain a trajectory across a host of mathematical spaces.

These high-tech digital debutantes are themselves likely to have been ousted by organisms that utilized RNA-based digital technology. Finally, RNA digital information storage technology may itself have been supplanted by organisms that employed a postmodern digital information technology known as DNA. In DNA-based information storage, a negative copy of the information-encoding strand is used to determine the accuracy of replication events and the fidelity of information transfer. If inaccuracies are introduced during replication, or if the DNA is damaged, the negative strand is able to supply the information necessary for their detection and correction.

But how can we conjure up these putative analog creatures from the depths of possible prehistory and peer back to a time long before the origin of genes? If there was a time when organisms were geneless, then how might the inhabitants of this *Geneless World* have been structured and what preceded them? If life on Earth originated from non-living matter, then how did its organization arise from a disordered sea of inanimate chemicals? What, indeed, precipitated the first moment of life on Earth, the creation of living from non-living?

9

Life Without Genes

We are now at the very beginning of life, at the interface between the physical and biological, between chemistry and biochemistry. A period of prehistory whose exact structure is irretrievable, but which we will nevertheless attempt to reconstruct by piecing together an assorted collection of intellectual fragments. But when we have completed this painstaking task, we are chastened by the knowledge that we are still only at the very beginning. For the true test of the plausibility of our reconstruction can only come from experiments. We must thus apologize for making speculations about an era about which we know so little, whilst at the same time acknowledging that our tentative suggestions as to how things might have been may help form the cornerstone of a more penetrating insight.

Let us first examine some of the intellectual fragments which will help us piece together our reconstruction. Only when we have done this, and breathed life into the *Geneless World*, will we attempt a logical dissection of its analog, geneless inhabitants. It is not, however, to the microscope, dissection table and scalpel that we will then turn, but to powerful modern computers. Perhaps the most important of these fragments is the notion that the complexity we observe in the structure and organization of living things is both embedded within and organically part of the physico-chemical world. It is not confined to living systems, but a potential aspect of the behaviour of certain types of systems. We should thus be able to explain the origins of the first living things without recourse to elements which lie outside the framework of modern experimental science.

If we accept that the causal factors underlying the generation of life are firmly rooted in physics and chemistry, then we should be happy to accept that living organisms have no more purpose than the moonlight which illuminates a dark night, or a cliff face exposed to the wind and sea. The universe is indifferent to particular things and events, whether living or non-living. No entity, whether purely physico-chemical or biological, has any more validity or imprimatur than any other. Once the universe has come into existence, the potential for everything else is inevitable. It might nevertheless come as a surprise that the order of living things is so firmly rooted in the physico-chemical world. For surely living things are very different to the pebbles that we find on a beach, the glass of wine in our hand, or the books stacked so untidily on our shelves?

Any notion that order is confined exclusively to the domain of living things should have been dispelled when, as schoolchildren, we watched copper sulphate crystals grow at the end of the piece of cotton to which we had attached a seed crystal before suspending it in a saturated copper sulphate solution. This observation might also have suggested that a system may have the potential to offer up several different aspects of itself, depending on the circumstances it encounters. Copper sulphate molecules can, for example, exist as a simple, information-free solution, or spontaneously form the iterative unit cells from which their crystals are constructed. We might also have marvelled at the geometry of a snowflake, which only moments before had been a simple, formless and information-free vapour, or watched oil drop onto a sheet of water and self-assemble into spherical globules. But any insight that we might have had into the natural sources of non-biological order that permeate the physico-chemical world was likely to have been overshadowed by the second law of thermodynamics, which states that order is an inherently unstable and improbable aspect of the physical world and that physico-chemical systems tend to destroy order spontaneously and decay into a state of maximal disorder.

Despite the fact that both can generate information, living things are quite different from non-living things. Capturing these differences in an unambiguous way is not, however, easy. We can, however, throw a hoop into the ring by suggesting that life may only be generated and sustained within the *complex* domain of the *Space of All Possible Behaviours of All Possible Physico-Chemical Systems*. But this will also contain self-organizing systems such as snowflakes, oil drops and copper sulphate crystals which, although able to exhibit complex behaviour, are not considered to be living. It is clear, then, that behavioural complexity alone cannot be the defining feature of life. It is nevertheless an essential prerequisite for life, and one of the minimal requirements for a living system.

Exactly what we mean by 'complexity' in this context will become apparent later on, but it is worth pausing to give it a little consideration. If I were to ask how much salt you had eaten in your life, with a little thought you may be able to remember how often you buy a packet and its approximate weight. If you found that you used a kilogram each year and you were thirty years old, you might estimate that you had consumed thirty kilograms of salt in your lifetime. On reflection, you might want to make some adjustments to this figure. It is unlikely, for example, that your consumption at the age of three months was the same as it was age three. Furthermore, much of the salt that you use at home is thrown out with the cooking water. Salt is also present in differing amounts in virtually everything that we eat and drink.

You might choose to write a computer program to help generate a more plausible estimate. If you were to do this, then instead of protesting about the difficulty of the problem in a general way, you could obtain a more specific measure of its difficulty in terms of the number of lines of computer programming necessary to compute its solution. If you were used to writing programs twenty lines long and the solution required 200 lines of programming, you might conclude that the problem was ten times more complex than the problems with which you were routinely familiar. However if I now asked you to calculate the amount of salt that had been consumed in human history, you might find that your program required more than 2,000 lines. This problem would thus be ten times more complex than the problems with which you were familiar. We should not, however, be misled into thinking that a problem has to sound complicated in order to be complex. In fact it is often the case that problems – such as that encountered by a salesman who wishes to visit a large number of towns by the shortest possible route – which at first sight appear simple, can turn out to be extremely complex.

This type of algorithmic approach that equates the complexity of a problem with the number of instructions needed to compute its solution was pioneered independently by the mathematicians Kolmogorov and Chiatin. By determining the number of algorithms necessary to compute a problem, we can obtain an objective measure of its algorithmic complexity. This approach to the definition of complexity is not, however, restricted to the type of problem we have just considered. It is equally applicable to the problem of describing physicochemical systems. It should be possible to specify the state of a system by writing a program that can faithfully reproduce it. If we were to calculate the number of computations necessary to produce this description, we would have produced a measurement of the system's algorithmic complexity.

We might, for example, wish to compare the complexity of a rockpool with that of a concrete block, a whirlpool, or that of a painting by Van Gogh with one by Seurat or Manet. One can imagine the algorithmic form of the program needed to specify Van Gogh's *Sunflowers* as being something like: **Take** one molecule of pigment x, **place** it at position BZE240 on the canvas, **then take** one molecule of pigment q and **place** it at position BZE241 and so on. The number of instructions needed to specify *Sunflowers* would provide an objective measurement of its algorithmic complexity. This could then be compared with that of Seurat's *Bathers at Asnières* or Manet's *Le Déjeuner sur l'Herbe*. We might, similarly, choose to compare the complexity of *Sunflowers* with that of a concrete block.

The notion that algorithmic complexity is itself a measure of anything useful is, however, absurd – if only because of the theoretical constraints that preclude us from making the measurements necessary to specify the exact quantum mechanical state of a system. What is more, the number of algorithms needed to compute a problem gives no indication of how long the program would take to run. If two different programs of identical length had significantly different running times, then surely this would influence our assessment of their relative complexities? We could certainly not ignore the running time if one of the programs completed its operations in five minutes and the other ran for eternity without giving the slightest indication that it would ever compute a solution and halt. This narrow algorithmic definition of complexity also fails to capture aspects of the ideas that an artist might wish to convey, and the associations, references and resonances contained within a painting.

Complexity is also a function of the observer's ability to respond to information, the structure of the information being defined transactionally, as a consequence of its interaction with the observer. An object may also contain information that only some observers are able to perceive; a bat, polar bear or fruit fly are, for example, unlikely to perceive *Sunflowers* in the same way as us. And to a blind person forbidden from touching or smelling the painting, it may convey no information at all. One is thus obliged to enquire, information for what, or for whom? The possibility that some aspects of particular systems are not computable, even in principle, must also be considered.

For these and other reasons, it seems unlikely that art historians will one day re-evaluate the work of great painters on the basis of algorithmic complexity. This is probably just as well, as a herd of elephants could doubtless produce a work far more algorithmically complex than anything Van Gogh ever produced. Given the limitations of this *static* notion of algorithmic complexity, might it be more rewarding to focus on *dynamic* aspects of algorithmic complexity? This would focus on the pattern of behaviour that unfolds in a system over a given period of time. With this emphasis on dynamic behaviour as opposed to static structure, we can define the dynamic algorithmic complexity of a system as the minimal set of algorithmic instructions needed for a blindfolded observer to be able to produce a facsimile of the behaviour we observed. Dynamic algorithmic complexity is thus the rate of change of static algorithmic information with time.

Evaluated from the perspective of dynamic algorithmic complexity, the paintings, rockpool and block of concrete appear relatively simple, as their structures are essentially frozen in time. Thus, although such systems appear complex when considered from the perspective of a static snapshot, when observed over an extended period of time they lack behaviour. Their information does not change appreciably with time and, from the perspective of dynamic algorithmic complexity, is impoverished. Although at the atomic level none of these systems is static, in macroscopic terms the block of concrete remains a block of concrete, *Sunflowers*, give or take a bit of sunlight and thermal agitation, remains *Sunflowers* and a rockpool, give or take a few ripples and temperature fluctuations, remains a rockpool. Unless *Sunflowers* were to spontaneously transmute into *Bathers at Asnières* or some other painting, we would be able to telephone the blindfolded observer each day, to inform them that it was not necessary to communicate any instructions to them, as *Sunflowers* had, yet again, exhibited no behaviour. *Sunflowers* and all similar systems thus lack dynamic complexity. A whirlpool, by contrast, does have a dynamic dimension to its complexity, as its structure changes unremittingly with time. We would need to supply the blindfolded observer with a continuously updated stream of information in order to describe the system's pattern as it unfolded across time. As each piece of information arrived, they could update their model.

Although the concept of dynamic algorithmic complexity as outlined above is a considerable improvement on static algorithmic complexity and captures at least one important aspect of living systems, it does not constitute a sufficiently rich or complete definition of complexity in its more formal sense. This is illustrated by the fact that despite being the antithesis of what we mean by complexity, a jar of diffusing bromine gas molecules is both statically and dynamically complex. Any complete description of the complexity of a living system must consequently transcend formulations couched exclusively in the language of algorithmic complexity.

The missing element may be defined by focusing on the nature of the behaviour that a system displays. Despite being dynamically algorithmically complex, the behaviour of a system may be devoid of order. Only truly complex systems are able to generate coherent behaviour at the molecular level. In the case of living organisms, this information is self-sustaining and persists across time. Although some physical and physico-chemical systems are able to generate ordered behaviour visible at the macroscopic level, their information is not generally self-sustaining and depends on the presence of an external constraint such as a heat gradient.

Although a specification of the microscopic state of a jar of diffusing bromine gas would, if indeed this were possible, require a formidable number of algorithms to compute, once the system had reached thermodynamic equilibrium it would cease to generate macroscopically visible behaviour. It would instead be fixed in a microscopic state of maximal disorder which is incapable of generating macroscopic behaviour. An observer would at this point notice that the gas had distributed itself across the jar. If they were then to watch the jar until the end of time, it is unlikely that they would witness any change in the system's macroscopic behaviour. This macroscopic monotony results from the lack of order at the microscopic level. Thus, despite having considerable dynamic algorithmic complexity at the microscopic level, the dynamic complexity in this type of system is unable to generate information at the microscopic level. This results in a lack of macroscopic behaviour.

The presence of dynamic behaviour which can generate microscopic information constitutes the third tier of our definition of complexity. Dynamically complex systems are able to generate information by constraining the range of their possible behaviours. This is achieved by selecting a limited but choice set of performances from the repertoire of possible alternatives. Rather than being free to explore their available state space, systems that exhibit dynamic microscopic order behave as if they were constrained by a set of rules that determine which states can be visited. Much like a grammar which orders regions of the *Space of All Possible Combinations of Words* into sets of rationalized utterances, the dynamics of some systems impose constraints on the processes by which their potential state space is explored. This sharp localization of a system within a confined region of its vast potential state space is the hallmark of true complexity. It is only by limiting its state space occupancy to a restricted set of possibilities that a system acquires the capacity to generate information. The minimal structural requirement necessary for this is that the system is open to the exchange of matter and energy and located far from thermodynamic equilibrium.

In summary, the three tiers of structure necessary for the generation of complex behaviour are: (1) static algorithmic complexity, (2) dynamic algorithmic complexity and (3) the ability of a system to constrain the way in which it explores its potential state space. Several purely physical and physico-chemical systems fulfil these requirements, but are not nevertheless considered to be living. So whilst being required for life, complexity as defined above is clearly not adequate for life. It appears that we need to hunt for additional elements to bring us closer to a formal description of the structural requirements for life.

One essential difference between complex systems that we call living and those that we do not is found not in the nature of the components from which the system is constructed, but rather in the way that they are organized. One of the things which distinguishes potentially living things from non-living things is, thus, their logical form. The pattern of organization that appears to be a prerequisite for life is known as a network structure. In networks, the individual components of the system are highly interconnected. The strength of these interactions is considerably greater than those between the components of algorithmically complex but random, information-free systems such as a jar of bromine. As a consequence of their connectivity, individual components within complex dynamic networks are able to feed information back to one another across large distances. This enables signals originating locally to be amplified and propagated to distant corners of the network.

In all of the living things with which we are familiar, networks are constructed from sets of interacting chemicals and cells. As a consequence of their organization, these and all other classes of network are tunable. This means that by adjusting the network's structure, the number of states that it can exist in and the frequency with which these are visited, can be modified. We may thus add a fourth tier to our definition of complexity, relating to the subset of complex systems that have the potential to generate life. They should (4) have a network structure, and this should be (5) 'tunable', so that the system can condense down into a tiny region of its potential state space. In addition to this, the dynamics of the network should be such that (6) it is able to generate stability which is relatively immune to random fluctuations in the environment. Finally, the network should be (7) self-sustaining, (8) self-replicating and (9) constructed from modular components which, in the words of Gabriel Dover, can be 'historically processed' by the 'incessant evolutionary process of molecular Lego'.

This random or semi-random molecular reshuffling, addition, deletion and modification of the components of chemical networks coupled to a process of evolution by natural selection, is the physical basis of network tuning. The repertoire of network structures is assembled by blind Darwinian engineering, which explores the theoretical repertoire of all possible network structures by bolting new molecules onto the network and removing or modifying others. However, the repertoire of available structures may be constrained by the self-organizational biases of the components from which the network is constructed.

The potential for structural flexibility confers networks with the capacity for generating behaviour. Thus rather than being immutable and hard-wired, the behaviour of a network can be modified by natural selection. The coupling of Darwinian evolutionary processes with the structural bias introduced by self-organizational processes provides the opportunity for competition between differently configured networks and adaptation to specific micro-environments. There is no inherent drive towards increasing complexity, but if an increase in complexity results in an increased efficiency of reproduction, then the more complex structures will tend to be over-represented and eventually replace their less complex competitors. A decrease in complexity, however, might also provide a more competitive phenotype.

In what follows I will suggest that analog, geneless life might have arisen naturally and spontaneously as a consequence of the complex sets of interactions and dynamics arising within self-organizing, highly interconnected and tunable chemical networks. Although these types of interactions might have originated either within or outside compartmentalized structures, structureless compartment-free networks would have had only a limited capacity to evolve. Thus, even if the first living things lacked discreet boundaries, compartmentalization represented an essential step towards modern complexity and all of the advantages this entails. Once enclosed within a non-informational, self-replicating structure such as a fatty micelle, such self-bounded, 'autopoietic' chemical networks would constitute a repertoire of variants on which natural selection could act, and the genotype could be connected to its phenotype. The word autopoietic, incidentally, refers to a collection of entities assembled into a network in which each component helps to form or transform the other components of the network.

The emergence of analog, geneless life in such tweakable and perhaps initially unbounded chemical networks may be no more surprising or unnatural than the crystallization of water into ice, or the attraction of iron filings to a magnet. Although, as I have already indicated, we may never be able to recreate our actual putative geneless ancestors, we can nevertheless model the logical structure and behaviour of abstract networks using computers. We can in this way obtain some insight into the types of molecular architectures that may have formed the skeletons, flesh and blood of our hypothetical geneless and, perhaps initially compartment-free ancestors.

In exploring the detailed structure of geneless organisms, we will consider the possibility that genes are neither required nor adequate for the generation of life. Unlike the encoded metabolisms of digital organisms whose essential logic and pattern is dictated by programs written in DNA microcode and housed in a central information centre, the non-encoded metabolisms of analog geneless organisms may have emerged as a consequence of the cooperative, self-organizing and tweakable dynamics of its components, without needing internal or external programs. Naked genes which lack the self-organizing, self-regulating network structure and dynamics which are a likely prerequisite for life are thus, from the perspective of complex living systems, informationally bankrupt. So, although the digital combinatorial nature of genes enables them to encode information, in the absence of a dynamic, self-regulating, self-sustaining and self-organizing network structure this information cannot contribute to a process of life. The ability of a system to self-regulate is known as homeostasis.

It is possible that it was only by the acquisition and embodiment of the network pattern of organization and homeostatic dynamics which characterize living systems and by becoming integrated into an open, far-from-equilibrium context, that genes were able to become informationally empowered. In the words of Freeman Dyson, one of 'the primal characteristics of life' is 'homeostasis rather than replication'. According to this definition, a hypothetical self-replicating and autonomous naked molecule, be it a string of digital RNA sequence, DNA sequence, a peptide or a prion protein, is not alive. In the absence of a homeostatic metabolism to generate a tightly controlled micro-environment in which they can operate, all of these would be lifeless. We will return to address the question of whether it is reasonable to draw a clear-cut line between potentially living and non-living structures in this way, at a later point. For now, however, we will stick to the homeostatic 'life pattern' criterion to define the boundary between the structures which we consider to be living and those which we do not.

Thus instead of being present right from the very beginning, genes may only have hopped onto the bus of life at a later date. In his 1985 Gifford lectures, Freeman Dyson excluded replication from the beginnings of life and relegated genetic replication to the status of 'an alien parasitic intrusion' which was only introduced at a later date. This implies that rather than providing the essential and inviolable informational foundations of life, genes were a later and therefore dispensable addition. We might speculate that self-replicating but nevertheless lifeless genes constructed from RNA or some other digital polymer, acted swiftly to capitalize on the opportunities that geneless life had on offer, and at some point parasitized these clumsy, primordial, analog creatures. Having carried history out of the domain of the inanimate, and in so doing discovering biology, the pioneering geneless organisms may eventually have been infected by genes or their precursors and established the informational foundations of modern digital life. It is possible that the hypothetical transition from geneless life to life which harnessed the informational capacities of both self-assembling and self-organizing chemical networks and digital sequences, was precipitated by a genetic plague resulting from an unprecedented increase in the local concentrations of digital self-replicating or non-self-replicating molecules. This would have increased the competition between molecules with similar or identical sequences and increased the chances of a geneless organism being infected with digital inoculum.

Under different circumstances, life might have acquired immunity to the genetic plague and remained in its geneless state indefinitely. The historically contingent acquisition of such immunity might, just as surely as the extinction of the dinosaurs at the end of the Cretaceous paved the way for the proliferation and diversification of mammals, have severely constrained the nature of subsequent life. It does not seem inevitable that genes, as previously defined, should have become a structural or informational component of living things, any more than it was inevitable that digital RNA gene technology should have been superseded by digital DNA technology. What does appear likely, however, is that given a situation in which individuals or populations of geneless organisms had to compete for limited resources, some of the infected lineages might have acquired a competitive advantage over their digitally sterile counterparts by acquiring a new structural catalytic function such as a chaperoning activity, as a consequence of their digital disease. Similarly, in a situation where populations of lifeless, self-replicating, non-symbolic 'pre-genes' were in competition for limited resources, any pre-gene able to harness the power of life through infection would have acquired a considerable advantage over its lifeless competitors.

A pre-gene is defined here as a digital polymer which like a gene encodes information within its combinatorial digital sequences, but unlike a gene is incapable of codification. Although a pre-gene encodes information within its digital sequence and self-replicates using a template-based mechanism, this information does not constitute an abstract representation and is not translated into different molecular languages analogous to mRNAese or amino acidese. The mechanism of template-based replication in pre-genes is furthermore imagined to be self-complementary, in contrast with the complementary mechanism found in genes. The assumption, however, is that some pre-genes had the potential to acquire the capacity for codification and complementary template-based replication and to evolve into fully fledged genes. Following a period of chemical evolution within stagnant pools of non-self-replicating digital pre-gene polymers, some of the pre-genes may have infected susceptible geneless hosts, after which these parasites may eventually have evolved into true genes. In this case, genes may, quite literally, have been born inside the metabolic networks of their geneless hosts. If this were the case, then rather than invoking a genetic plague, it would be more appropriate to invoke a non-genetic, digital plague. For in this model, it was only as a consequence of infecting geneless organisms that the digital information encoded within the sequences of the primordial digital pre-gene polymers was able to acquire the accoutrements of genes.

These included, amongst the other things we have mentioned, a capacity for symbolic codification, a complementary template-based mechanism of self-replication of the type found in modern genes, and mechanisms to ensure high-fidelity replication.

Prior to being infected with digital inoculum, the biological information of geneless creatures would have been stored in an exclusively analog manner. However, once infection had occurred, the distinction between the analog hosts and their digital parasites would have become blurred, and the eventual children of this unholy alliance, the first proto-genes and eventually genes, would have set about redefining the architecture of life. The formidable capacity for information storage inherent within the digital sequences of proto-genes is likely to have provided the new digitalized organisms with a chemical network-tuning capacity which had hitherto been unattainable. Like an out-of-tune pianoforte, the structure and connections of the previously analog and geneless organisms could now be reconditioned, forced, tuned and tweaked, to increase their capacity for generating ordered behaviour.

For the first time in history, life, in its new digital proto-genetic clothing, acquired the potential for producing protracted lineages, as it was now able to sustain a degree of informational continuity which had hitherto been unobtainable. Unlike their geneless precursors, which had only a limited capacity for heredity and explored finite and limited state spaces, the new proto-genetic organisms held the key to infinity. Fitted out with its new digital technology, life was now free to move from the finite to the infinite and to make its first tentative explorations of the newly accessible expanses of *Gene Space*. Only later would it wander into the small corner of *Gene Space* that we have called *DNA Space* and discover regions which specified amongst many other things, unicellular organisms, ammonites, trilobites, dinosaurs, mountain lions and man.

The digital takeover or 'informational symbiosis' hypothesis which I have outlined suggests that the information of modern life originated from a primordial fusion event in which information written into the digital sequences of pre-genes was united with analog information stored in the dynamics of chemical networks. Although there is no direct evidence to support this hypothesis, its general logic and predictions are falsifiable. Furthermore, the precedent and paradigm for the parasitism of metabolic structures by naked genetic material has been established by viruses.

It should be possible to model the plausibility of fusion processes in which information from two distinct sources cooperates and eventually integrates to provide a new structure incorporating features from both systems. If mathematical modelling and computer simulations suggest that this type of informational symbiosis is plausible, then it would be reasonable to perform experiments using biological components. We should not assume, though, that informational fusion occurred simultaneously with infection. It may have taken a very long time for the two sources of information to cooperate and integrate. The probability of such a fusion event occurring might also have been very low. Repeated independent digital infection events may have been necessary to produce structurally stable and informationally viable fusion mutants. However, until this occurred, the pre-gene digital polymers most likely functioned as catalysts which accelerated the rate of one or more of the reactions within the chemical networks of their geneless hosts and influenced their dynamics in a neutral or favourable manner. In conferring an adaptive advantage to their hosts, the digital parasites would have increased the likelihood of their own survival. Many of the clonal genetic infections would have undermined the geneless dynamics and been deleted. Only those parasitized clones in which the effects of the digital infection were phenotypically neutral or conferred some adaptive advantage would have persisted.

In addition to examining the plausibility of the informational symbiosis hypothesis and the mechanisms by which dynamically and digitally represented information might fuse, it is also of interest to examine the limitations imposed on geneless systems by their lack of digital codification competence. It is almost certainly the case that the complexity of the potential creatures housed within the *Space of All Possible Geneless Organisms* was severely constrained by limitations intrinsic to their non-digital and non-symbolic information representation strategies. What is more, the lack of combinatorial information representation strategies would have excluded geneless organisms from the infinite combinatorial spaces that such technologies make available, making their capacity for heredity extremely limited. It would be interesting to define these logical constraints more precisely and in so doing draw a formal limit to the level of complexity that analog organisms lacking digitally-encoded information might attain.

It is also important to address the question of how dynamic, self-organizing and geneless chemical networks might have mutated, replicated, repaired themselves and evolved as the result of natural selection. Could they initially have existed as unbounded entities in free solution, or was it necessary for them to become enclosed within the geometrically defined boundaries of fat droplets, micelles or other membrane-like structures? There are several other points that we might want to consider. How long, for example, might individual geneless organisms have lived for? Nanoseconds, seconds, minutes, hours, days, weeks, months or years? Had they not involuntarily been bitten by the sharp, infective jaws of lifeless digital parasites, would geneless life have prevailed? Would it have been possible for geneless structures to approximate to the biological complexity present in even the most modest of contemporary organisms? There is one thing, however, of which we can be certain, and that is that once it had tasted and greedily devoured the digital pre-genetic apple, life would never be the same again.

If experimental evidence corroborates the plausibility of the informational symbiosis hypothesis, we will have to acknowledge that the informational foundations of modern life may have had a dual origin. This consists first of the information that arises spontaneously in certain classes of self-organizing, open and far-from-equilibrium chemical networks which, when tuned appropriately, are able to generate, perpetuate and modify information as a result of their dynamic behaviour. This information itself arises from the constraints imposed on both the manner and extent to which the networks are able to explore their vast potential state spaces. Random reshuffling, addition, deletion and modification of the modular components from which they are composed would presumably have generated a repertoire of structural and dynamical variation on which natural selection could act. Those structures able to generate evolvable dynamics with the optimal balance of stability and flexibility and which were adapted to their micro-environments, were likely to have had a significant competitive advantage over those which were not. The second source of information issues directly from the combinatorial digital sequences of non-self-replicating lifeless pre-genes, or self-replicating lifeless pre-genes, which were assembled by natural selection.

The informational symbiosis hypothesis suggests that following digital infection, the information of individual lifeless pre-genes may have interacted and eventually integrated with the information stored in the complex dynamics of geneless networks. This hypothetical fusion event is itself likely to have been preceded by a period of chemical evolution within the internal environment of the geneless organisms. During this time, randomly generated alternative versions of the founder digital parasites would have competed for survival with one another and with foreign sequences introduced as a result of super-infection. Some of the parental information would presumably have been jettisoned following informational fusion, and in the process a new coherent informational hybrid may have formed. By functioning as symbols that specified the chemical structure of the geneless components of organisms, self-replicating proto-genes derived from pre-gene precursors, would eventually have replaced the primitive geneless machinery they inherited. One may imagine the dynamics of the pioneering lifeless organisms as constituting an informational stepping stone which provided the rudimentary benefits of life, whilst proto-genes began to experiment with their own symbolic strategies of information representation. The non-encoded metabolism of geneless organisms thus eventually acquired an abstract symbolic strategy of information codification in the form of the combinatorial digital sequences of genes and their associated genetic code, and the digital genetic paradigm for life was established.

Although I have suggested that the information of the first geneless organisms was represented using a non-codified analog strategy, we should not, as I have intimated, be surprised if these organisms were constructed from digital but nevertheless non-genetic components such as peptides. Furthermore, randomly produced digital chemical components, although sharing a common physical structure with pre-genes, would not, in the absence of self-replicating mechanisms operating at a reasonable level of fidelity and a potential mechanism for information codification, qualify as pre-genes. Although pre-genes, harbouring the potential to evolve into proto-genes and eventually true genes by capturing the means for codification and the ability to perpetuate the information stored within their digital sequences, in the absence of such mechanisms these components would not be able to encode digital information which could contribute to a process of life.

The combinatorial complexity of diverse populations of non-genetic digital molecules may, nevertheless, have provided the potential for dynamic means of analog information storage within chemical networks. The chemical complexity inherent in vast collections of randomly assembled combinatorial digital polymers may thus have constituted one of the necessary prerequisites for the creation of life. Chemically heterogeneous populations of such molecules harbour the potential for evolving into lifeless pre-genes and geneless forms of life. We cannot assume, though, that combinatorial complexity provided the historical means by which sufficiently diverse mixtures of chemicals initially chanced upon life. The formidable power of combinatorial chemistry may only have been discovered by life at a later date.

Now that we have discussed the possibility that life might have begun naked and embarked on its first perambulations without the benefits of its modern genetic clothing, it is time to address the question of how our hypothetical geneless ancestors might have originated in the first place. How, after all, could information have been generated from non-information? And once generated, how was this new biological information perpetuated? In order to answer these questions and to prepare the ground for a discussion of the possible architecture of the hypothetical geneless organisms, it will be necessary to return to the second law of thermodynamics. For it is only by understanding the statistical mechanisms that underlie nature's inherent tendency to destroy information, that we will be able to fully appreciate the mechanisms that generate and perpetuate the information of living things.

10

The Origins of Geneless Information

It is a common intuition that life stands somewhat mysteriously in defiance of the physical and chemical laws of the universe. Indeed, the second law of thermodynamics informs us that the universe tends towards a state of maximal disorder. We have all watched an ink blot dispersing across a piece of filter paper, or bromine diffusing in a jar. But who has ever seen a splattering of ink spontaneously reassemble into a blot, or a process of diffusion running in reverse? So what is the driving force behind this irreversible decay of order into disorder and given nature's inherent tendency to destroy information, how might the information of life have originated? An investigation of these issues will carry us to the heart of the *Geneless World* and provide the remaining intellectual fragments that we will need to flesh out the anatomical details of its inhabitants. However before we begin to explore the mechanism by which information is degraded, we will first examine the basic structure of systems which are able to generate information. There are fundamental structural differences between complex information-generating systems and those devoid of information. It is these differences which underpin the complexity of living things. A consideration of both types of system will demonstrate how information can be generated and maintained without violating the second law of thermodynamics.

The minimal architectural requirements for information-generating systems are that they should be open to the exchange of matter and energy and located far from thermodynamic equilibrium. Equilibrium corresponds to the state in which any gradients within a system have been dissipated. The gradients themselves may be of a chemical or thermal nature, or may represent other types of inhomogeneity, for example differences in pressure. When the homogeneous state of equilibrium has been attained, the system is symmetrical and all of the regions within it are indistinguishable. Systems may be divided into those that are at or near equilibrium and those located far from equilibrium.

The concept of equilibrium may be illustrated by imagining a bathtub filled with water. If the water is well mixed, pre-existing heat gradients will be neutralized and the temperature at the top end of the bath will be identical to that at the bottom end. At this point the system is in thermal equilibrium. Let us now imagine that you are able to miniaturize yourself so that, relative to your diminished presence, the bathtub seems as large as the Atlantic Ocean. If in this telescoped state you were to commence a transatlantic crossing, you would notice that irrespective of the direction in which you travelled, everything would look and feel the same. As a result of the homogeneity of your surroundings and your inability to distinguish one location from another, you would have no concept of having crossed a volume of space. You would, furthermore, have only a poor concept of time, as no external events would have occurred to help you mark its passage. However, if a hole was drilled at the bottom end of the bathtub and water allowed to drain out whilst a tap containing boiling water was opened at the top and the bottom end cooled with ice, a heat gradient would be established from the top of the bathtub across to the bottom end. The temperature of the water would now no longer be uniform and the symmetry of the system would be broken. The system is at this point displaced from thermodynamic equilibrium. The word 'thermodynamic' emphasizes that thermal equilibrium is a dynamic phenomenon. If you were to miniaturize yourself once again and attempt another crossing, you would notice that if you moved in one direction the temperature of the water increases, whereas in another it stays the same, or decreases. The existence of local temperature differences would enable you to derive a concept of direction, a sense of movement, regional heterogeneity, space and time. Thus while equilibrium systems are symmetrical and devoid of information, systems displaced from equilibrium have a broken symmetry and are replete with self-generated information.

The state of non-equilibrium is maintained by the continuous consumption of matter and energy. This translates into the flow of water through the bathtub and the energy needed to boil it to 100°C at one end and freeze it to 0°C at the other. If the hole in the bathtub was repaired and the tap switched off, the system would be closed to the exchange of matter with the surroundings. Similarly if the electricity supply was disconnected, the energy needed to boil and freeze the water would not be available. The system would thus be closed to the exchange of energy with the surroundings. If this were the case, then the heat gradient across the bathtub would be dissipated, all regions would attain the same temperature and the system would be in thermodynamic equilibrium.

The extent to which a system is displaced from equilibrium can be estimated from the extent of the gradient across the system. This gives an indication of the ease with which equilibrium might be attained. If the temperature of the water was 100°C at one end and 0°C at the other, the temperature gradient would be severe and the system described as being far from equilibrium. If, on the other hand, the temperature of the water was 40°C at one end and 37°C at the other, the gradient would be negligible. In this case, equilibrium would be relatively easy to attain and the system is near equilibrium.

Unlike non-equilibrium systems, which are open to the exchange of matter and energy, a jar of bromine is a near-equilibrium system that is closed to the exchange of matter and energy with the environment. The flux of matter and energy through living organisms generates and maintains the gradients that underlie their order. The inability of closed systems to dissipate matter and energy makes them inept at information generation. A jar of bromine is a low-energy system consisting of a huge number of identical molecular components, each of which interacts with its companions only weakly because of the short-range nature of their interactions. Each molecule moves independently of its companions and there is no cooperation. The behaviour of these types of closed, near-equilibrium systems can be modelled using linear mathematical equations. Linear equations describe relationships between variables that are related in a predictable, linear fashion, where, for example, if the first variable is increased by one unit, the second may increase by two. The predictable nature of linear equations means their solutions can be calculated with great accuracy and certainty.

In contrast to equilibrium systems, living creatures and certain non-living structures are high energy systems constructed from non-identical components, many of which interact strongly. This results in the emergence of coherent, higher-order behaviour. The behaviour of each component correlates with many others in the network over considerable distances, which means that local behaviour can have far-reaching consequences. Highly correlated non-equilibrium systems of this sort cannot be accurately modelled using simple linear equations and instead require non-linear equations. Unlike linear equations, these are very sensitive to the starting conditions, as a small increase or decrease in the value of a variable can be hugely amplified by the network of interconnected components. Whereas in linear equations variables are related to one another in a simple additive manner, in non-linear equations they may be squared or raised to higher powers. If plotted in the form of a graph, instead of producing the straight lines of linear equations, non-linear equations generate complex shapes that are very much harder to solve. Powerful modern computers have, however, enabled these previously intractable equations to be solved with relative ease. Even so, the sensitivity of these equations to details of the initial conditions means that even a small inaccuracy in measurement will be greatly amplified. Our inability to define initial conditions to an infinite degree of precision means that the behaviour of non-linear equations will always contains elements of unpredictability. The greater the inaccuracy of measurement, the more significant these will be.

Systems that are closed to the exchange of matter and energy can readily equilibrate. In a jar of bromine, this corresponds to the situation in which the molecules are maximally disordered. Open systems, on the other hand, are driven by a continuous stream of matter and energy derived from their surroundings. This dissipation of matter and energy enables energy-charged systems to generate and perpetuate information. One need only think of a sandcastle which is progressively eroded by the tide. If left alone, it would rapidly decay and return to the disordered state of maximal probability represented by a featureless beach. However if we were to repair any damage as it occurred, the structure of the castle would be maintained. The individual grains of sand in the founder would not be the same as those in subsequent castles, but the general pattern would be indefinitely conserved.

The origin of life must consequently be closely related to the origin of non-linear systems open to the exchange of matter and energy and located far from equilibrium. Unlike their equilibrium counterparts, these systems are able to generate information from complete randomness. If we can demonstrate that these types of systems are able to arise naturally and spontaneously, we have established a tentative mechanism by which the skeletons of the most primitive forms of life might have originated. Whereas equilibrium systems are steamrollered by the second law of thermodynamics, living systems preserve their informational structures by destroying the information which surrounds them. However, as we will see, only certain types of non-equilibrium systems are able to generate processes of life.

The second law of thermodynamics was formulated with a specific sort of model system in mind; that is a closed, near-equilibrium system not dissimilar to the jar of bromine discussed above. Open, far-from-equilibrium systems are however more complex than closed, near-equilibrium systems. The differences between them are of central importance to understanding the order of living systems. Although open, far-from-equilibrium systems are able to generate and perpetuate information, their focal order occurs at the expense of an overall increase in the disorder of their environment. Open, far-from-equilibrium systems thus parasitize the order of their surroundings and generate their own order by avariciously degrading the informational content of their environment. In the words of Ilya Prigogine, the order of living things 'floats on disorder'. The second law of thermodynamics is thus at no point violated. The apparent paradox of the existence of information in a world governed by the second law of thermodynamics, is thus a straw man. There is in fact no paradox at all.

Before examining the structure of information-generating systems in more detail, we will take a little time to consider the structure of closed, near-equilibrium systems, which lack microscopic order. The first account of why closed, near-equilibrium systems like a jar of diffusing bromine tend inexorably toward a disordered state in which their information is destroyed, was provided by the theory of statistical thermodynamics which emerged from the work of Sadi Carnot, Ludwig Boltzmann and James Clerk Maxwell. This stated that the instability of information results from the fact that information-rich states are intrinsically improbable compared with their disordered alternatives. The principles underlying statistical thermodynamics may be understood in the following way.

Imagine a robotic schoolboy set loose on a giant scramble net similar to those found on army assault courses, but which stretches high up into the sky. The robot represents a bromine molecule, and the criss-crossing rope of the scramble net a playground which it can explore. Much like its biological counterparts, the robotic schoolboy has a tendency to wander aimlessly and impetuously from place to place. Bromine molecules are similarly driven into motion as a result of the heat energy transferred to them from collisions with air molecules. Unlike real schoolboys, however, robotic schoolboys lack volition. The higher the ambient temperature, the faster the robot moves and the only way to immobilize the robot, is to freeze it to the coldest possible temperature, which is absolute zero (–273°C). This may be more practically achieved by removing its batteries. As the robot lacks free will, he is indifferent to his trajectory across the scramble net. He might consequently climb up, down, diagonally, in a straight line, or zigzag across the scramble net in any number of different ways. Being impartial to his surroundings, he is as happy in one place as in any other.

The robot is free to explore all of the scramble net grids, which constitute the state space of the system. Let us now imagine that he is left to explore for several days. We will assume that there are 1000 grid positions on the scramble net, each of which is formed by the intersecting strands of rope. The two-dimensional *Scramble Net World* which the robot inhabits thus consists of 1000 possible states. As all of the regions in the scramble net are identical, and the robot has no preference for any particular coordinate on the scramble net map, no state is any more likely to be realized than another. If we, the schoolmasters in search of our recalcitrant mechanical pupil, could switch him off by remote control and then conduct a search for him, we would have a 1 in 1000 chance of locating him first go.

Now imagine placing two robotic schoolboys within adjacent squares in the corner of the scramble net and removing their batteries. Each is indifferent both to its position and to the presence of its companion. Having the robots positioned side by side in the corner of *Scramble Net World* is an ordered configuration that represents a highly improbable state of affairs. If we were to replace their batteries and allow them to wander around the scramble net for several days, the probability of finding any particular robot would remain 1 in 1000. As probabilities are multiplicative, the probability of finding the robots in their initial starting positions would be 1 in 1000 multiplied by 1 in 1000, which is one in a million (1 in 1,000,000). It is clear then, that even in a relatively cramped and sparsely populated *Scramble Net World*, the ordered arrangement is vastly less probable than arrangements in which the robots are randomly arrayed. The antics of just two robotic schoolboys might consequently tax even the most patient schoolmaster.

The larger the scramble net and the greater the number of robots, the more improbable an ordered configuration becomes and the harder it is to locate a specific robot. If, for example, we were to increase the size of *Scramble Net World* to 100,000 grids, the probability of finding a robot becomes 1 in 100,000. The probability of finding two robots in their starting positions would, however, be ten billion (10,000,000,000) to one against. Given the intrinsic improbability of ordered, information-rich configurations, it should now be easy to see why they have a tendency to collapse into one of their vastly more probable disordered alternatives.

I would now like to imagine a scramble net which stretches from the Earth to the Moon and contains a million quadrillion or 1,000,000,000,000,000,000,000 (10^{21}) grids. It is necessarily large, as it belongs to a school which accommodates ten million (10^7) robotic pupils. The schoolmaster left in charge of this class understands that any attempt to keep tabs on individual pupils would be futile. His first go chance of locating any particular pupil is after all one sextillion (10^{21}) to one against. In a *Scramble Net World* of this size inhabited by ten billion robots engaged in independent ramblings, the number of distributions corresponding to a disordered configuration vastly exceeds the infinitesimally smaller set which correspond to an ordered configuration. The collapse of ordered configurations of robotic schoolboys into random, disordered configurations occurs as an inevitable consequence of this discrepancy. It depends on the fact that every position in *Scramble Net World* is, as a consequence both of the robots' indifference and the absence of any attractors which draw them to given regions, equally probable.

Despite his being denied information about the location of any individual pupil, the schoolmaster wonders whether he might nevertheless predict the behaviour of the whole class. He thus dreams up the following experiment. All ten billion robots are placed in adjacent grids at the corner of the scramble net. Their batteries are then removed and all of the robots painted red. Their batteries are then replaced and they are left to perambulate in the usual manner. Using a huge and specially constructed wide-angle telescope, he then focuses on the entire scramble net and, as if stalking a tiger, sits back in his armchair, patiently waiting for events to unfold.

Although unable to visualize individual robots, what he now sees is a slow but regular dispersal of red moving out from the corner and radiating slowly across the scramble net, which eventually appears uniformly red. The schoolmaster repeats the experiment several times and finds that the same result is obtained on each occasion. It seems that when treated *en masse*, robotic schoolboys behave in a predictable manner and global order emerges from local disorder. Thus despite their individually erratic movements, vast numbers of robotic schoolboys inadvertently cooperate to generate a higher order behaviour, which manifests as the generation of maximal disorder. So when left to their own devices, robotic schoolboys will destroy information in a unique, but nevertheless entirely predictable way.

Having obtained this result, the schoolmaster decides to investigate whether the global regularity of the robots' behaviour is discernible at the individual level. He thus performs one further experiment. A random selection of ten thousand robots are painted blue. They are then placed in one corner of the scramble net with the others and set loose. Despite the inevitable attainment of a uniform sea of red at the end of each experiment, the pointillistic blue dots floating amidst the sea of red repeatedly change their pattern of distribution and never appear in the same place at any one moment, or on any separate occasion. The ability to predict that the ordered structure of ten billion robots clustered together will decay into a uniformly disordered distribution is thus independent of local knowledge of the position of individuals. Indeed, the statistical behaviour of very large numbers of things is resistant to such details. Although helping him to draw general conclusions about his robotic class, the schoolmaster's discoveries are unable to furnish insights that help resolve the indeterminacy at the local level. Indeed the position and behaviour of any particular robot will always elude him.

Gases, like bromine, consist of around ten billion billion molecules per cubic centimetre that move randomly across vast scramble net-like state spaces that exist not just in two dimensions, but in three. Bromine molecules behave like robotic schoolboys and once released from their starting position will tend to randomly explore all of the possible locations in their own type of three-dimensional scramble net-like space. But rather than being powered by batteries, they are instead driven to relentless movement by environmental heat energy. As the ordered state in which the molecules cluster together in one region of the potential state space is hugely improbable, the system will inevitably decay into the state of maximal probability. This corresponds to the disordered situation in which all regions of the state space are visited with an approximately equal frequency. The macroscopic consequence of this continuous sampling of the state space is that the bromine colours the jar uniformly brown. The probability of the bromine molecules reverting to their information-rich state is so small that it is effectively non-existent.

The most probable arrangement of large numbers of indifferent things arrayed across a space is thus a random and constantly changing distribution in which different regions of the space are sampled with an approximately equal frequency. The statistical laws of physics rest upon this fundamental tendency of ordered matter powered by thermal agitation to decay into its most probable state. And yet despite this, life, which is the very embodiment of ordered complexity, prevails. So, does the phenomenon of life implore us to postulate an extraneous agent located beyond the laws of physics and chemistry? As we have already said, the answer is an emphatic no. Unlike bromine molecules diffusing across a closed jar, or robotic schoolboys meandering across a *Scramble Net World*, living systems are located far from equilibrium and are open to the exchange of matter and energy with their environment. There is no mystery surrounding the generation and perpetuation of the information of life, as it is not obtained without exacting its own price, which is a proportional destruction of the information of the environment, whose disorder consequently increases. An eye for an eye, an informational tooth for an informational tooth.

In living systems, the units of structure are non-identical and interact strongly. The robots in the *Scramble Net World* analogy are consequently no longer homogeneous antisocial clones, but co-operate to form interconnected networks. As a consequence of this and their open, far-from-equilibrium structure, they are able to restrict their exploration to a fraction of all the possible states, instead of visiting every region of their state space with equal frequency. The nature and extent to which they are able to compress their state space occupancy is influenced by the architecture and dynamics of the components from which they are constructed. The condition of life thus corresponds to a situation in which immensely powerful magnets are scattered across the *Scramble Net World*. Once released from their starting position, the robots no longer explore their state space in a random manner, but fall under the attracting force of the magnets instead. Each magnet has its own region of influence, and any robot entering its basin of attraction is drawn towards it. Although the extent of the potential state space is unchanged, the number of states available for the robots to visit is significantly reduced. The structures of life are absolutely dependent on this dramatic biasing of probabilities. The dice of life are thus heavily weighted.

In living systems, the robots remain in an ordered configuration, instead of randomly dispersing across the scramble net as the laws of probability predict. They are thus able to anchor themselves and maintain their pattern, despite the ever present threat of thermalization. The absence of life, on the other hand, corresponds to the situation in which the cluster of robots lose their pattern and collapse into its most probable state, which is a uniform random distribution. Life thus mimics the situation in which the robots have been cooled down to absolute zero. At this temperature, the environmental heat energy motor which drives thermalization is switched off and statistical phenomena are no longer relevant. Thus, although life does not undermine the laws of chance, it modifies their outcome and living things behave *as if* they have withdrawn from the world of statistical phenomena.

In his book *What is Life*, published in 1944, Erwin Schrödinger compared our attempts at understanding the mechanics of life to the confusion that an engineer familiar with only steam engines might experience when presented with an electric motor. But the confusion would be short-lived, as when given an opportunity to examine the motor, he would soon realize that its structure reflected an underlying functional principle. Schrödinger speculated that 'while not eluding the laws of physics as established up to date', living matter 'is likely to involve other laws of physics hitherto unknown, which, however, once they have been revealed, will form just as integral a part of this science as the former'. Furthermore the regular and lawful unfolding of the events of life, is 'guided by a mechanism entirely different from the probability mechanism of physics' and constitutes a state of affairs which is 'unknown anywhere else except in living matter. The physicist and the chemist, investigating inanimate matter, have never witnessed phenomena which they had to interpret in this way. The case did not arise and so our theory does not cover it . . .'

It is reasonable to speculate that the unknown laws of physics that Schrödinger was alluding to are what have come to be known as the laws of complexity; fledgling laws that are only just beginning to be formulated, but which might provide fundamental insights into the dynamical behaviour of far-from-equilibrium, open and highly interconnected systems. The laws of complexity suggest how non-equilibrium systems are able to counteract the thermalization pressures which attempt to drive them into their state spaces in an entirely random manner. As a consequence of the natural dynamics of highly interconnected non-equilibrium systems, vast regions of their state spaces become forbidden no-go zones, and state space exploration is confined to a narrow subset of all of the possible regions. This contrasts dramatically with random, equilibrium systems, which are free to explore their state spaces with impunity, but are consequently unable to generate information.

But how is this restriction of state space exploration achieved and what is the source of life's order? Max Planck made a distinction between two types of order. The first is the order which arises from disorder. This alludes to the statistical laws that describe the collective behaviour of huge numbers of independent entities. The second is the order which arises from pre-existing order. An example of this 'order from order' principle is the configuration of the planets in the solar system which is, at any moment, a direct consequence of all its previous states. The principle that determines the clock-like motions of planets does not invoke statistical phenomena and is based exclusively on Newton's law of gravitational attraction. Given a starting constellation, the regular movements of the planets will continue indefinitely. Unlike statistical systems, the motion of planets can be understood by the application of purely mechanical laws, which ensure that aspects of the order of the past are imprinted onto the future.

The 'order from disorder' principle, on the other hand, rests upon the statistical foundations of phenomena involving huge numbers of entities. Schrödinger argued that it was most unlikely that living things could derive their order from a source of this kind, precisely because of the large numbers of elements that would need to be involved. He thus inferred that the order of living things is derived from the order from order principle. In making this inference, Schrödinger established the foundations for the central paradigm of modern biology. This is that the information of contemporary life stems from the execution of a digital program written in the DNAese language of genes and that each genetic program is itself programmed by a pre-existing precursor program.

In 1944 Oswald Avery, Colin MacLeod and Maclyn McCarty showed that the genetic order from order principle of modern organisms is embodied in DNA. The elucidation of the three-dimensional structure of DNA by Watson and Crick suggested how the fundamental information of life can be represented digitally in linear combinations of the **C**, **T**, **G** and **A** bases from which DNA is made. It also provided a simple explanation of how this information could be replicated. DNA constitutes the chemical basis of virtually all contemporary strategies of biological information storage. Genes which are constructed from DNA or other digital chemistries are thought to be the essential precursors from which the information of all life originated. It follows from this genetic paradigm that life began with a primordial grandmother gene, the root source of what Richard Dawkins has called the 'digital river' of genetic information. A river flowing across the expanses of time from its source, deep inside the womb of a metaphorical Eden.

We have already considered the possibility that the digital DNA river may not have issued directly from Eden. Indeed it may have been a tributary of a digital RNA river, which may itself have been a tributary of a river filled with primordial digital genes, pre-genes or proto-genes of an unknown chemical nature. In making the suggestion that life began without genes and that genes were only acquired by analog, geneless organisms at a later date following a process of digital infection, we are faced with the possibility that the ultimate source of the DNA digital river may not in fact lie within a digital Eden at all. Although it defines the informational structures of modern life, it is possible that the order from order principle may not have underpinned the informational structures of the very first living things. Eden may in fact have been an analog, gene-free haven. If this was the case, then it was upon the geneless planes of an ancient analog world that the first organisms walked tall. The origin of lifeless digital pre-genes may have predated the discovery of the analog Eden, but it was in this gene-free laboratory that the forces of chance, drift, contingency, complexity and evolution by natural selection discovered the precious and fragile gift of life.

Of course Eden was not necessarily a discreet place. It is likely to have existed simultaneously and degenerately in several places and incarnations. In fact, an Eden existed in any nook or underwater cranny where a successful life experiment had taken place. However, although the history of life might have begun with multiple Edens, it is unlikely that there is more than one historical Eden whose lineages persisted and were the antecedents of modern life. It is the life discoveries within this charmed garden that generated the descendants whose information was to seed all future life on Earth. A single, geneless clone, unravelling its covert potential for complexity by means of a diverse spectrum of digitally infected descendants, randomly dispersing across time and the expanses of several multi-dimensional information spaces.

But the river of biological information issuing from Eden, the mother source of all life on Earth, may not have been digital at all. If life did begin without genes, then it is only as we sail gently down the analog river linking the past to the present that the analog and geneless source of all information is eventually joined by a digital tributary. This represents the small subset of primordial lifeless digital pre-gene parasites which, as the result of a process of infection, were able to mingle their digital information with the analog information of their geneless hosts. As we continue to sail down the metaphorical river of biological information, we may have to travel a considerable distance before the waters begin to pulsate and fill with the digital information of RNA and eventually DNA. Still further down the river, which bifurcates into countless rivulets of private information, we notice the appearance of a new tributary. History repeating itself and turning full circle as extra-genetic cultural information mingles with the digital information of DNA. If we were to continue our journey into the future, we should not be surprised if at some point the digital DNA river began to foam and swirl once again, as information from some unknown source comes to mingle, dilute and eventually displace the information encoded within the ancient and jaded sequences of DNA. Cast aside like the out-dated technologies that preceded them, DNA symbols, now hundreds of millions of years past their sell-by date, might themselves finally be ousted and returned to the archives of possible history.

We have discussed how living organisms behave as if they are largely immune to the potential informational disturbances resulting from thermal agitation. Unlike closed equilibrium systems which are subject to the tyranny of statistical law, living systems appear to behave in a more mechanical way. In the case of the organisms with which we are familiar, the ability to generate and perpetuate information depends on both genetic order from order principles and on the metabolic principles which non-living systems lack. We will shortly consider a very different type of strategy for generating transmissible order that is based on non-digital, analog principles. However before examining the anatomy of these structures which may have formed the flesh of our hypothetical geneless ancestors, we will first examine how genes are able to escape from the chaos of statistical law and mimic mechanical behaviour.

The transient immunity of modern life from an inevitable collapse into formless noise is grounded in the properties of digital information technologies, which enable essential information to be conserved across time. These include, perhaps most importantly, the way in which individual digital symbols may only be altered in a quantum manner. In the case of DNA, any one of the **A**, **C**, **T**, or **G** base symbols may suffer a quantum mutation and be replaced with one of the remaining symbols. Symbols may also be inserted into or deleted from a sequence. But there is no opportunity for continuous, analog change. An **A** cannot, for example, be changed into an **A/////C**, which is a hypothetical intermediate between an **A** and a **C**. It cannot similarly be changed into an **A////////////////T**, which is a hypothetical hybrid of an **A** and a **T**. This requirement for discreet, quantum mutations and the lack of any possibility for continuous change, is essential for maintaining the structural integrity of digital order from order principles.

The relative stability of different types of digital genetic principles is influenced by the nature of the chemical bonds within and between each molecular symbol. It also depends on the context in which a given gene finds itself. The robustness of the information stored in modern genes is achieved in many different ways. These include: the proofreading and editing capacities of DNA and RNA polymerases, checkpoint proteins which examine replication events, apoptopic programs which destroy cells whose DNA database is irreparably damaged, and repair mechanisms which monitor DNA synthesis and replication. Such direct methods are supplemented by a host of indirect methods, which include homeostatic mechanisms that regulate the chemical environment surrounding genes and physical factors, such as the way in which DNA is packaged. Dynamic factors such as the speed of DNA synthesis and transcription are also likely to play a role, as the unwinding of the two strands increases their susceptibility to physical and chemical damage.

The grammar of the genetic code, furthermore, appears to have been selected to minimize the impact of mutations. Individual amino acids are, for example, encoded by more than one codon, so that whilst changing the information of the DNA sequence, a point mutation does not necessarily alter the information encoded by the gene. If a mutation does result in the production of a codon that encodes a different amino acid, the code is configured such that the new amino acid is likely to be chemically similar to the one it replaces. Some genes demonstrate what is known as codon bias, and use a particular codon with a higher frequency than equally good alternatives. In addition, with the exception of their binding and active site regions which are of critical functional importance and exquisitely sensitive to change, and other regions of structural importance, the tertiary structures of proteins are remarkably tolerant to amino acid substitutions. Indeed the alpha helices and beta sheets which are the essential modular elements from which tertiary structure is generated, are very robust.

Naama Barkai and Stanislaws Leibler have suggested that the metabolic networks which form the computational devices of living cells have themselves been selected for their insensitivity to change, indicating that robustness at the level of collective gene function may add yet another level of stability to the genetic order from order principle. If order is equated with the global information contained within the modular units of biochemical function, then genetic circuits and the information of the proteins they encode may suffer extensive damage without substantially affecting the higher-level structures which define the computational function of the circuit. Evidence from gene knockout studies suggests that there is also a considerable degree of functional degeneracy in living systems. Thus if one circuit is compromised, the function may often be restored by an alternative circuit.

The digital DNA order from order principle is thus underpinned by a hierarchy of order-preserving principles; several of which are unlikely to be available to genes made from different building materials. RNA, for example, can form local double-stranded regions, but is unable to adopt the globally double-helical structure characteristic of DNA. Without a complementary strand to serve as a reference negative copy of the coding information, a hypothetical RNA repair enzyme would have no objective record of the core information and is consequently unlikely to be able to detect and repair damage.

Quantum mutations occur with a finite frequency in all digital genetic sequences. A stretch of DNA, or indeed any other digital sequence, may thus, much like a lump of uranium, be considered as having a half-life. This is the time that it takes for half of the DNA symbols in a given sequence to decay into different symbols. Although it is possible to broadly predict that a certain number of atoms within a lump of uranium will decay within a given period of time, the laws of physics do not allow us to predict which particular atoms will do so. Sequences of DNA also tend to decay in this random, stochastic manner. Although the decay is inevitable, its pattern is unlikely to be the same on any two separate occasions; furthermore unlike uranium and other radioactive atoms that have a fixed rate of decay, the rate at which a stretch of DNA decays is influenced by the physico-chemical environment in which it finds itself.

Evidence suggests that in some instances, the manner in which DNA decays is non-random. Biasing of mutational events manifests both as 'hot spot' and 'cold spot' regions which have respectively higher and lower probabilities of suffering a mutation and as biases which influence the direction of change once a position has been targeted. Thus, although the global mutation rate is essentially sequence-independent, there are instances in which some regions or motifs are more mutable than others. The fact that certain patterns of decay are seen more commonly than others adds yet another tier of complexity to the DNA order from order principle. Although as a consequence of the random nature of DNA decay it is impossible to predict which particular nucleotide will mutate over a fixed duration, once a position has been hit the direction of change may be non-random. If, for example, an **A** is targeted, it has the potential to change either to a **C**, **G** or a **T**. Directional bias occurs when the probability of changing an **A** to a **C**, for example, is greater than it changing to a **G** or a **T**.

Although it might be of interest to assign an informational half-life to a sequence, this is difficult, because the informational impact of a mutation depends on several factors. These include the location of the mutation, whether or not a premature termination signal is generated, and whether the mutation is silent, in which case the sequence of the encoded protein is unaffected. A single mutation in a DNA sequence may, as a result of changing a single amino acid, destroy or profoundly modify the structure and/or the function of the protein it encodes. So despite having only a modest impact on the information of the DNA sequence itself, the overall informational impact can be profound. A single or double insertion or deletion near the beginning of a gene has a devastating informational impact, as it introduces a shift into the reading frame and results in the production of an entirely different protein sequence.

The half-life of a gene is also sensitive to environmental factors such as the presence of ultraviolet radiation, or chemical mutagens that increase the rate at which DNA decays. The reactive oxygen molecules that enable cells to respond to signals at their surface and are generated as toxic products of several metabolic pathways, for example, produce several types of DNA damage. Even the water that is essential for the maintenance of DNA structure may attack it and undermine its informational content. In view of its continuous exposure to such dangers, DNA, as has already been mentioned, is critically dependent on strategies that minimize the informational impact of any damage that is incurred. Mutation detection and repair mechanisms constitute one of the most important ways of modifying the half-life of genetic sequences. These include base excision repair (BER) mechanisms that excise and repair mismatched or chemically damaged bases, and nucleotide excision repair (NER) mechanisms that cut out and repair whole nucleotides, double-stranded break mechanisms (DSBR), and mismatch repair (MSR) that corrects errors introduced while DNA is being copied.

The DNA polymerase enzymes involved in DNA synthesis also have an important role in influencing the half-life of a sequence. DNA polymerase enzymes isolated from different organisms have different error rates. Whereas the fidelity of DNA synthesis is high in some species, in others it is more error-prone. An identical piece of DNA will consequently have a different half-life if it is replicated using one type of polymerase, as opposed to another. Individual species may even utilize several different polymerases, each with a different error rate. Polymerases can also differ in the frequency with which they make particular types of errors. DNA polymerases could presumably be designed to have error biases of one type or another. This would influence the pool of genetic variability available to a species by biasing its exploration of *DNA Space*. Its repertoire of possible futures might then be fundamentally constrained.

In some genetic systems, the overall baseline rate of mutation can, under certain circumstances, be dramatically increased. Hypermutation mechanisms rapidly furnish certain types of bacteria with the genetic variability needed to provide a hasty response to adverse conditions. Hypermutation also occurs in antibody genes of the human immune system. In this case, however, the mutations are not introduced into the germ line and so cannot be passed on to subsequent generations. The hypermutation seen in antibody genes provides an excellent example of how mutations can be focused onto discreet hot spot regions of a sequence, as the hypermutable sequences encode the regions of the antibody molecule that are of key structural importance. In focusing mutations principally, but not exclusively onto these regions, the immune system achieves great economy in the deployment of its hypermutational resource.

Most damage to DNA is faithfully repaired, but occasionally a mutation will not be repaired and becomes fixed into the genetic sequence. This occasional fixation of random quantum mutations ensures that the genetic order from order principle is able to balance the requirement for high-fidelity information transfer with the equally important requirement for informational flexibility. When coupled to a process of selection, the fixation of occasional mutations ensures that the population of informational structures is able to update its representation of the environment in which it is evolving. As we have noted, it is important that the rate of mutation should be high enough to produce occasional variants, but not so high that the core information of the system is undermined and melts away into noise.

In addition to mutation, other processes generate informational diversity and contribute to the DNA order from order principle. Recombination generates genetic diversity by reshuffling genes like a pack of cards. This is achieved by allowing corresponding regions on different chromosomes to exchange with one another. By recombining the starting material in a novel way, recombination makes best use of the finite resource of available genetic material. Although ultimately dependent on the variation generated by mutations, it maximizes their informational impact. The order of genetic information thus lies both in the sequences of individual genes and in the genetic context in which alternative examples of each gene find themselves. The success of recombination as a strategy for generating new genetic information relies on the fact that different combinations of alternative forms of the same set of genes can behave quite differently. By generating information in this way, it is likely that recombination allows error-correction rates to be set at a higher level of efficiency than might be possible if mutation was the only generator of variation.

The existence of different versions of the same gene within a population is known as polymorphism. The efficiency of recombination as a process for information generation is dependent on the degree of polymorphism in a population. If the gene pool is highly polymorphic, then the potential number of different combinations is huge. The degree of polymorphism in any particular gene varies enormously. Some genes for example, have literally hundreds of alternative versions. In the case of the MHC genes of the immune system, for example, these alleles confer resistance or increased susceptibility to various diseases. The genes encoding the venom of some poisonous snakes are also highly polymorphic, whereas others have little or no polymorphism.

In genetic systems that utilize sex, variation is thus, amongst other important mechanisms such as gene duplication, accomplished by a mixture of mutation and recombination. Like mutation, however, recombination is not a completely random process. Although all regions in a genome might in principle recombine with the same frequency, local rates of recombination are often non-uniform. Some hot spot regions undergo recombination at much higher frequencies than others, whereas cold spot regions recombine only rarely. Nature then, like a crooked dealer in a Las Vegas casino, does not shuffle her genetic cards randomly, but goes out of her way to keep certain combinations together, whilst at the same time disrupting others. What gambler after all, would risk a radical reshuffle if they had been dealt the essentials of a royal flush?

Having considered the disordered behaviour of closed, near-equilibrium systems and the genetic order from order principle that rationalizes random statistical behaviour and in so doing generates biological information, we are faced, once again, with the problem arising from the fact that the order from order principle of genes presumes a source of pre-existing order. A mother program which programs its daughters. No matter how far back we choose to look, we are forced to infer the existence of a precursor mother, grandmother, or great-grandmother program *ad infinitum*. This situation of infinite regress is clearly absurd and it seems reasonable to hypothesize that there was a time when Earth was devoid of programs. The challenge, then, is to explain how digital genetic programs might have originated in a program-free context. Is it possible, even in principle, for digital programs to emerge spontaneously from information-free noise? If we conclude that this is impossible or unlikely, then might some other intermediate have functioned as a stopgap or stepping stone technology? How indeed was the first genetic program programmed?

In a more general sense, and without making assumptions about the nature of the mechanism used by the very first living organisms to represent their essential information, what we appear to need is a process which can generate information from a lack of information, or 'noise'. However, although an order from disorder principle might initially generate biological information from noise, an order from order principle would be mandatory for the heritable transmission of this information. The principle need not necessarily be genetic however. If the newly-created biological information was to persist, it would have been essential that it incorporated a capacity to evolve. The order from order principle should thus be able to secure the core information which defines the structure, whilst at the same time allowing a limited fraction of the information to mature. In so doing the system is conferred with the potential both for heredity and evolvability. Whereas a limited heredity system is confined to the exploration of a relatively small and well-circumscribed potential state space, systems with an unlimited capacity for inheritance are able to explore infinite state spaces.

Given these considerations, is there any experimental evidence to suggest that digital genetic programs of any length and constructed from any of the historically plausible chemical building materials are able to spontaneously self-assemble from a random soup of information-free precursors? Using the terminology of Steen Rasmussen, Carsten Knudsen and Rasmus Feldberg, is there any class of matter which, in the absence of a pre-existing informational template, is able to spontaneously generate information from non-information and is consequently 'self-programmable'? If self-programmable matter does, or has existed, then once a program has self-assembled from informationless noise, would it be able to self-replicate? Evidence suggests that the nucleic acid components from which DNA and RNA are made are not self-programmable. Modern DNA and RNA is only able to replicate in the context of highly evolved enzymes. They appear at the very best to have only a rudimentary capacity for non-enzymatic replication. Furthermore, the type of replication that occurs in the absence of enzymes is not the same as that occurring when enzymes are present.

The formation of a complete and natural cycle of replication requires both high-fidelity complementary template-based copying, release of the newly synthesized molecule from the template and the initiation of further rounds of replication. Both enzyme-free replication and template-based copying have been independently achieved to a limited extent. The self-complementary replication which has been achieved in experimental systems is not, however, equivalent to the natural complementary mode of replication characteristic of nucleic acids. In the case of DNA and RNA, replication involves the production of a newly-synthesized daughter strand which is complementary to the mother template. In artificial systems on the other hand, the daughter strand is identical to the mother template and thus self-complementary. The combination of complementary template-based copying coupled with the initiation of a new round of replication has not yet been realized. There are furthermore good theoretical reasons related to the efficiency, rate, fidelity and generality of complementary synthesis reactions to suggest that enzyme-free complementary copying and the subsequent initiation of self-replication is not realizable even in principle. Experiments using artificial systems have, for example, demonstrated that only a very restricted range of sequences are able to act as efficient templates for complementary synthesis.

Julius Rebek and Günter von Kiedrowski have independently

demonstrated the feasibility of self-replication in organic polymers that are chemically very different to DNA and RNA. These artificial replicators are not self-programmable, but the fact that they can in some instances undergo complementary self-replication and are able to mutate and evolve by natural selection emphasizes once again that the search for primordial self-replicating molecules must be cast widely. The possibility that life could have emerged from regions of *Chemistry Space* with which modern life is unfamiliar must thus be taken seriously. A variety of informational molecules present themselves as candidate primordial genetic materials. These include molecules that share the sugar backbone common to DNA and RNA and others such as peptide nucleic acid (PNA), which do not. Leslie Orgel and Wojciech Zielinski have shown that modified nucleic acids are more replication-friendly than naturally occurring ones, indicating that the earliest genes may have been chemically modified relatives of DNA or RNA. The repertoire of structures that combinatorial libraries of polymers composed of unusual bases might have been capable of generating may have been more enzyme-friendly than their DNA or RNA counterparts. Until the chemistry of all of the plausible candidates has been fully explored, we must acknowledge the possibility that at least one of these chemical systems might be self-programmable. The ancestral programs that may have templated programs written in DNAese or RNAese may thus have been chemically quite distinct from their descendants.

In 1996 James Ferris and Leslie Orgel demonstrated how RNA and small peptide 'mini-genes' of up to fifty-five amino acids long could have spontaneously assembled on one of the catalytic surfaces provided by minerals such as clay, or hydroxylapatite which were present on the primordial Earth. If the same reactions occur in the absence of a catalytic surface, the maximum length of the mini-genes obtained is only ten amino acids, as hydrolysis competes with polymerization which is less efficient in the absence of a catalyst. It is possible that, under favourable conditions, mini-genes of an even greater length might self-assemble in a similar manner. Random libraries of RNA sequences of approximately this order of magnitude have been shown to contain ribozyme species that mimic the action of enzymes. Mineral surfaces could thus have functioned as factories for synthesizing the ribozymic machinery necessary for the generation of rudimentary life. Unfortunately, however, in this system the newly synthesized mini-genes stick tenaciously to the mineral surface on which they are made and are unable to initiate a new round of synthesis to complete a full cycle of replication.

Although the matter in the Ferris and Orgel experiment is not self-programmable, it constitutes a significant step in that direction. The experiment also indicates that rudimentary life may have been baked on mineral surfaces like a primordial pancake, rather than being cooked in the primordial ocean like a minestrone soup. It also raises the possibility that the first living organisms inhabited a two-dimensional *Flatland* and proliferated by sprawling out across surfaces. No experimental system has yet generated self-assembling mini-genes that are able to self-replicate by functioning as complementary templates for the construction of daughter mini-genes. So despite having the potential to unlock digital combinatorial spaces of a size dependent on both the length of the mini-gene and the number of elements incorporated into it, these miniature digital pancakes do not satisfy our definition of a gene.

In 1974 Manfred Sumper demonstrated that RNA could be synthesized without a template. However, template-free synthesis requires the presence of a highly evolved RNA replicase enzyme. It is unclear whether this helps catalyze the formation of the founder template, or whether the RNA nucleotides self-assemble as the result of an enzyme-independent process and are subsequently amplified by the replicase. Nevertheless, this experiment indicates the feasibility of complementary synthesis coupled with replication, and raises the possibility that RNA might be able to catalyze its own synthesis and replication using a ribozyme that mimics the activity of an RNA replicase. But despite an intensive search, no such ribozyme has yet been discovered.

In 1996 David Lee and Reza Ghadiri showed that a thirty-two-residue peptide can self-replicate by autocatalytically templating its own synthesis. This is achieved by accelerating the ligation of pre-existing fifteen- and seventeen-residue precursor fragments. The parent peptide functions as a structural template on which the fifteen- and seventeen-residue fragments are organized. The existence of this peptide replicator raises the possibility that the information of primordial organisms may have been encoded by self-replicating peptide pre-genes. If this was the case, then the software/hardware distinction may have been blurred in the earliest examples of life. These pre-gene replicator molecules do not qualify as true genes as they do not codify information using a microcode. The information in the amino acid sequence of the peptide does not stand as an abstract symbol for another piece of information in the way that a codon in a DNA gene symbolically represents the information of the amino acid it specifies. What is more, the mechanism of templated-based replication is different and inherently more complex than that found in DNA, as it depends both on the primary structure of the amino acid sequence and on its tertiary structure. The replication process also depends on the presence of pre-existing precursor peptides that can be ligated together, unlike DNA replication in which the template allows *de novo* 'unitary' synthesis of a daughter molecule from individual non-polymerized precursor units. And unlike the complementary template-based self-replication found in genes, this peptide system replicates using a self-complementary mechanism of replication. A gene of sequence **AAAAA** will thus as a consequence of complementary template-based replication produce the *complementary* sequence **TTTTT**, whereas a peptide **YYYYY** replicating by a self-complementary template-based mechanism, will produce the *self-complementary* daughter sequence **YYYYY**.

Although both spontaneous synthesis and self-replication of peptide sequences have been demonstrated independently, there is as yet no known situation in which synthesis and self-replication are coupled. Thus, like the nucleotide constituents of DNA and RNA, the amino acid constituents of peptides do not appear to be self-programmable. However, the demonstration that a peptide sequence can template its own synthesis raises the possibility that the sequence specific complementary molecular recognition seen in DNA and RNA replication may occur in other chemical systems. Complementary molecular interactions between peptides are, however, likely to be far more complex than the simple Watson-Crick base-pairing seen in DNA and RNA.

In addition to the practical difficulties associated with realizing enzyme-independent informational replication, a major theoretical problem constitutes another stumbling block for any theory of the origin of life that places self-programming and template-based replication at its centre. Manfred Eigen, who first described this paradox in 1971, demonstrated that the amount of information which can be encoded in a digital genetic sequence is limited by the copying fidelity of each digital symbol. The numerical value of the copying fidelity associated with a stretch of digital sequence thus determines its error threshold. If this is exceeded, the information will melt away into meaningless noise. The error threshold thus limits the maximal length that a digital sequence can attain and consequently the amount of information it can encode. Without an external agent to minimize error propagation, the maximal sustainable length of a piece of digital information is around 100 bits. This amount of information is unlikely to be adequate to code for error-correcting enzymes and so it appears that an informational crisis threshold precludes the possibility of any free-living genetic organisms with a genome of a size greater than around 100 digital symbols. It is important to remember, though, that an RNA sequence of 100 nucleotides would be sufficient to allow the formation of the structural elements and motifs such as bulges, pseudoknots and stem loops, that are essential for ribozyme function. It is thus quite possible that the library of all possible RNA sequences of length 100 or fewer nucleotides contains molecules able to function as rudimentary error-preventing and correcting machines. No such ribozyme has as yet, however, been discovered and although an intensive search of artificially constructed combinatorial RNA libraries is currently underway, it is by no means certain that such ribozymes exist. But if enough error-correcting and synthetic competence could be found to allow the formation of mini-genes that were 101 or more nucleotides units long, the *Space of All Possible Nucleotides of Length 101* or more units may contain error-correcting RNA polymerase-mimicking ribozymes with greater efficiency than those present in the library containing mini-genes that are only 100 units long.

Given that we lack a precedent for self-programmable matter in any candidate genetic materials and that in the absence of an external error-correcting mechanism the Eigen paradox limits the amount of information that can be stored in genetic systems, are there any non-digital means by which information could have been spontaneously derived from noise? And if information can be stored non-genetically, is the information-storing capacity of such systems sufficient to surmount the information crisis suggested by the Eigen paradox? It should be remembered that we have defined a gene in a very specific way, so as to include only those information-encoding entities that are digital, replicate using a complementary template-based copying mechanism and which utilize some type of codification. However, is it possible that there were geneless informational systems that were able to contribute to a process of life, but which had non-digital architectures, non-encoded metabolisms and did not replicate their information using a template-based strategy? Might such systems have mimicked the capacity for heredity inherent to genes and mutated their information, competed with one another and evolved as a consequence of a process of natural selection? If this was the case, then a non-digital, extra-genetic principle of order may have laid the informational foundations for all subsequent life on Earth.

11

Patterns Without Programs

In 1810 the Swiss entomologist Pierre Huber observed that many species of ants supplement the manpower of their colonies by capturing and subjugating slaves from neighbouring colonies. Thus although outlawed amongst humans, in the ant world slavery continues to flourish. In procuring slaves, a colony acquires an extra supply of workers, some of which have skills their captors lack. Once captured, the slaves are integrated into the day-to-day life of the colony, but things do not always run smoothly and there are several documented instances of slave revolts. Slaves of the species *Leptothorax curvispinosus*, for example, have been known to take bites at the head and thorax of their surrogate Queen. Young slaves have been observed showing a similar disrespect for royalty. One, for example, was seen dragging a Queen backwards and forwards whilst repeatedly stinging her.

Two types of slavery are found in ants: true slavery in which ants of the same species are enslaved and far commoner slavery-like situations, in which ants of a different species are captured and subjugated. Some ant species are occasional slave holders, with some colonies using slaves whilst others do not. In populations of *Formica sanguinea* ants for example, the number of colonies holding slaves varies between 2% and 98%. Other species only rarely take slaves and even then do so only by accident. A range of strategies are employed to procure slaves. *Formica subintegra* ants, for example, use chemical warfare to subjugate their victims and when raiders attack a colony they spray propaganda substances at the defenders. These disseminate disinformation by mimicking the defenders' alarm chemicals and inducing a state of confusion.

The honeypot ant *Myremecocystus mimicus* practises true slavery.

The fact that the foraging grounds of neighbouring colonies usually overlap frequently leads to territorial conflagrations. But battles between rival colonies are more akin to symbolic tournaments than true conflicts, and participants are seldom injured. Jousting matches in which individuals extend their legs and drum their antennae onto the abdomens of their enemies often continue for days. If, however, a colony is unable to deploy sufficient workers to the site of the tournament, it is likely to be overrun. The defeated Queen is killed or driven off and larvae, pupae and workers are carried back to the conqueror's nest. By capturing slaves, the size of the victor's work force is much increased, but no new skills are acquired. This is not, however, the case with the amazon ant *Polyergus*. These large, shiny red or jet black beasts are formidable opponents and slaughter defenders that resist attack. But although skilled in the art of warfare, they are good at little else. Their sickle-shaped mandibles are adept at piercing the body armour of their opponents, but are unsuited for essential tasks such as digging, or nursing their larvae and pupae. They are not even able to procure their own food, and are completely dependent on the housekeeping skills of their captives. The indolence of their parasitic lifestyle is punctuated only by occasional slave-hunting sorties into neighbouring colonies belonging to ants of different species.

We have already discussed the possibility that modern life may have arisen as a consequence of the parasitism of analog geneless organisms by digital pre-genes. These were defined as digital polymers that encode information within their combinatorial sequences but which do not self-replicate using the complementary template-based strategy of genes, and which lack mechanisms for the symbolic codification and translation of encoded biological information. Like *Polyergus* ants which enslave aspects of the complexity of foreign ant communities, digital pre-genes may have parasitized and enslaved the metabolic networks of analog geneless organisms. Such self-organizing analog, geneless creatures may thus have provided the informational context for the evolution of modern genes. In this scenario, pre-genes are imagined as having manipulated the spontaneous order and dynamics of self-organizing analog systems with their digital propaganda. In this light, modern genes may in one sense be viewed as parasites which ensnare and constrain the natural order of self-organizing systems. The foundations of modern life may thus be rooted in the exploitation of the spontaneously forming order of ancient geneless creatures.

But how might the informational structures of analog, geneless organisms have originated without genes? For surely all biological organization must be specified by genes? Furthermore, given that such hypothetical geneless organisms are not envisaged as containing a centralized symbolic specification of themselves, is it reasonable to call them living at all? The weather is after all a complex geneless system, but lacks any internal codified representation of its structure that is physically distinguishable from the components of the weather itself.

The candidate non-genetic principle that may have been responsible for generating the first geneless creatures is the spontaneous order which arises in certain types of highly interconnected, open and far-from-equilibrium systems. The tentative principles thought to underlie the self-organizational or self-programmable properties of the matter from which such systems are composed have come to be known as laws of complexity. When considering how information is represented in these types of systems, we must dispense with the linear, digital notion of a program, borrowed from computer science, and consider the information of the program as being stored in a distributed, dynamic, non-digital or analog manner across the entire system. In such dynamic, program-free informational systems, there is no formal distinction between hardware and software and no centralized information database. The program is no longer a discreet and central set of coded instructions, but rather a dynamic and emergent property of the whole system.

Unlike the order perpetuated by genes, which depends on the communication of a pre-existing source of order from one digital system to another, the order that emerges in complex self-organizing systems arises spontaneously in the absence of a precursor. Natural order of this kind, which Heinz von Foerster has called 'order from noise', represents a dimension of the statistical behaviour of molecules which is unique to certain types of open, far-from-equilibrium, heterogeneous and highly interconnected systems. The matter contained within a system that has the type of organization necessary to generate order from noise may, with the proviso that we are using the word program in the sense discussed above, be described as self-programmable.

The apparent paradox of the emergence of spontaneous order in the context of a law that appears to demand the exact opposite is, as we have already seen, a straw man. This is because the classic model system which the second law of thermodynamics describes is an idealized, closed, low-energy, near-equilibrium system unable to exchange matter and energy with the outside world and which consists of homogeneous and only weakly interconnected components. Living things, on the other hand, are high energy systems located far from thermodynamic equilibrium. They are constructed from heterogeneous and highly interconnected elements which dynamically exchange matter and energy with the environment. This flux and dissipation of matter and energy through living systems is the root source of their order. Like light focused through a magnifying glass, this flux of matter and energy allows order to be concentrated onto the system at the expense of the increasing disorder of the surrounding environment. When an open system and its surroundings are considered as a single system, the equations balance up and it becomes apparent that the second law of thermodynamics is not violated. However, because of their open structure and their ability to siphon order from the surroundings and translate it into an increase in their own order, systems located far from thermodynamic equilibrium are able to behave *as if* they were immune to the second law of thermodynamics. The information contained within repertoires of alternative states of self-programmed natural order should, in principle, be mutable and thus amenable to a process of Darwinian evolution by natural selection.

It follows then, that the order of living things may have originated from two independent sources. First, the spontaneous self-programmable order from noise principle, which operates in certain classes of open, far-from-equilibrium systems, and second, the order from order principle, which originated with the inception of the first self-replicating digital program. The elaboration of these sources of information was dependent on the iterative generate-test-retest algorithm of evolution by natural selection. But is it possible that the order from order principle of genes was only able to contribute to processes of life after rudimentary geneless organisms utilizing the order from noise principle had been established? An analog order from noise principle may indeed have been necessary to kick-start digital life into existence. Once generated, these hypothetical geneless organisms may have constituted chemical micro-laboratories in which the first genes evolved and established the digital genetic foundations of modern life.

In addition to providing the context and ground rules that allow matter to self-program spontaneously, the laws of complexity are likely to influence the manner in which information that arising both from noise, and from digital programs, is able to evolve. The laws of complexity thus relate both to the *de novo* generation of information and to its subsequent elaboration and optimization; and its machinations are consequently likely both to complement and constrain the process of evolution by natural selection. These intertwined principles, tempered by historically contingent events, are likely to represent the principal causal factors responsible for shaping the information of modern life. Whereas the so called 'modern synthesis' of Darwinian theory focuses on the molecular genetic events which underlie the mechanics of evolutionary processes, the broader 'postmodern' synthesis focuses on the interplay between natural selection, historical contingency, as yet poorly defined laws of complexity, and physicochemical factors that shape and constrain the nature of living things.

Although natural selection may ultimately take the upper hand, the laws of complexity appear to be important in helping maintain the homeostatic stability of living things and ensuring that the living structures amenable to Darwinian processes are built to be maximally evolvable. This may be achieved by influencing and constraining the dynamic behaviour of genetic systems. The information of life is not consequently exclusively dependent on the frozen accidents of chance events, but may opportunistically freeze and exploit the seams of available program-free information inherent to natural self-organizational processes. Life is not therefore the sole result of a string of highly improbable chance events and time is not the only victor in the historical game of life-creation.

The laws of complexity, or complex self-organization, offer life a guiding hand and help make the improbable possible. They are the hypothetical principles which govern the spontaneous information-generating behaviour of certain classes of open and highly inter-connected systems located far from thermodynamic equilibrium. These laws are likely to be as fundamental as the statistical laws which govern the behaviour of classic near-equilibrium systems. If such laws are indeed important for the construction of living things, then instead of being marginalized as an intrinsically improbable phenomenon, life becomes a natural and integral part of the behaviour of certain classes of open, far-from-equilibrium, network systems. Rather than requiring an extraordinary miracle of chance to get it started, life may in fact be an expected, inevitable and inexorable property of matter located within the context of certain types of open and far-from-equilibrium systems.

But what exactly do we mean by the terms self-organization and spontaneously emerging program-free order? Before considering more complex examples of self-organization which may have been responsible for the generation of analog geneless life from non-living matter, we will examine some simpler examples. One need only think of a droplet of fat which spontaneously assembles into a spherical micelle as it contacts water. There is clearly no internal or external digital program instructing the droplets and specifying their architecture. The information needed to assemble the droplet is instead inherent to the physics and chemistry of the whole system. Thus, although the fat molecules lack a program, they behave as if they were being controlled by instructions encoded in an internal program and are in this sense self-programmable. In 1992 Pascale Bachmann and Jacques Lang demonstrated that aqueous micelles of sodium caprylate are able to self-organize and self-replicate. It seems, then, that, at least in this simple case, a system that lacks internal genetic instructions can spontaneously generate the information needed for its own self-assembly and replication. It thus also behaves as if were under the influence of instructions encoded within a centralized internal digital program. Systems that can self-replicate without the execution of a set of internal, digitally encoded instructions belong to a class of systems which Leslie Orgel has called 'non-informational'. Even a system as simple as a layer of sand placed on the bottom of a rigid evacuated container and which is oscillated at different frequencies, is able to spontaneously generate a host of complex geometrical patterns that include squares, stripes and hexagons.

The mathematician Alan Turing devised a simple test to determine whether a machine can think, but concluded that the question itself is absurd, because if a machine behaves as if it can think in a manner indistinguishable from that of a human, then the question of whether or not it actually *is* thinking is irrelevant. We might choose to apply a sort of modified Turing test to network systems, in order to determine whether their informational structures are established by the implementation of a genetic program, or whether they emerge spontaneously as a consequence of a self-organizational process. Although some tests might not be able to distinguish between programmed and program-free systems, one assumes that each type of system will generate a unique profile of behaviour and have its own strengths and weaknesses. These would only become apparent, however, when the systems were tested in the appropriate manner.

A further example of program-free self-organization can be found in the structure of proteins. In an earlier chapter we discussed the secondary, tertiary and quarternary structures of haemoglobin in some detail. When appreciating the remarkable complexity of haemoglobin's architecture, we should remember that the higher order components of protein structure self-organize as a consequence of their primary structure. It is almost as if we had disassembled St Paul's Cathedral, moved it to the Sahara Desert and then sat back and watched as it spontaneously rebuilt itself without the intervention of a single architect, builder, foreman or crane. Although some proteins require the assistance of chaperone proteins and foldase enzymes to help them fold into their higher-dimensional structures, the complete informational potential for the higher order structures of naturally occurring proteins is inherent within their primary structure. Primary structures thus function as highly compressed informational ambassadors for their folded derivatives, and although primary structure is monodimensional, it contains all the information necessary to generate a protein's complex array in three-dimensional space. A protein sequence thus constitutes simultaneously both the hardware and software of a self-assembling protein nano-machine.

The process by which linear amino acid sequences are transformed into higher order structures is doubtless lawful and governed by a set of complex rules. However, despite some highly successful attempts at structural prediction, the rules which govern the spontaneous assembly of higher order protein structures and enable the needle of a native fold to be found in the haystack of the *Space of All Possible Alternative Protein Folds* have yet to be elucidated. Nevertheless, the increase in order represented by the folded protein may be broadly be envisaged as offset both by the increase in the disorder of the solvent molecules which surround the protein and as a result of the fact that the folded protein has a lower energy than its unfolded counterpart. Thus despite the apparent focal increase in order represented by the realization of a folded structure, there is a net increase in the overall disorder of the system, which includes the solvent molecules surrounding the folded protein.

The structure of open, non-equilibrium systems is very different to the closed, near-equilibrium systems modelled using the linear mathematical equations of classical thermodynamics. Classical linear equations are thus unable to capture the complexity of open, far-from-equilibrium systems. Ilya Prigogine devised a new non-linear thermodynamics to describe non-equilibrium systems. As the ordered structure of such systems is derived from the flux of energy and matter through them, Prigogine called them 'dissipative structures'. The non-linear structure of dissipative systems makes it impossible to derive the properties of the whole system from any one of its components. This is because at the transition from equilibrium to non-equilibrium, long-range interactions creep into the system and undermine its previously linear behaviour. At this point the system begins to function as a whole, which is far more complex than the sum of its parts. These new behaviours emerge as a result of the system's interconnectivity.

The behaviour of closed, near-equilibrium systems is predictable and independent of the starting conditions. Thus irrespective of whether the bromine is initially positioned at the side or in the middle of the jar, the gas molecules will diffuse and attain a uniform distribution representing the state of maximal disorder and probability. History in these systems is marginalized and irrelevant. Imagine, for example, a metal ball which on being released from the edge of a funnel, may roll to the bottom along a number of different routes. The end result is the same regardless of the route by which it arrived there. The final low-energy resting state at the bottom of the funnel acts like a magnet, attracting the ball towards it in an inevitable manner. The state of maximal disorder observed in a jar of bromine will similarly be attained irrespective of the pathway taken by each molecule.

The situation in open, non-equilibrium systems is quite different. Instead of having just a single attracting point, the ball may instead be siphoned off into any number of different sub-funnels. Each of these contains its own magnet and associated region of influence, or 'basin of attraction'. If the route the ball takes happens to lead into the domain of influence of a particular magnet, the ball will be drawn into the sub-funnel controlled by that magnet. In this case history assumes great importance, as the particular pathway taken by the ball determines the sub-funnel that it is captured by, and so its final state. There is no inevitability in a non-equilibrium system and chance historical events have a significant impact on the system's future. This lack of inevitability has the consequence that the success of any non-linear mathematics of complexity in predicting specifics will always be undermined by our inability to define the initial conditions to a sufficient degree of precision. The sensitivity of such systems to small differences in the starting conditions, and the difficulties associated with determining these to an infinite degree of precision, means that although we may be able to model the coarse patterns of behaviour that emerge in complex systems, we are precluded, even in principle, from making more detailed predictions.

A simple and easily visualized example of a non-living dissipative system is the whirlpool that forms in a bathtub when the plug is removed. If the tap is left running, the familiar vortex structure of the whirlpool with its spirals and tapering funnel may be maintained indefinitely. The structure of the vortex is conserved even though the water molecules from which it is composed change from moment to moment. The structure of the vortex arises spontaneously and is not programmed by an internal information command centre. If all of the atoms in your body were labelled with a fluorescent pen, they would all, under the appropriate conditions fluoresce. If your fluorescence was monitored on a daily basis, a progressive decay in the strength of the signal would be observed as atoms present at the time of labelling were replaced by unlabelled atoms. Much like the whirlpool that forms in a bathtub, living organisms are dissipative, non-equilibrium systems whose structures are preserved across time as a result of the continuous dissipation of energy and matter.

The order of living things is not acquired for free but, as we have seen, at the expense of the increasing disorder of the environment. One has only to think of the disorder generated by a single lunchtime snack to be reassured that the second law of thermodynamics is in no way contravened. The fish on your plate had to be caught, which involved the destruction of the chemical order of the diesel in the trawler used to catch the fish, the destruction of the order of the food eaten by the fishermen to give them the energy they needed to catch the fish, the destruction of the order of the trees used to make the newspaper in which your fish was wrapped and of course the destruction of the information in the newsprint. It also involves destroying the order of the fish itself and possibly a transient destruction of the order of your personal life, if for example, your partner, would have preferred to dine in a fashionable Soho bistro!

If order can emerge spontaneously within self-organizing dissipative systems, might the information of life have originated within a self-organizing non-equilibrium system of some sort? In 1900 Henri Bénard observed the emergence of complex patterns in a thin layer of oil heated from below. This has come to stand as a general model of how patterns similar to those underlying the order of biological machines can arise spontaneously in the absence of codified digital instructions. The observation was studied more systematically by Lord Rayleigh in 1916. The patterns and pattern switchings in time and space which Bénard and Rayleigh observed were found to arise exclusively from the dynamics of the system and were not dependent on the implementation of an internal or external program. Their observations may be understood in the following way.

A thin layer of water placed on the surface of a large metal plate is uniform, symmetrical and featureless. If we were to miniaturize ourselves, every region of the fluid would appear identical and we would consequently have no way of perceiving the concept of space. This is because the fluid at this point is completely homogeneous and lacks gradients, as it is in thermodynamic equilibrium. All regions are at the same temperature and the water is motionless. The system is stable and if by chance it is perturbed, for example, by someone sticking their finger onto its surface, it readily returns to its former state once the stimulus is removed and any memory of the perturbation is erased. The stability of the equilibrium state means that all instants are indistinguishable from one another, and we are consequently unable to develop a concept of time. The system at this point is devoid of behaviour.

If the fluid is now gently heated from below, a heat gradient is established and the system is displaced from the stable state of equilibrium. The extent to which this occurs depends on the magnitude of the heat gradient. As long as it does not exceed a critical threshold value, heat will be transferred from the warmer lower regions to the colder upper regions by conduction. The heat gradient established across the fluid is linear and the system remains stable. As before, any memory of perturbation events is erased and the system continues to lack behaviour. However if the value of the heat gradient across the fluid exceeds a critical value, hexagonal honeycomb-like patterns appear at the surface of the fluid. These patterns, which are best observed by sprinkling aluminium dust onto the fluid's surface, are known as Rayleigh-Bénard cells. As the system is taken progressively further from equilibrium, several different types of pattern may emerge. These include: parallel 'rolls', triangles, spirals, 'targets' and 'patchwork quilts'.

Whereas before the heat energy was dissipated by a process of conduction, the honeycomb cells appear as a consequence of the emergence of a mechanism of heat dissipation which involves the bulk movement of fluid, known as convection. Below the threshold value, bulk movement of the fluid is prevented by its viscosity and the dissipative effects of conduction. However once the threshold is transgressed, these stabilizing effects are neutralized and the system becomes unstable. This results from the fact that the hot water nearest the heat source is less dense and thus lighter than the cooler water above it. There is consequently a tendency for the hot, lighter volumes of water at the bottom of the fluid to move to the surface and for the cooler, heavier volumes of water at the surface to move down to the bottom. If a small volume of water near the bottom happens to be displaced slightly upwards by a perturbation, it will find itself in a region of higher density and will experience an Archimedes force which amplifies the upward movement. If, on the other hand, water near the surface experiences a perturbation which pushes it down, it will find itself in a less dense region and the Archimedes force will amplify its descent. Below the critical threshold value, friction within the fluid dampens these effects and prevents the formation of the circulating currents which generate the honeycomb patterns. Hot water rises through the centre of each honeycomb cell and cooler fluid descends at its edges. The honeycomb structures increase the efficiency with which heat is transferred to the environment and help dissipate the heat gradient. Interestingly, if the honeycomb cells are examined closely, adjacent cells are found to have convection currents which turn in opposite directions. If we were to skip from cell to cell, we might encounter a pattern of cells which rotates in either a clockwise, anticlockwise, clockwise (CW-**AW**-CW) direction, or an anticlockwise, clockwise, anticlockwise (**AW**-CW-**AW**) direction. Either pattern may be found, but the system has to 'decide' which pattern will be realized. It is impossible, however, to predict which pattern will emerge on any given occasion. Furthermore if the system is taken above and below its threshold value on several occasions, it is likely to make different 'choices' at each independent transgression event.

The sudden appearance of Rayleigh-Bénard cells once the threshold has been transgressed shatters the featureless and timeless world that our miniaturized incarnations have become accustomed to. Whereas previously the fluid was homogeneous and symmetrical, it is now heterogeneous and has a broken symmetry. By skipping from one cell to another we could now acquire a notion of distance and space simply by counting the cells that we passed. Furthermore, when we are stationary, we could tell where we were by observing the direction of rotation in the cell we occupied. The astonishing feature of the Rayleigh-Bénard cell phenomenon is that, whereas below the threshold of instability each molecule of water operates as an independent, disconnected and randomly moving agent, above the threshold each molecule cooperates with huge numbers of other molecules to generate a complex, higher order pattern. Each cell arises as a consequence of the coordinated activity of around one hundred billion billion (10^{20}) water molecules. To revert to our robotic schoolboy analogy, a situation is achieved in which each robot explores only a tiny region of its huge *Scramble Net World* state space. The sudden change from simplicity to complexity is reminiscent of the type of phase transition which occurs when water turns into ice.

In addition to local cooperation, distant regions of the fluid become correlated and participate in the global long-range pattern of alternating cells. Thus as a consequence of an energy flux which displaces it from equilibrium, order can be generated spontaneously in a system previously devoid of information. The heat gradient is non-specific and does not contain a code or program that instructs the assembly of the patterns. The information of the patterns arises exclusively as the result of a self-organizational process that originates spontaneously in response to the heat gradient constraint. The honeycomb patterns thus provide a paradigm for the origin of order in the absence of a program, as well as a glimpse of the type of anatomy that may have underpinned the elementary forms of hypothetical geneless life. They also provide a suggestive and interesting example of how a rudimentary but nevertheless complex structure can be generated suddenly and discontinuously without the agency of a classic process of natural selection.

The information that arises within the Rayleigh-Bénard system represents a level of complexity well below that of even the simplest living thing. It is important though, because it illustrates how life-like features, normally associated with systems instructed by information encoded in genes, can emerge in a simple system without genes or a genome. The abstract structure of the Rayleigh-Bénard system constitutes a good model for all complex self-organizing systems. The difference between these types of simple self-organizing but non-living systems and the simplest types of living creatures may thus be one of amount, rather than of kind. It is possible to imagine a continuous scale of self-organized complexity encompassing a range of systems of varying complexity that include: oil drops, snowflakes, crystals, folded proteins, vortices and Rayleigh-Bénard cells at one end and complex multicellular creatures at the other.

There is no reason to believe that the patterns of living things are generated by different sorts of principles to those found in purely physical or physico-chemical systems. It is more likely that in living systems the complexity of inanimate non-equilibrium systems is augmented by the injection of codified digital information that enslaves the natural order emerging spontaneously within them. Eric Schneider and James Kay have suggested that genes act as 'informational databases of self-organizing strategies that work'. In acquiring the ability to codify information and to tune and tweak self-generated order, genes have, amongst other things, attained the capacity to reproduce self-assembling patterns across time. The Rayleigh-Bénard system provides an insight into how this might occur and indicates how genes might, in principle, force or channel the spontaneously generated information that arises within such systems, in particular directions.

We have seen that once the threshold is transgressed, the system is faced with two possibilities. It may generate either a CW-**AW**-CW pattern, or an **AW**-CW-**AW** pattern. Although the same system may select different patterns on different occasions, it is not possible to select both outcomes simultaneously. Both are equally probable. So what determines which pattern is selected? Once its threshold has been exceeded, the destabilized system is sensitive to even the slightest perturbations. The first perturbation that the system encounters as it crosses the threshold value determines whether the CW-**AW**-CW or the **AW**-CW-**AW** pattern is realized. Unlike the sub-threshold situation in which the system erases memories of chance perturbations, the supra-threshold system is imprinted by chance events and acquires a historical dimension. The subsequent evolution of the system is influenced by historically contingent events that cannot be erased. The evolution of any particular Rayleigh-Bénard system on any particular occasion thus reflects a complex interplay between chance and determinism.

Let us imagine that the alternative CW-**AW**-CW and **AW**-CW-**AW** patterns have different outcomes for the supra-threshold system. The CW-**AW**-CW pattern might be more aesthetically pleasing than the **AW**-CW-**AW** pattern, so that any system realizing the **AW**-CW-**AW** pattern is immediately discarded by a fastidious observer. In this case the two alternative patterns compete for survival in a process of aesthetic selection. Any factor which biases the selection process so that the probability of one outcome exceeds that of the other confers a selective advantage over a system which continues to make its decisions on a random basis. We may imagine then that one of the earliest functions of pre-genes, was to bias the evolution of complex systems that explore vast alternative *Solution Spaces* and which, in the absence of biasing factors, would do so in a purely random manner. Pre-genes may initially have functioned to minimize the impact of chance events on self-organizational processes by guiding them towards optimal, historically tried and tested solutions.

The critical thresholds of instability at which alternative solutions present themselves to non-equilibrium systems are known as bifurcation points. Each solution represents a route by which the system might stabilize itself. Under unbiased conditions each solution is equally probable and the outcome is determined by the fixation of chance perturbations. But one could imagine situations in which the outcome at a bifurcation point is biased, so that the system consistently flows along one trajectory, rather than along equally probable alternatives. By stabilizing one particular outcome, the system becomes a historical object. In the metal ball analogy discussed earlier, factors that bias the outcome at bifurcation points are the magnets which attract the ball down one particular sub-funnel as opposed to another. The breaking of symmetry at bifurcation points by chance perturbations thus appears to play an essential role in the generation of information within complex systems.

As the value of the constraint across a system is increased beyond the first bifurcation-generating instability, it passes through new sets of instabilities and their accompanying bifurcations. These are associated with increasingly complex patterns. Each system is defined by a hierarchy of potential instabilities. Successive tiers of possible solutions, however, only reveal themselves as each successive instability is transgressed. Each island of stability which a system might in principle be funnelled into is known as an attractor. The set of initial conditions that channel the system into a given attractor is known as its basin of attraction.

The rate at which a system bifurcates increases as it is displaced further and further from its first instability in the direction of non-equilibrium. The distance to each successive bifurcation point in fact halves progressively. If a system is driven too far from equilibrium, the repertoire of possible solutions that satisfy any given point of bifurcation becomes so large, that its behaviour appears chaotic. If, for example, the heat gradient across a Rayleigh-Bénard system is too severe, the orderly patterns disappear and the system becomes turbulent. This apparently chaotic behaviour reflects the fact that the system is no longer confined to selecting a single attractor solution, but is instead free to sample a vast repertoire of alternative attractors. It occurs as a result of the excessive amount of information generated when a system is displaced very far from equilibrium.

The types of program-free, instructionless systems able to generate information spontaneously are united by the fact that they are held in a state of non-equilibrium by an externally applied constraint which drives them to bifurcation-generating instabilities. The bifurcations within a system are satisfied by repertoires of different attractor solutions which are 'scanned' before a single solution is selected and fixed by a chance perturbation. The outcome at each bifurcation point might, in principle, be forced so as to influence the trajectory along which the system proceeds. I suggested earlier that pre-genes, and later proto-genes and genes may have functioned to generate a programmed, non-random pattern of selection at bifurcation points. Grégoire Nicolis and Ilya Prigogine have in fact drawn an analogy between the chance perturbations that influence the choices made at bifurcation points and the mutations introduced into genes. They have also suggested that the fixation of a chance perturbation by the selection of one attractor outcome at a bifurcation point rather than another, is analogous to the fixation of a mutation by natural selection. The random fixation of a chance event may thus, as a consequence of its influence on the dynamics of a system, be amplified and have profound consequences for its subsequent evolution.

The significance of the statement that genes are 'informational databases of self-organizing strategies that work' is now apparent. For although self-organizing systems that have been driven beyond their threshold of instability are deeply historical and carry the imprint of multiple random perturbations, the likelihood of the sequence of choices being repeated in another unbiased system is very small in any situation other than one where a very small number of attractor solutions are available at a limited number of bifurcation points. Thus although individual systems are historical, the collection of systems lacks history as each new system is a *tabula rasa* upon which new history is written. In the absence of a record that documents the choices made from the repertoire of possibilities at each bifurcation point, it is not possible for a system to reproduce its information by instructing another system and in this way to establish a historical lineage. Any particular historically imprinted system thus represents a hugely improbable state of affairs. The extent of the improbability of any particular structure to some extent reflects the number of independent bifurcation points it has transgressed and the size of the available repertoire of possible solutions that it encountered at each bifurcation point.

By encoding and thus dictating the broad pattern of choices that are made at critical bifurcation points, genes might in principle condense a complex system down into a discreet and sharply defined region of its potential *Bifurcation Solution Space*. Coded instructions may thus enslave the natural but unstructured order of complex self-organizing systems and guide its decision-making process along historically prescribed pathways that lead to tried and tested attractor solutions. Genes may be envisaged as generating the macro-complexity of biological systems by using coded historical knowledge to reduce the immense potential for micro-complexity. In the absence of coded instructions, the decision taken at each bifurcation point is random and non-deterministic. However, once enslaved by genes, such systems would behave more deterministically.

This model of genes as agents that modify the manner in which the mathematical patterns of complex self-organizing systems unfold in time and space appears generally reasonable. It assumes, however, that complex systems are not able to introduce bias into the decisions made at bifurcation points without the intervention of genes. Genes are envisaged as acting opportunistically on the unstructured mathematical edifices of otherwise free-running self-organizing systems. Although these systems are able to generate information, its quality is low as the information is structural rather than symbolic. It turns out, however, that this is not a reasonable assumption, as geneless systems are able to generate their own biases and to behave as if they were under the influence of genetically encoded instructions. Rudimentary informational continuity is, in principle, attainable without genes. Before we discuss the types of mechanisms that may have enabled geneless organisms to shape and perpetuate the spontaneous information they generated, it is worth discussing what we mean by the term information.

The word information is derived from what the ancient Greeks called *eidos*, or form, and is a measure of the amount of order, or structure in a system. Carl Friedrich von Weizsäcker has defined information as 'that which is understood'. This implies that the concept is inextricably bound up with the process of communication between a sender and recipient. Information may be considered as consisting of two principal components which we may call structural and instructional, or syntactic and semantic. The structural content describes the relationship between the symbols that encode the information. The quantity of structural information in the word '**man**' is, for example, identical to that of the alternative syntactic combinations '**mna**', '**nam**', '**nma**', '**amn**' and '**anm**', as each sequence contains the same number of characters, or bits of structural information. Only the word '**man**,' however encodes semantic information. If read by a person conversant with the English language, this string would be recognized as a hypersymbol representing the generic concept of a man. Whereas the symbol string '**man**' produces an image of a man in the mind of the observer, an alternative string '**amn**' does not. The identification of the word '**man**' as a hypersymbol with a particular meaning presupposes a tacit understanding or agreement based on shared knowledge between the sender and recipient. The nature of this agreement might change, and the nonsensical string '**anm**' could be assigned the meaning of the former word '**man**'. The word '**man**' could itself be reassigned to represent the concept of a 'tree', or remain unassigned and thus nonsensical. As long as this evolution of meaning occurs within the context of an agreement between the participants that utilize the communication channel, the nature of the syntactic unit chosen to represent any particular concept is semantically inconsequential.

Information theory as originally formulated by Shannon and Weaver in 1949 addresses the communication of structural information between a source and receiver, but does not encompass its semantic dimension. As far as information theory is concerned, the informational content of a genome is equivalent to that of a random sequence of an equivalent length. It does not account for the fact that some DNA sequences act as binding sites for regulatory proteins or, in the case of genes, actually encode abstract semantic information which might be translated into a protein sequence or functional RNA molecules, while random sequences lack a semantic dimension. The order that is spontaneously generated in some non-biological systems would not, at first glance, appear to fulfil the criterion of von Weizsäcker's definition, as it is apparently neither codified nor communicated. But if we think of the honeycomb cells as the sender of the message and the surrounding water molecules as the receivers, the structure of a honeycomb cell may be envisaged as informing the water molecules and communicating the information which determines their spatial location. Without this information, their positions would be random and the structure of the cells would not be dynamically perpetuated across time.

The geneless information that emerges spontaneously in Rayleigh-Bénard systems driven far from equilibrium is exclusively structural. At no point do we witness the birth of the type of abstract, codified, semantic or symbolic information found in genes. The information is instead stored within the structure of the system itself, in the analog manner reminiscent of gramophone records, and there is no formal distinction between the information and the machinery in which it is represented. Thus although the water molecule receivers are sensitive to the syntactic component of the system's information, there is no semantic dimension. The information generated within a Rayleigh-Bénard system nevertheless constitutes a potential source of semantic information, despite the fact that there is no semantically sensitive 'receiver' to decode the message. The situation is analogous to one in which the BBC World Service broadcasts to a non-English-speaking public that have no radio sets. Although the Rayleigh-Bénard system may in one sense be imagined as a generator of potential semantic information, there is no mechanism for communicating this and no receiver to discern grammar or meaning. In what sense then, is it meaningful to suggest that the information generated in such systems includes a potential semantic dimension at all?

If the system were repeatedly taken above and below its critical point of instability, the honeycomb patterns would appear and disappear at each transgression event. Each time the patterns were generated, either the clockwise-first or anticlockwise-first pattern of rotation would be realized. If the former is donated by 1 and the latter by 0, then over a period of time the system's behaviour could be represented by a string of 1s and 0s which might read: 1000101001000100101010000111111100100010101. A string of symbols of this sort constitutes an abstract record, or 'fingerprint', of the system's exploration of its potential state space over time. If this string were communicated to a receiver that was aware that the 1s and 0s represented the combination of patterns that had been realized within the system, then it could in principle be used to artificially reproduce the history of the system. The fact that there is no observer or mechanism for doing this, diminishes the status of this component of the information to potential semantic information. If, incidentally, we were to observe a non-random pattern such as 1111111111111111111111111111111111111111, 0000000000-00000000000000000000000000000 or 11111110000000011111 1110000000001111111, then we might suspect the presence of an underlying biasing factor.

The structural information content of genes is considerably less than that of the systems they encode. Thus rather than constituting an exact database of the system's structural information content, genes encode a far more economical representation of this information that incorporates only the essential aspects necessary for recreating the system. The great economy of the biological codification process may be illustrated with two examples.

In an earlier chapter we discussed how the higher dimensions of protein structure are able to self-assemble from a minimal specification at the level of their primary amino acid sequence. Although many enzymes are assembled from a single protein, others are more complex, and constructed from a number of different protein components, each of which is encoded by a separate gene. The enzyme

F_1-ATP synthetase, for example, which generates the ATP molecules that provide the energy which powers reactions in cells, is made up of a stalk and a head assembled from ten to twelve different proteins. These self-assemble to generate a biological motor of considerable architectural complexity. The relatively minimal structural information contained within the ten to twelve genes that encode the primary amino acid sequences of the sub-components of the F_1-ATP synthetase head is able to generate not only the structural information necessary to form the tertiary structures of each of the subunit components, but also that needed to assemble the subunit components into the molecular complex which forms the functional architecture of the machine. The relatively small structural information content of the genetic specification for the protein components that make up the synthetase thus stands as a highly compressed informational ambassador for the structural informational content of the fully assembled multi-component enzyme complex. Most of the information needed to specify the final architecture of the synthetase does not consequently need to be encoded. It is just 'out there'; a fundamental aspect of the physico-chemistry of proteins and their interactions in the context of a particular solvent and micro-environment. The compression of the structural information of the F_1-ATP synthetase machine is only possible because its genetic representation makes historical assumptions about the nature of proteins and the types of environments in which their higher order patterns of organization are likely to emerge. It is most unlikely, for example, that the F_1-ATP synthetase would self-assemble in the presence of a detergent, or inside a bottle of oil. There is thus no explicit genetic representation of the fully assembled F_1-ATP synthetase within the genome. The assembled structure of the functional multi-component enzyme is instead computed from the genetic instructions which specify the structure of each individual sub-component.

However a metabolism, which may, incidentally, be defined as a complex collection of proteins that cooperate to transmit signals or to function as structural or catalytic nano-machines and which form the essential functional circuitry of all known living things, constitutes an even higher order level of emergent, self-assembling structural complexity. The F_1-ATP synthetase is itself integrated into a metabolic network. Rather than existing at a poorly defined position in the cell, the enzyme is targeted to the cell membrane. The genes encoding the enzymes that form and regulate metabolic circuits thus also contain a highly compressed representation of the structural information associated with the spatial address of each protein within the cell and the set of chemical interactions they are likely to participate in. In addition to this, the control systems that regulate the concentration of each component of a metabolic pathway are not specified by genes but are instead computed by the pathways themselves.

The self-organizing control loops which regulate metabolic pathways constitute an even higher level of dynamic structural organization, which is not explicitly represented within the genome, but is instead inferred indirectly from it. The individual protein components which comprise a given metabolic pathway and the genes that encode them, stand as highly compressed informational ambassadors for a hierarchy of higher-dimensional self-organizing structures ranging from the secondary, tertiary and quarternary structure of individual proteins, to the modular assembly of protein complexes, the architecture of complex metabolic pathways and the regulatory mechanisms by which these pathways are controlled. All of these self-organizing higher-level phenomena rest, crucially, on the primary structure of the protein and therefore on the sequence of the gene and the DNA sequences that regulate its expression. We have once again demonstrated that genes are highly compressed informational ambassadors for a host of higher order self-organizing strategies which work. There is also a sense in which the protein components of a metabolism may be thought of as miniature analog computing devices, able to compute both their own higher-order structures and those of the metabolisms and control loops in which they participate. Biological structures are thus composed of multiple, hierarchically organized levels of self-organization that are ultimately rooted in or indirectly programmed by lower-level, genetically specified 'scaffolding' protein structures, which act as seeds for a host of higher-order levels of self-organizing structural organization.

Given the highly compressed nature of genetic information, it is possible that it will never be possible to fully compute the higher-level components of structural information from the sequences of genes and DNA regulatory elements alone. What is more, DNA is not the only database of compressed biological information. The potential structural information encoded within the set of genes and regulatory sequences generated as a result of the fusion of a sperm and an egg can only be made explicit in the presence of the mRNA sequence information in the cytoplasm of the egg and the correct pattern of induction signals delivered by the follicular cells that surround the fertilized egg. The egg and follicular induction signals are themselves a machine which supplies the information necessary to specify how the potential structural information stored within the new genome should be realized. It is for this reason, as Lewis Wolpert has argued, that the scenario of the movie *Jurassic Park* is implausible. Although we may eventually be able to recover or infer a dinosaur genome, it is unlikely that we will ever be able to recover or correctly infer the structural information contained both within the dinosaur eggs and in the follicular cells of pregnant females. Without this, we could not guarantee that the potential structural information stored within the dinosaur genes would be realized in a historically authentic manner. Having said this, we might, as a result of acquiring a detailed understanding of the mechanisms found in other types of closely related species, eventually be able to make educated guesses about the possible informational content of dinosaur eggs and follicular induction signals. This might then enable us to construct plausible dinosaur-like creatures, whose historical validity although not falsifiable, is nevertheless likely to be plausible.

The extent to which the copious structural information that defines the structure and dynamics of a metabolism can be compressed, illustrates how much free structural information can be generated spontaneously by self-organizing, geneless processes. Given a concentrated source of the primary protein sequences necessary for their construction and a specific non-equilibrium context, the informational structures of living things appear to self-assemble into coherent biological micro-machines and biochemical pathways. The encoded information content of individual genes is superimposed upon this non-encoded and instruction-free tapestry of emergent structural information. The economy of this strategy of information compression is remarkable. The type of structural information compression achieved by the abstract digital microcode of genes does not, however, occur in Rayleigh-Bénard systems. Although these systems are able to generate structural information, there is no economy in their representation. Each system stands as an exact analog representation of itself. All of the structural information is explicit and none is inferred, which has the consequence that the Rayleigh-Bénard system is amnesic and readily forgets its past. Thus, although the dynamic behaviour of the system might in principle be represented digitally, as the alternative solutions that present at each bifurcation point represent binary decisions, these choices are recorded in an exclusively analog manner and there is no codification of the historical pattern of decisions. There is, furthermore, no abstract reference text or master template with which the current state of the system's structural information can be compared. The analog nature of the information generated by Rayleigh-Bénard systems makes them subject to all the limitations of analog machines.

We have established two important points about the type of information that simple Rayleigh-Bénard-like systems can generate. First, it is structural not semantic, and second, it is analog not digital and has not been compressed. However, all contemporary creatures utilize codified structural information that is digital, incorporates a semantic dimension and is highly compressed into an abstract symbolic notation. So although we have observed the spontaneous emergence of structural information in the Rayleigh-Bénard system, it appears to lack the informational features that we associate with contemporary life. But these features of historically mature biological systems have taken billions of years to evolve and it is clearly unrealistic to expect the very first living creatures to have incorporated them in anything other than an extremely rudimentary manner.

It seems reasonable then, as was suggested earlier, to invoke a situation in which there is a continuum between what we might call pre-life, proto-life and life. Information-generating but nevertheless non-living complex systems, like that described by Rayleigh and Bénard, clearly lie at the pre-life end of the scale. They are non-equilibrium dissipative structures which spontaneously generate structural information, but which are unable to compress structural information and to generate abstract, codified representations. The first proto-organisms, on the other hand, may eventually have incorporated a very rudimentary capacity for codification and abstract representation. The semantic dimension in the information content of such proto-organisms would, however, have been a much attenuated version of that found in the creatures that eventually supplanted them and with which we are familiar.

We might now wonder whether it is reasonable to argue that digital codification and the capacity to represent semantic information are necessary features of life? If we were able to demonstrate the presence of intrinsic organizing principles within complex self-assembling systems that were able to introduce bias into the decisions made at bifurcation points but which were nevertheless non-digital and lacked a semantic component, might such systems in some sense be alive? Might the proto-life ancestors from which we have all descended have generated attractor states and bias within the dynamics of their self-assembling metabolisms without the agency of genes? The answer to each of these questions may well be yes and, as we will see, experiments using computer programs which simulate the abstracted properties of networks of highly interconnected and tightly coupled clusters of interacting non-equilibrium elements indicate that under certain conditions, attractors which introduce significant bias into the behaviour of a system can emerge spontaneously and effortlessly. The study of the generalized abstracted logic of living things using computers has come, in its broadest sense, to be known as artificial life, or 'A life'.

The central role of feedback processes in generating complexity in systems composed of multiple interacting components has been illustrated by Stuart Kauffman. He imagines 10,000 buttons scattered on the ground. Two of these are randomly picked up and connected by a thread. This connection allows the behaviour of each button in the pair to feed back to its partner. If the connected pair is thrown back on the floor and one button in the pair is picked up, its partner will be carried along passively as a passenger. The buttons are then thrown back onto the floor and two more buttons are picked randomly and connected. If this process of random picking and connecting is repeated many times, we eventually find that most of the buttons we select are already connected to another button. When an additional button is then connected to a pre-existing pair, a small cluster of three or more interconnected buttons is formed. If we continue to pick, connect and throw buttons back in this manner, the system eventually and quite suddenly reaches a critical point of interconnectedness. At this point if a single button is picked up, we find that all or a large number of the remaining 9,999 buttons are pulled along with it. The collection of buttons, which originally behaved as independent agents, is now a single interconnected cluster, and the behaviour of each button within the cluster feeds information back to all of the other buttons within the cluster. The slightest tug on any button will send out ripples of consequence that will be felt by all of the other buttons in the network. A highly interconnected network structure thus spontaneously and dramatically 'crystallizes' out and local behaviour is now able to exert action at a distance and to have profound global consequences.

This simple example illustrates some important aspects of how information can arise spontaneously in systems that have the potential to generate highly interconnected networks. The complexity of the system's behaviour is dependent on the fact that, once a critical level of interconnectedness has been attained, each component of the system is able to feed back information about its current state to all of the other components. The complexity of these interactions explains why non-linear equations are needed to model the behaviour of these systems. The behaviour of any button at any moment is computed as a function of what all the other 9,999 buttons are doing at that exact instant. The influence that each button exerts on all the others depends on several factors, including how close they are to one another, how many connections exist between them and the strength of the connections. The transition from a situation in which all of the buttons behave as autonomous agents, to one where they form a huge interconnected web of interactions, occurs suddenly as a 'phase transition'. This is reminiscent of the type of abrupt change seen, for example, when water turns into ice. Above a certain minimal level of interconnectivity, the network structure crystallizes out spontaneously. Computer simulations show that the greater the number of components in the system, the more sudden is the transition from simple to complex behaviour. Erdos and Renyi demonstrated in 1959 that the phase transition occurs when the ratio of connections to buttons reaches a critical threshold value of 0.5. As the number of threads begins to exceed the number of buttons, the network precipitates out suddenly and spontaneously.

This type of phase transition from simple to complex behaviour is not restricted to clusters of buttons, or indeed to Rayleigh-Bénard systems. Chemical systems are also able to form interconnected networks of metabolic reactions that feed back to one another and which, in so doing, are able to generate complex behaviour and structural information reminiscent of that seen in Rayleigh-Bénard systems. Whereas, in Rayleigh-Bénard cells, the structural basis of the feedback is the convection currents that circulate in either a clockwise or anticlockwise direction, the basis of the positive and negative feedback found in chemical reaction systems is autocatalysis and cross-catalysis. An autocatalytic chemical reaction is one in which the products of a reaction are able to increase the rate of their own production by forming a positive feedback loop. Let us imagine a simple chemical reaction in which the reactants A and B combine to form the products C and D. The reaction is autocatalytic if either the C or D products feeds back to increase the rate at which the A and B reactants are converted into C and D. The positive feedback loop so formed is known as an autocatalytic loop.

If the reactants A and B are mixed together in the presence of an appropriate environment and catalyst, they will interact to form the products C and D. But however long the reaction is left, some of the A and B reactants will always remain. This is because all chemical reactions are reversible. Thus whilst the A and B reactants are being converted into C and D, the C and D products are at the same time being converted back into A and B. The forward and backward reactions thus compete with one another. If left for a sufficient period of time in an appropriate chemical context, the system will eventually reach equilibrium. At equilibrium the ratio of products to reactants reaches a fixed value, known as the reaction's equilibrium constant. At this point the forward and backward reactions are exactly balanced. The value of the equilibrium constant varies from one to reaction to another. If high, the ratio of products to reactants is high and the majority of the reactants will be converted into products. If, however, it has a low value, the ratio of products to reactants will be low and the reaction will contain more reactants than products. Although a catalyst is able to increase the rate at which equilibrium is attained, the value of the equilibrium constant, and thus the final equilibrium concentration of reactants and products, remain unchanged. A chemical reaction that has attained the stable state of equilibrium is analogous to a Rayleigh-Bénard system which is below the critical threshold of instability and dissipates excess heat energy by conduction. The chemical system is symmetrical, featureless and devoid of behaviour. Any chance perturbation is rapidly obliterated and no memory of any perturbation events is retained.

In 1951 Boris Belousov demonstrated that if a simple autocatalytic chemical reaction of the type that we have just discussed is displaced from equilibrium, it will, much like a Rayleigh-Bénard system, suddenly and spontaneously undergo a phase transition and generate complex behaviour. This self-programmable property of autocatalytic chemical reactions can be systematically studied by artificially modifying the concentrations of reactants and products in a stirred flask and thus opening the system to the flow of matter. In doing this, the distance of the chemical reaction system from equilibrium can be tuned by adjusting the rates at which reactants and products are pumped in and out of the flask. The numerical distance of the artificially induced ratio of products to reactants from the value of the natural equilibrium constant is analogous to the magnitude of the heat gradient across the Rayleigh-Bénard system. Belousov and, later, Anatoly Zhabotinski observed that once a critical artificial ratio of products to reactants is reached, the system begins to exhibit complex behaviour. In the case of the particular chemicals used by Belousov and Zhabotinski, the behaviour manifests, amongst other things, as a regular periodic change in the colour of the reaction mixture from yellow (**Y**) to colourless (C). If observed over a fixed duration, the following pattern of colour changes might for example be observed in the system: **Y**, C, **Y**, C, **Y**, C, **Y**, C, **Y**, C, **Y**, C. These chemical oscillations are analogous to the convection cells of the Rayleigh-Bénard system, with the feedback of the convection currents being replaced by autocatalytic chemical feedback loops that cooperate and become correlated over large distances. Much like the convection currents of Rayleigh-Bénard systems, the autocatalytic chemical loops amplify any chance perturbations that occur within the system. While at equilibrium every instant was equivalent, once the system has transgressed the critical threshold of instability and is held in a state of non-equilibrium, the symmetry of time is broken and the system may behave very differently at different instants. The so-called Belousov-Zhabotinski (BZ) reaction illustrates how relatively simple chemical reaction systems can self-organize and program themselves under non-equilibrium conditions to generate structural information and complex behaviour.

Autocatalytic, chemical reaction loops appear to underpin the behavioural complexity of at least some chemical systems. Although the complexity of the BZ reaction results from the somewhat artificial situation in which the flux of reactants and products is modified using an external pump, complex non-equilibrium behaviour can be generated using more plausible mechanisms. One such method is to couple autocatalytic reactions which generate excess heat with ones that do not. In this way, the energy needed to drive reactions away from equilibrium can be internally generated within the metabolism itself. If the BZ reaction is not stirred, the concentration of reactants and products is not spatially uniform. Under these conditions, spiral or concentric waves are generated within the system. The focus from which these waves issue functions as a source that transmits a 'message' to the rest of the chemicals in the system, which act as 'receivers'. Oliver Steinbock has demonstrated that the chemical networks within a BZ reaction can be harnessed and made to function as a parallel processing chemical computer which can be used to solve mathematical problems. This suggests that a geneless, self-organizing chemical reaction system might, at least in principle, provide the rudiments of a computational device that could be used to help generate an elementary manifestation of life.

The metabolisms of modern organisms consist of huge numbers of interconnected autocatalytic loops. Although the individual components of these loops are encoded by genes, the functional anatomy of the autocatalytic circuits themselves is not represented genetically. Instead, the biochemical circuits which comprise the computational machinery of living things is left to self-assemble. Although the logic of genetic programming incorporates compressed cybernetic information that enables metabolic circuits to be coordinated and modified to meet environmental challenges, the essential pattern and nature of the connections between the components of the circuits is not genetically programmed. Is it possible, then, that the first living organisms were constructed from self-assembling biochemical networks which lacked a genetic specification? If so, then life may have begun without genes and the first living things may have been autocatalytic geneless metabolisms that did not incorporate the principle of template-based information storage and replication, but derived their order from non-linear chemical feedback processes. In such systems, information is not stored digitally in a centralized information database, but is represented in an analog manner in the network of chemical interactions itself; there is no distinction between the informational database and the thing itself. Hypothetical geneless systems of this sort would not have any type of program or codification whatsoever. If this were the case and the first living things were self-assembling biochemical networks, then life began complex, not simple, and free-living self-replicating digital molecules may only have been incorporated into the structure of living things at a later point. The order of life may thus have originated not from pre-existing order, but from program-free noise. Although there is no direct experimental evidence to support this somewhat heretical geneless-life hypothesis, there are compelling theoretical and suggestive but nevertheless indirect, experimental reasons for taking it seriously.

But how might such biochemical networks have self-assembled to generate a set of chemical reactions organized into a coherent interconnected metabolism? One possibility is that, instead of being constructed from autocatalytic reactions where the C and D products of a reaction (in which the reactants A and B are converted into the products C and D) feed back to increase the rate at which the reactants A and B are converted into C and D to form an autocatalytic loop *within* the same chemical reaction, the catalytic feedback loops are formed *between* different chemical reactions in a process known as cross-catalysis. In this situation, although no individual reaction is autocatalytic, the set of interconnected reactions that feed back and cross-catalyze one another is *collectively* autocatalytic. This results from the fact that every reaction within the cross-catalytic network is indirectly connected to every other reaction within the network, including itself. Every member of the reaction set thus contributes to the maintenance and perpetuation of the whole set. So although the products C and D do not catalyze their own formation, they may catalyze the conversion of the reactants E and F into the products G and H. Similarly although G and H may not catalyze their own formation from the reactants E and F, they may be able to catalyze the formation of the products C and D from the reactants A and B. The two reactions thus have the potential to become locked into a collectively autocatalytic reaction set. Such sets of collectively autocatalytic reactions may incorporate large numbers of different chemical reactions. The only criterion for becoming a member of the set is that each product has its formation catalyzed by at least one other member of the set. This makes each member critically dependent on the existence of the whole network of reactions which, by providing the appropriate catalytic activities, are necessary for its synthesis from a primary set of 'food' molecules.

Although in Kauffman's model none of the components of the collectively autocatalytic reaction sets can self-replicate, and rely instead on the organization of the rest of the network for their replication, some of the components might participate simultaneously in both cross-catalytic and autocatalytic self-replication. A product C or D might thus *autocatalytically* increase the rate at which it is produced from the reactants A and B (and thus self-replicate) whilst simultaneously *cross-catalyzing* the conversion of the reactants E and F of a separate chemical reaction, into the products G and H. It is participation in cross-catalysis, however, that gives this *individually* autocatalytic or self-replicating reaction the potential to become a member of a *collectively* autocatalytic reaction set. The incorporation of autocatalytically self-replicating components into the collectively autocatalytic set would be expected to occur only very infrequently, as if any component received both autocatalytic and cross-catalytic inputs, it might take over the network and destroy its structure. Although this tendency to outgrow the rest of the components would initially select against collectively autocatalytic networks that had incorporated autocatalytic self-replicating components, it is nevertheless possible to imagine situations in which such elements might be held in check by inhibitory influences originating from within the collectively autocatalytic set itself.

The notion that the very first living things may have been constructed from collectively autocatalytic, chemical networks was advanced independently by Rossler, Eigen and Kauffman in 1971. One of the most convincing arguments in support of this hypothesis, however, has come from Stuart Kauffman, who has used computer modelling to demonstrate how complex, collectively autocatalytic networks might self-assemble in the absence of an informational precursor. Richard Bagley, Doyne Farmer and Walter Fontana have also used computer simulations of artificial, collectively autocatalytic networks to show that these types of gene-free metabolisms are dynamically stable and able to store and mutate their information, to reproduce and repair themselves, and to evolve. Computer simulations of randomly constructed, collectively autocatalytic metabolisms have, furthermore, indicated that these types of networks have powerful self-organizing properties that enable them to generate immense order and homeostatic robustness without the need for any internal or external pre-programmed genetic instructions. This type of homeostatic resistance to the potentially chaotic effects of chance perturbations is an essential property of all living things. Chris Langton, Stuart Kauffman and others have indeed demonstrated that randomly constructed networks of collectively autocatalytic chemical reactions may be tuned by altering fundamental parameters such as the degree of connectivity between the elements from which the chemical circuits of the network are constructed. An appropriately tuned network possesses great order and stability, whilst at the same time retaining the plasticity necessary for evolutionary processes. A poorly tuned network, on the other hand, may be hopelessly chaotic and devoid of order and robustness.

Freeman Dyson and Stuart Kauffman have independently argued that the first metabolisms crystallized out from a complex sea of unconnected or partially connected molecules to form an interconnected set of collectively autocatalytic chemical reactions. This conversion from a disconnected to an interconnected state is envisaged as occurring suddenly as a 'phase transition', in a manner analogous to that seen in the button and thread system discussed earlier. As a critical level of connectivity and chemical diversity is transgressed, the collectively autocatalytic metabolism spontaneously self-assembles. Although containment of the reaction mixture within a spatial compartment to generate a 'proto-cell' would provide reactants at a concentration sufficient for this to happen, other methods of concentration that do not involve compartmentalization are also plausible. The very first living creatures may thus have been acellular 'liquid' organisms; formless, spontaneously generated, digitally infected, collectively autocatalytic metabolisms that were confined to a well-defined region of space such as the two-dimensional surface of a rock or a concentrated volume of free solution, but which were not in the very first instance, compartmentalized.

12

A Journey Through the Geneless Zoo

And now I am falling again. But this time more like a feather than a stone. Or maybe a bit of both. As if I myself were now the unwilling participant in someone else's strange and apparently unnatural experiment. At first I see vistas of Italy. Pisa with its crooked tower and Tuscan hills with their ochre colours and walled mediaeval towns. We pick grapes in the sunshine and there is laughter by the swimming pool. Then there is Torre Guinigi with its awkward leafy tree, the open air cinema and the restaurant by the railway track. I carried you on my shoulders you know, all the way to the top! And you told me that my painting was 'pretty'. For such things we circumnavigate the globe. However these images soon fade and I am left to anticipate the next moment. For a while this does not appear to arrive. And the funnel leads me through *Gene Space* with all of its strange numbers and combinations. This will surely be a long journey. Down and down, deeper and deeper. And I am reminded once again of previous events, the endless rows of faceless boxes and of the herb garden and vegetable patch. We should not be fearful. For we have been here before. We have *all* been here before. Down to the roots. To the beginning. But choose the right path. Please ensure that you at the very least do that. Select the right lineages, whatever or whomever they may be. The ones of course that will lead us home.

The screams from *RNA Space* are muffled by the speed of our descent and as we are flung still deeper into the void, the funnel progressively narrows. Still more spaces are sliced one by one like the pieces of a birthday cake. And each cross section appears more curious than the next. Natural. Artificial. Beauty. Beast. Tinker, tailor, soldier, spy. We cut another slice of foreign space and shudder with the impact. 'Deeper, still deeper!' I hear you cry. And so we do. Indeed we will continue this descent despite our better judgment. And soon we will arrive at our strange destination. The landscape with the ultimate power to constrain. The seeds of untold and perhaps limitless complexity of shape, colour and form. And, notwithstanding, the capacity to transform.

I hear the voices in the playground. The folded bud, the leaking tap, the kitchen sink, the city sleaze. For from this place all things sprang. And then the unwanted, but apparently necessary and therefore inevitable dialogue.

'You are too small, I cannot see you'.

'Then we shall make you smaller'.

'But I do not necessarily wish to be smaller'.

'But you will be smaller nevertheless.' For in the *Sea of Funnels* you have little choice and you will be unable to change your direction. 'We will transform you as we require. We will invent and then reinvent your shape. You are powerless. We will redefine your boundaries.' And you were wrong to think that you had any choice. That was the one thing that you *never* had.

We are through the gates. Safely home. Well almost, or so we think. And then the voices. 'We will allow you to see the way things might have been.' And I am able, momentarily, to crawl from one funnel into another. The sound of a horn. And in a painful swirl which enables me to take in both the past, present and the future in a single turn, I search relentlessly and pointlessly for the ziggurats. There are no walled Tuscan towns and I am unable to carry you up the tower. There are no grapes or swimming pools and the dusky olive groves disappear one by one, like defunct pieces on a chess board.

Turquoise blue once again and now the sunlight makes flickering patterns on the smooth concrete surface beneath the water. For a moment I am part of that pattern, organically tracing its sinuous route across this globular and sometimes impossibly granular terrain. The mallard struts confidently across the grass and we laugh as the Canada geese splash down on the water. And so into the *Geneless Zoo*. For perhaps this is the way in. Unable to appreciate that this *is* life, although in a guise or incarnation with which we are unable to relate.

And you lead me past the cages, expectant of spots and stripes and teeth and fur, only to be disappointed by the absence of even a silent roar. The colours of this rainbow are invisible. There is no pot of gold. But there is indeed a pot. Clear, colourless and yet not entirely empty. No tusks, no sliding slimy sideways stagger, no pink forked tongues, or orange beaks. But a zoo nevertheless.

As we open the lid of the *Geneless Zoo* and turn it upside down, the creatures that fall out are quite unfamiliar. For they lack genes. Indeed some of them lack even structure. Shapeless and opaque, they are as fragile as a piece of tissue paper. Some of them have no precursors. They inhabit the wilderness at the borders of life-lines and before them there was almost certainly nothing. There can indeed have been nothing, as they exist at the edge of a mathematical precipice.

We have suggested that the hypothetical creatures which inhabited the *Geneless Zoo* did not rely on digitally encoded instructions and that in some cases they had no need for history as they might, in their very simplest incarnations, have been created spontaneously; the parentless children of a self-organizing alliance between physics and chemistry. These analog, self-programming biomachines which lacked genes, symbolism and indeed any type of encoded instructions, were clearly not as sophisticated as modern creatures. Their capacity for heredity would have been severely constrained and each of their multiple lineages are likely to have been extinguished and liquidated at the whim of chance events. Their watering holes were the darkest shadows of extinction. But living creatures nevertheless they were; limited heredity and limited lineages, but heredity and lineages none the less.

If the very first living creatures did lack genes and if these hypothetical geneless creatures were able to spontaneously self-assemble in the absence of parent precursor structures, then we have addressed the paradox which overcasts any attempts to explain the origin of life and may offer a reasonable retort to the question 'what programmed the first program?' The first program was not programmed by a pre-existing program. Some classes of chemical systems may be able to program themselves without the need for either internal or externally applied instructions.

We have already discussed the implications of using analog instead of digital machinery for information representation and perpetuation. We should consequently be aware of the very severe constraints that analog information-storage technology would have imposed upon the first living creatures. The nature and extent of their existence would have been rudimentary compared even with the very simplest forms of living things with which we are familiar. It was not until the precursors of genes were cooked and baked within one of the geneless lineages that life discovered the digital secret of unlimited heredity and the mathematical landscapes associated with the complexity we witness today became visible and accessible for the first time.

The origin of rudimentary existence in such hypothetical geneless systems constitutes a paradigm for the creation of analog life in the absence of pre-existing, historically acquired, digitally encoded information. These hypothetical proto-creatures represent the minimal level of complexity beneath which existence is not possible. But how might such creatures and the metabolic machinery from which they were constructed have spontaneously self-assembled in the absence of a *bauplan* or building instructions? How, indeed, might a complex collectively autocatalytic self-replicating network structure have been generated from a host of disjointed components, each of which were themselves incapable of template-based or indeed any other mode of self-replication?

The case for the spontaneous self-assembly of complex, collectively autocatalytic metabolisms rests upon the notion that, within a random library of chemicals of any sort, a fixed proportion of the components will, by chance, be able to catalyze chemical reactions which might occur between different members of the library. Let us imagine that the buttons of the previous chapter represent individual chemical species, and the threads between them the set of catalyzed chemical reactions in which they might participate. Computer simulations indicate that if the number of catalyzed reactions is roughly equivalent to or greater than the number of different types of chemicals within the system, a collectively autocatalytic network may spontaneously emerge from a previously informationally sterile chemical broth. This result rests on the assumption that there is a finite probability of finding a catalyst for a particular chemical reaction within a random polymer library of any sort.

Critics such as Leslie Orgel have argued that the capacity of random chemical libraries to furnish a supply of mutually compatible catalysts sufficient for the spontaneous self-assembly of a collectively autocatalytic set is much over-estimated, but indirect experimental evidence from combinatorial chemistry suggests that the hypothesis may not be unreasonable. However, the fact that it is still unclear whether the spontaneous emergence of collectively autocatalytic sets is chemically plausible, remains a significant criticism of the hypothesis. It is, nevertheless, worth observing that even if it were possible to demonstrate that no plausible natural chemistry could generate such a system, we need not rule out the possibility that analog, geneless life could be coaxed into existence by some ahistorical process, articulated within the confines and context of some as yet undefined artificial chemical system.

Stuart Kauffman has made the point that, as the length of a chemical polymer such as a peptide, single-stranded DNA or RNA sequence increases, the number of different types of molecules that can potentially be synthesized increases more slowly than the number of chemical reactions in which each of the molecules might participate. This purely theoretical consideration suggests that the crystallization of a collectively autocatalytic network becomes almost inevitable, if the probability of finding compatible cross-catalytic activities within the polymer system in question is sufficient for this to occur.

Our discussion of random RNA sequence libraries provided a concrete example of a combinatorial library which is replete with potential catalytic activities. Surveys of other types of randomly constructed chemical libraries such as those composed from peptides, single-stranded DNA or antibody molecules, yield similar results. One would therefore not be surprised if these properties were found to generalize to other types of chemical libraries. Furthermore, the frequency with which catalytic activities may be found within libraries appears compatible with the spontaneous emergence of collectively autocatalytic sets. There has not, as yet, however, been a systematic study of the frequency, distribution, efficacy, stability, kinetics and mutual compatibility of the repertoire of potential catalytic activities which exist within different types of combinatorial libraries. Such studies are critically important if the plausibility of the hypothesis is to be tested.

The probability of finding appropriate catalytic activities within any particular library of chemical polymers is related, amongst other things, to the ability of the polymers within the library to map onto the abstract shape space which we have previously referred to as *Catalytic Shape Space*. This abstract space takes into account both the shape of the enzyme's binding site, which determines its specificity, and its active site, which defines its catalytic efficacy. A random library of double-stranded DNA molecules would, for example, contain few if any catalytic activities, as double-stranded DNA molecules have a very limited capacity to form different three-dimensional molecular shapes and to bind and configure the metal ions that form the essential components of many active sites. RNA, on the other hand, for reasons that we have already discussed, is able to mimic many of the binding sites of proteins and to organize metal ions into arrays which are compatible with the generation of active sites. The same appears, to a greater or lesser extent, to be true of peptides, antibody binding sites and single-stranded DNA molecule 'aptamers'. There are presumably many other types of chemical polymers which are able to map onto *Catalytic Shape Space* and which might consequently furnish hardware components for the construction of a collectively autocatalytic set.

Although any complex collection of chemicals could provide the components necessary for the spontaneous self-assembly of a collectively autocatalytic set, collections of polymers such as RNA sequences, single-stranded DNA sequences or peptides are statistically more likely to be replete with catalytic activities. This arises as a consequence of the intrinsic power of combinatorial modes of *Catalytic Shape Space* exploration. It seems reasonable, however, to conjecture that some types of chemical libraries are more likely to contain mutually compatible sets of cross-catalytic activities than others.

The size of the library of chemicals required for the spontaneous self-assembly of a collectively autocatalytic set is proportional both to the probability that the library will harbour catalytic activities of appropriate specificities, competence and mutual compatibility and to the ratio of the number of different chemical entities within the library to the number of possible reactions that might in principle occur between them. The greater the number of possible reactions between the components of the network, the higher the probability that one of them will catalyze a reaction within the network. The more structurally complex the chemical entities from which the potential catalytic repertoire is constructed, the greater the number of chemical reactions in which they are likely to be able to participate. Thus, if one wanted to search for candidate chemistries which might be used to create artificial collectively autocatalytic geneless metabolisms or indeed for historically plausible candidates, we would need to identify a set of molecules that offered both a high frequency of potential catalytic activities and a high ratio of potential chemical reactions to molecular components. Such compact libraries would offer the greatest potential for generating a collectively autocatalytic set, whilst at the same time incorporating the smallest diversity of components. These types of chemical libraries represent candidate self-programmable technologies.

If life did begin in this way, then the basic complexity of the first living, cross-catalytic network systems arose suddenly and spontaneously as a consequence of a process of self-organization and not through a sequential and incremental process of classic Darwinian evolution by natural selection. Although such systems, once formed, would be substrates for natural selection in the usual way, the genesis of the founding structure would not necessarily involve the historical processing routinely associated with systems constructed by Darwinian processes. Because such structures represent the minimal complexity compatible with life, they could not, by definition, have been the products of a process of heredity. Even if the network itself was assembled sequentially and incrementally by the modular addition of discreet layers of chemical circuitry and was preceded by a sustained period of chemical evolution, this could not be regarded as a classic process of heredity. It would not consequently be reasonable to say that the newly created proto-organism had at the very moment of its inception evolved from a precursor informational structure.

Life may thus be both a natural and expected consequence of the behaviour which emerges spontaneously within certain types of complex 'self-programmable' chemical reaction systems. And although once assembled, a cross-catalytic metabolism would be historically processed by classic Darwinian mechanisms, its initial order does not appear to originate as a consequence of a logically interconnected series of selection events on a host of simple components, each of which can be successively more 'programmed'.

Assuming that such collectively autocatalytic chemical reaction networks are indeed chemically plausible, in what sense might they be considered living? There is no generally accepted definition of a living thing. However, if attempting to compile a list of requirements, one might amongst other things assert that living things are: (1) discreet, well-defined, compartmentalized entities that have a clearly defined boundary and which are (2) composed of interlocking networks or metabolisms of cooperating and mutually compatible chemical reactions. They should furthermore be able to (3) store and retrieve information, to (4) self-replicate or reproduce their informational structures and thus possess at the very least a rudimentary or limited capacity for heredity. They should also be able to (5) mutate their core informational structures so as to produce a population of variants which are (6) amenable to positive and negative selection by classic Darwinian processes and which are consequently able to update and optimize their representations of the outside world and thus to evolve. They may consequently produce informational lineages which incorporate structural changes that result from chance mutations and which have been selected or fixed as a result of contingent historical events or structural constraints, and which could be traced back to an ancestral founder structure.

These properties should originate from within the system itself so that, in the presence of an adequate supply of food and energy, the system is able to maintain its informational structure across time. Living systems should thus (7) incorporate mechanisms for furnishing and delivering the matter and energy necessary to realize the chemical transformations embodied within their metabolic circuits and to ensure that the system remains displaced from equilibrium. The system should also be able to (8) repair its informational structures if they are damaged or misreplicated and display a degree of (9) homeostasis and dynamic stability. This should make it relatively immune to the effects of chance perturbations and confer it with the property of robustness. The system should not, however, be too stable, as this would preclude the possibility of evolutionary change. It should thus be (10) optimally poised to be stable and robust whilst retaining a potential for informational flexibility or evolvability. That is to say the ability to update its representations of the causal texture of the outside world at discreet intervals. This enables living things to interact dynamically and interactively with the environment in which they find themselves, to generate appropriate responses to specific contingencies that they encounter, and to process environmental information so as to provide optimal behavioural responses to enable their continued survival. This implies that one may (11) attribute at least some degree of computational competence to living things. Although their computational style is clearly different to that of artificial computing devices, the notion that living things perform computations which enable them to dynamically respond to events within their environment follows as a logical consequence of their ability to process information. Indeed the accuracy and relevance of their representations, and consequently their ability for continued survival, are critically dependent on their capacity to dynamically reconcile their internal state with that of their environment.

As a living system can broadly be defined by its overall organization as opposed to its specific material structure, it should not be necessary for our definition to include an account of the physical and chemical nature of the system's components. The material and formal aspects of living systems may thus be considered independently. Given their important informational and effecter functions in contemporary living things, the lack of other clear-cut candidates and their apparent availability in the prebiotic environment, it seems reasonable to suggest that the hypothetical collectively autocatalytic metabolisms that may have been the precursors of modern creatures were constructed from single-stranded DNA, RNA or peptide molecules. In principle, any of these technologies might have sufficed. The hypothetical geneless metabolisms may even have self-assembled from a heterogeneous collection of these components. The criterion for their inclusion into the collectively autocatalytic set being principally their availability and their binding site and active site compatibilities, which would enable them both to cross-catalyze other members of the set and to form part of a reaction that could itself be catalyzed by another member of the set. However one or several technological 'takeover' processes may have occurred in the history of proto-life. Indeed a founding set of hypothetical, 'low-tech', non-encoded catalytic hardwares may at some point have been supplanted by other types of encoded, 'high-tech', chemical technologies. The chemical components of the very first geneless metabolisms may therefore no longer be represented in the molecular repertoires of modern digitalized creatures.

One significant omission from this list of essential requirements is any exact requirement for the manner or style in which the putative living system should store and retrieve its information. There is also no mention of the mechanism by which the system becomes informed in the first place. If one were exclusively concerned with modern life, one would almost certainly insist that the information should be encoded using a symbolic digital microcode written in the language of DNA. One of the most compelling characteristics of the metabolic machinery of contemporary living things is that chemistry is transcended; the bases of DNA molecules are not selected on the basis of their physical and chemical properties, but rather on their symbolic or informational properties. Furthermore when information is retrieved from these genetic micro-databases, it is obligatory that the symbolically encoded information is decoded and translated into a protein effecter language. This implies the existence of a grammar that specifies the rules which define the translation process. The codified information embodied within the digital DNA sequences of genes thus contains highly compressed instructions for an organism's construction and operation.

However symbolism and its associated codes and translation machinery are unlikely to have been an option for the very first living things, so it seems reasonable to dispense with this requirement. One might nevertheless argue that any candidate living thing that does not embody some type of rudimentary codification process or program, and is not itself generated by the systematic execution of a set of instructions issuing from a central command centre, must be confined to the category of 'quasi-life'. We should not forget, however, that a significant number of DNA sequences which are critically important in regulating the execution of genetically encoded programs and that have a clear computational interpretation, are never translated into protein sequences and function exclusively as analog logic devices. These provide an example of semantically meaningful biological information which is not decoded in the usual manner, but nevertheless subserves a critically important biological function.

If one unconditionally accepts the list of minimal requirements outlined above, and if a putative self-assembling metabolism was found to satisfy them, then one could, quite legitimately, at least with reference to this particular definition, refer to it as a living creature. If it turned out though that one or more of the requirements were satisfied on only a rudimentary or conditional basis, we might decide to forgo the term 'living'. However in recognizing that such systems may possess many of the properties normally associated with living things, we might decide that it was inappropriate to categorically rule out the possibility that at least a small subset of the complete set of all possible such systems might in some sense be living. We may prefer to use the term 'quasi-living' when referring to these creatures, in order to indicate that they inhabit a hypothetical grey zone or transitional region between the somewhat arbitrarily defined states of life and non-life. This does not seem intrinsically unreasonable, as there does not appear to be any reason why the states of living and non-living should be sharply delineated. Indeed the precursors of living things may have existed in a state somewhat intermediate between that of non-living and living and which falls outside any narrowly circumscribed definition of life formulated within the framework of contemporary living things. The first quasi-living creatures might also have existed only transiently, perhaps 'quasi-living' for only a few seconds, minutes or hours before being destroyed.

We will now look at each of the requirements for life more closely, in order to establish the extent to which they might be satisfied by the hypothetical collectively autocatalytic geneless organisms that may have inhabited the geneless plains of the *Geneless World* and provided the informational scaffolding for the complex forms of contemporary life with which we are familiar. It should then be possible for us to decide whether they might reasonably be classified as either quasi-living or living things.

Let us start by examining the way in which exclusively analog metabolisms are able to store information without a genome or genes. This is clearly an important issue as, unlike the metabolisms of modern creatures whose protein components are precisely *encoded* by internal genetic specifications, the components from which the metabolisms of geneless creatures self-assembled are imagined as being *non-encoded*. Such analog, non-encoded networks contain no internal description of themselves that can be separated from the metabolism itself. There is no discreet, centralized, digital database, no microcode and no symbolic representation of the structure of each component as is the case with historically processed metabolisms informed by genes. The information of the network is instead held in an entirely analog fashion and there is no master template or reference point with which its current state can be compared. It is distributed across and contained within the precise organization and concentration profile of the chemical species from which it is composed and in the complex network of chemical transformations that each component of the network participates in. The specification of such putative geneless organisms is thus represented by the concentration landscape that its components generate at any particular moment.

The network itself thus stands as a continuous chemical analog record of its own structure, dynamics and history, and there is no distinction between the machine and its specification. This is in stark contrast to genetic systems which do not store a 'picture' of their core structural information, but rather a central set of minimal instructions for *recreating* it. There is also no information compression and thus none of the economy that characterizes genetic representations. Whereas genetic systems are able to capture and summarise the essential details of complex analog states of affairs and to subsequently use their compact, abstract, digital representations to recreate the analog machinery they encode, analog metabolisms lack mechanisms for distilling their information down into its essential modular components. If an encoded metabolism is destroyed, it may be recreated from its genetic description by the implementation of a process of 'instructed synthesis'. However if a non-encoded metabolism is destroyed, the analog means for recreating it are also destroyed and the information of the structure is irretrievably lost.

Much in the same way that the programmed elements of encoded metabolisms are able to self-assemble in the absence of an explicit representation of how this should occur, the non-programmed components of non-encoded metabolisms may also have self-assembled in the absence of a program. The information in such analog networks is not, however, stored and retrieved in the usual manner. Unlike genetically encoded information, which can be silenced and selectively accessed both at different times and in different spatial locations, in non-encoded metabolisms all of the information is operational all of the time. Furthermore, while genes in the words of Sydney Brenner 'do not contain the means, but rather a description of the means' for the execution of the genetic program that directs the construction and operation of the organism, and consequently require the machinery of the egg to decode this information and to install the means to do this into the daughter machine, geneless metabolisms contain both their own special type of non-digital program and the means for its execution.

Jeffrey Wicken has stated that 'information required for specification has little to do with information content'. From this perspective, in order to attribute an information content to a system one has to have a notion that the system is actually conveying something. This implies that there is a receiver to decode the information using specific rules of interpretation which define the structure of an appropriate grammar. It is true that for the core aspects of an informational structure to be faithfully perpetuated across time, an accurate communication device is needed. And there is no question that the symbolic codification strategy used by all known living things constitutes the most powerful mechanism for communicating information. But it seems unreasonable to insist that living things should have had a capacity for unlimited heredity right from the very beginning, and more reasonable to suggest that the very first living things had only a rudimentary capacity for transmitting their essential information across time.

We should not, however, vilify them for this reason, although we may, as has been suggested, prefer to acknowledge this shortcoming by using the term 'quasi-living' instead of 'living'. For given a sufficiently large time scale, one might expect the sequences of genes to diverge so far from their parent structures that one might well question the extent to which the information communicated by the founder sequence was in any meaningful sense represented in the structure of the final genetic 'receiver'. It is unclear, then, whether such distinctions can reasonably be made on formal as opposed to essentially arbitrary considerations.

We have discussed the static structural aspects of the information contained within collectively autocatalytic networks. But might there also be a dynamic component to the information they generate? That is to say, given that these self-assembling networks lack genes and digital genetic programs, might they nevertheless be able to program themselves and to spontaneously generate the type of information to which Wicken was referring? Is there anything to suggest that geneless metabolisms are capable of generating non-random dynamics? And if this is the case, would such information exhibit the type of homeostatic stability and robustness that characterizes the informational structures generated by the symbolically encoded instructions of genes? Furthermore, given that modern genes use their complementary strand as a reference master copy with which to compare and repair sequences that have been damaged or misreplicated and that non-template-based metabolisms lack such mechanisms, would they be able to repair themselves, or to conserve their information after a cycle of low-fidelity replication?

If analog networks are characterized by self-programmable dynamics, then we may subdivide the natural order that arises within them into two related but distinct components. The first of these concerns the self-assembly of the network elements, whereas the second pertains to the self-organizing dynamics of the assembled structure. The second component is not necessarily a consequence of the first. Any analog mutations introduced into the network may thus affect either its structure, dynamics, or both. But how might this type of dynamical information emerge spontaneously in the absence of the logic of historically programmed genetic instructions?

We have already discussed the way in which the relatively simple B-Z system is able to spontaneously generate complex dynamical behaviour as a consequence of autocatalytic chemical feedback loops. So we should not be surprised if more complex chemical feedback systems which may in some cases incorporate both autocatalytic and cross-catalytic components connected together into a collectively autocatalytic network, also exhibited similar or more complex types of self-organizational behaviour. However, given the immense structural complexity of these hypothetical geneless chemical networks and the ahistorical and relatively *ad hoc* manner in which they are likely to have self-assembled, we should equally well not be surprised if they were chaotic and devoid of dynamic order.

As there are no known natural or artificially constructed examples of collectively autocatalytic chemical reaction networks, it is not possible to study their dynamics directly. This may not always be the case, unless, as has already mentioned, these non-encoded analog metabolisms do not make chemical sense and are thus precluded from existing, even in principle. But what we can do is model the generic behaviour of artificial chemical networks using powerful computers. This provides us with a unique opportunity to take a glimpse at the possible anatomies and behaviours of the types of geneless creatures that may have inhabited the analog, instruction-free plains of the *Geneless World* and which may eventually have established the digital genetic lineages from which we have all descended.

The ability of computer simulations to model and mimic the dynamics of artificial metabolisms rests upon the fact that networks of highly interconnected molecular components have a clear computational interpretation. Each component of the metabolism may be seen as transforming an input signal into an output. Their input is the information they receive from other proteins in the network that interact with them and their output is the way in which they in turn interact with other components in the network. An enzyme may be imagined as 'reading' the concentration profile of its input reactants and translating this, by updating its own state of activity, into an output concentration of products. In the case of encoded protein metabolisms, each protein component may be viewed as an idealized digital switch which is either ON or OFF.

Proteins and the functions they confer, can be interconverted between ON and OFF states in a number of different ways. These include mechanisms for increasing or decreasing the concentration of the protein or its substrates, allosteric changes that occur in response to changes in the concentration of the reactants and convert the protein into an active or an inactive state, the presence of activators and inhibitors which modify enzyme activity by altering the availability of its binding site, and reversible chemical modifications such as phosphorylation or dephosphorylation, which induce a conformational change that alters accessibility of the binding site.

Any protein component may receive more than one input. When this occurs, the protein onto which the inputs converge needs to integrate this information in order to produce a coordinated response. The rules that determine the way in which such integration events take place are known as Boolean rules. A protein which is converted to the ON state only when both of its two inputs are ON, is, for example, governed by what is known as a Boolean AND function or 'gate'. That is to say an integrated ON response is only produced when both input one AND input two are ON. Another example of a Boolean logic rule, is what is known as an OR gate. A component that has two inputs and whose response is under the control of an OR gate, will produce an ON output if either its input one OR input two is in the ON state. There are many different types of Boolean functions which could, in principle, be implemented by chemical components. The chemical circuits and regulatory elements that implement these logical functions thus constitute the basic elements of computation in biochemical systems. In some cases the logical response of a gate is situation dependent, and 'fuzzy logic' is observed where under one set of conditions the gate may behave according to the logic of an AND function, and under another like an OR gate.

In genetically encoded metabolisms, non-coding DNA sequences play an important regulatory role and function as analog gating devices that integrate information about the concentrations of two or more proteins and produce an appropriate output response by altering gene expression. The computational complexity of such analog gates can be astonishing. The promoter region of the sea urchin gene *Endo 16*, for example, contains at least thirty different analog regulatory elements which are dispersed across about 2,300 bases of DNA. Chiou-Hwa Yuh and Eric Davidson have suggested that this collection of regulatory elements functions as a tiny analog computer. The program that runs on this computer is encoded in each of the regulatory elements, which are not themselves ever translated into protein sequences and thus perform an exclusively computational function.

The encoded metabolic circuits of modern organisms have been trained over billions of years by a sustained process of Darwinian evolution. The functional properties of each circuit have been programmed into its components by optimizing multiple parameters. These include such things as the rate constants of individual reactions, the affinity constant of enzymes for their substrates, the stability and folding efficiency of each component, the nature and specificity of their binding sites, the efficacy of their active site and their ability to form complexes both with themselves and other components. The function of non-translated DNA regulatory elements in regulating the dynamics of the metabolism is, as we have already seen, every every bit as critical as the genes that encode the protein components themselves. Like the effecter components, the logic of the program which determines the functional dynamics of the circuit has been hammered out by history. As each component and the programs in which it participates have been historically processed and optimized for a particular function within a specific environment, the network is restricted to modes of behaviour that have been successfully tried and tested.

Non-encoded collectively autocatalytic networks received no such education and lacked evolutionary 'intelligence'. The individual elements from which they self-assembled were historically naive and had not been taught by evolution how, for example, to interact optimally with their substrates, to fold rapidly and efficiently, or to stabilize themselves under both normal and extreme conditions. Unlike 'smart', encoded metabolisms, they would not have been able to benefit from the types of carefully honed logical rules that regulate the dynamics of contemporary metabolisms and which have been discovered by evolutionary processes and captured digitally in the sequences of genes and their control elements. Surely such randomly constructed chemical networks, composed of random, non-educated components that integrate their inputs according to the logic of randomly assigned Boolean rules and in the absence of historically assembled regulatory programs, would not stand even the slightest chance of generating orderly dynamics?

Remarkably, this is not the case. Certain types of randomly constructed networks of interconnected elements, each of which has been randomly assigned a Boolean function, can actually generate a huge amount of order. Although this cannot compare with the complexity of even the most rudimentary of encoded, historically processed metabolisms, it is nevertheless highly significant. Fortunately, the formal behaviour of interconnected collections of ON/OFF components can be modelled on computers without reference to the particular technology in which they are implemented. Such models are, in a very broad sense, able to capture the generic properties or 'logical skeletons' of all classes of biochemical networks. In fact the models are so general that they may be applied to any type of interconnected network system including, for example, the global economy, in which case the components might represent companies that define an economy and the connections between them local and global market forces. The fact that we do not know the identity of the molecular components from which such hypothetical geneless metabolisms may have been constructed does not thus prevent us from modelling their possible modes of behaviour. We should, however, be aware that such simplified models of abstracted, artificial chemistries ignore many of the parameters that are critical in defining the properties of real biochemistries. The computationally rich and historically defined features of instructed biological metabolisms are precisely the types of things that enable even the simplest genetically programmed single-celled organism to perform its essential functions.

Let us now imagine a collectively autocatalytic chemical network that self-assembles from a total of 100,000 different enzyme components, each of which can be idealized as existing in either an ON or an OFF state. The ON state corresponds to a high concentration of the component, whereas the OFF state corresponds to a low concentration. In the case where the concentration of the enzyme remains the same and only its activity is regulated, for example by an allosteric change or reversible chemical modification, the ON state corresponds to a situation in which the enzyme is active and the OFF situation to one in which it is inactive. As each enzyme may exist in one of two different states (ON or OFF), the metabolism explores a hyperastronomically large *Concentration Landscape World* containing a total of $2^{100,000}$ ($10^{30,000}$) different concentration states, in which the metabolism might at any moment reside. Each state represents a distinct pattern of enzyme concentrations. One of the possible states in the *Concentration Landscape World*, for example, corresponds to a situation in which all

100,000 of the enzymes are switched ON, whereas another corresponds to one in which they are all switched OFF. In yet another, the concentration profile may be such that 50,000 of the enzymes are ON and the remaining 50,000 are OFF.

Each enzyme within the network is able to determine its own ON or OFF status by integrating the profile of inputs it receives from the other components in the network to which it is connected by means of cross-catalytic or inhibitory loops. At discreet intervals each enzyme may be imagined as integrating its profile of inputs according to the logic of its Boolean gate by 'looking up' the appropriate output response and updating its own state according to the logic that the gate dictates. An enzyme controlled by an AND gate will, for example, switch to an ON state if both of the cross-catalytic inputs that it receives are ON. However an enzyme under the control of a NOT gate will only be switched ON if one of its inputs is ON and the other is OFF. As a result of each component synchronously or asynchronously updating its ON/OFF status at any given moment, the concentration profile of the network will change. At the next instant the updating process is repeated and the state of the network will change once again. The dynamics of the network thus emerge as a consequence of huge numbers of these computations being iterated in parallel.

The distinctive train of concentration patterns or 'states' generated as the system explores its *Concentration Landscape World* is known as the 'trajectory' of the system. Eventually the system will encounter a state that it has encountered previously and the same succession of states will commence once again and repeat indefinitely, much like the repetitive flickering patterns on the electronic advertising boards which adorn Piccadilly Circus. The pattern of successive states that any given network repeatedly cycles through is known as its 'state cycle'. If the state cycle is short, then the dynamics of the network are orderly and only a very small number of states are visited. The network might, however, visit all or a large number of its possible states, in which case the state cycle will be long and the associated dynamics chaotic. In this situation, each successive state is selected randomly from the repertoire of all possible states.

The state cycle of a network is known as an 'attractor' and the set of initial states that flow into the state cycle, its 'basin of attraction'.

Whereas closed, equilibrium systems are driven into their potential state spaces in a completely random manner and are thus under the influence of large chaotic attractors, open, far-from-equilibrium systems with a sparsely connected network structure are able to generate orderly, small attractors spontaneously and in the absence of programmed instructions. These small attractors draw the system into an infinitesimally small region of its potential state space and ensure that the trajectory of the system through its potential state space is non-random. Each network system has one or more attractors and each of these has its own unique basin of attraction and drains a distinct collection of initial states.

In an earlier chapter we discussed the *Scramble Net World* space occupied and explored by a collection of 1000 robotic schoolboys. Our metabolism consisting of up to and around 100,000 different, interconnected chemical components may, similarly, be imagined as exploring the *Concentration Landscape World* it inhabits. In the *Scramble Net World*, orderly dynamics were represented by the situation in which movement of the robots was restricted to tiny regions of their potential state space. Chaos, on the other hand, was represented by a situation in which their movement was unconstrained and the robots were free to explore the space with impunity. A metabolism with chaotic dynamics would be expected to explore its potential state space in an entirely random manner. The concentration landscape would change continuously, with the concentration of each component in the network flickering ON and OFF in a chaotic manner devoid of any discernible pattern. In an entirely ordered network however, the concentration of each component would be frozen into a fixed and unchanging pattern of ON and OFF values. Both types of behavioural dynamics are clearly unsuitable for living processes, one being too random and chaotic and the other too rigid and inflexible. So what type of dynamics are actually observed in a randomly constructed Boolean network which self-assembles from 100,000 different components?

Stuart Kauffman has studied the behaviour of idealized and highly simplified, randomly constructed, collectively autocatalytic artificial metabolisms using computer simulations. These have indicated that the principal features which influence their dynamics are:

(1) the number of inputs that each component receives, (2) the nature of the Boolean gate that controls the way in which each component integrates its inputs and computes its output, and (3) the bias that each gate introduces into its output response.

If each component receives only one input, simulations show that the network becomes frozen into a highly ordered and very simple mode of behaviour, cycling between only one or two of the $10^{30,000}$ possible states in its *Concentration Landscape World*. If every component receives an input from every other component, there is feedback overload and the behaviour of the network is hopelessly chaotic. The concentrations of each component flicker ON and OFF in a random, frenetic manner and the system cycles haphazardly through a vast number of the states in its potential concentration landscape.

The quite extraordinary result, however, is that if each component receives just two input connections, the system spontaneously generates a small attractor which ensures that it cycles repeatedly between only 317 of its $10^{30,000}$ possible states. This staggering degree of order is generated without the agency of digitally encoded instructions. Indeed the principal prerequisite for generating spontaneous, self-organizing dynamical order within this system is simply that it should be constructed so that its components are sparsely connected. When this condition is satisfied, the system will spontaneously generate a small and orderly attractor. It then ignores the majority of the states in its huge potential state space repertoire and visits only an infinitesimally small fraction.

The collectively autocatalytic network can, in this way, focus the food molecules with which it is fed into high concentrations of only a relatively small number of molecular species and may consequently function as a true metabolism. It is clear then, that even in a simulation where the structure and logic of this geneless network is thrown together in an entirely random fashion, immense order can be generated without the intervention of encoded digital programs. Although once generated, this natural order constitutes an excellent substrate for a process of evolution by natural selection, the order itself arises independently of any such process.

This remarkable condensation of the system down into a tiny region of its potential state space occurs naturally and spontaneously without the agency of any internally or externally applied instructions. Profound order emerges effortlessly in the absence of a precursor program, and the ahistorical program-free system behaves as if it had been programmed to avoid most of the states in its *Concentration Landscape World*, cycling through only a very small and highly select number. This staggering dynamical order arises purely from the way in which the network is configured. Unlike genetically instructed metabolisms in which the order of a historical program is imprinted into each of the 'smart' network components and superimposed onto any intrinsic self-organizational phenomena, in non-encoded metabolisms the observed order arises ahistorically and exclusively as the result of spontaneous auto-programming.

Of the large number of possible Boolean logic gates, it turns out that only some are able to generate and sustain orderly dynamics. The dynamics of a simulated network can consequently be made more orderly by assigning such order-generating functions to components, rather than employing a hotchpotch mixture of logic gates. Order can also be generated by using a set of logic gates that introduce bias into their output response. Let us consider, for example, a component that receives two inputs and whose output response is controlled by an OR gate. The component is able to receive four possible input patterns. Each input can be OFF (0, 0), or ON (1, 1), one can be ON whilst the other is OFF (1, 0), or vice versa (0, 1). In the case of an OR gate, three of the four possible input patterns (1, 1), (1, 0) and (0, 1) produce an ON response. An OR gate will thus produce an ON response 75% of the time and is biased towards an ON response. An AND gate on the other hand, whose logic demands that input one and two must be ON in order to produce an ON response, produces an OFF response 75% of the time. This is because only one of its four possible input patterns (1, 1) produces an ON response, whilst the others (1, 0), (0, 1) and (0, 0) do not. The overall bias of the network can be determined by averaging out the bias found at each gate. If the overall bias exceeds a threshold level unique to each network, the network is likely to generate orderly dynamics. The overall order in a network can in fact be tuned by systematically adjusting the bias value. This may be achieved by changing the set of logic gates deployed by the components in any given simulation. This effect is particularly pronounced in densely connected networks, which have a tendency to be characterized by feedback overload.

It appears then, that genes and the digital instructions they encode are not the only source of biological order, and that order can be generated *de novo* in the absence of pre-existing order. With this in mind, we can return to the question 'what programmed the first program?' and reiterate our suggestion that it may not have been necessary for it to have been programmed by the microcode of a pre-existing program. The first rudimentary systems of biological heredity may instead have emerged spontaneously within the complex self-organizing dynamics of sparsely connected, open and far-from-equilibrium collectively autocatalytic networks. The small, order-generating attractors which form spontaneously within such networks appear, at least in computer simulations, to be able to generate their own set of distributed analog instructions without the need for codification or digital symbolism. Analog attractors are thus able to arise spontaneously within the dynamics of appropriately constructed network systems and, like strange spectres, to 'trap' them within infinitesimally small volumes of their vast potential state spaces.

So, are self-organizing, sparsely connected chemical networks of this sort that are controlled by powerful small attractors able to generate homeostatic stability, to repair themselves and to evolve? In large networks, small attractors generally have a wide basin of attraction. Thus, in addition to the states that form part of the cycle, a large number of other initial states will flow back into the parent attractor state cycle. These basin states are often very similar to the states of the attractor state cycle they drain into, so any perturbation or damage to the network is likely to produce a new state which, although not part of the parent state cycle, will nevertheless drain back into it. The network thus has a tendency to 'flow home' to its small attractor, which enables the dynamics of the system to remain orderly and under the influence of its attractor, even in the presence of perturbations. This confers it with robustness. Mutations to the structure of the network are, for the same reason, unlikely to displace the system from its attractor. This allows the network to evolve smoothly across what we might call *Geneless Metabolism Space* without introducing catastrophic changes into its structure or dynamics.

Much like the *RNA Space* landscape that we explored in our discussion of the *RNA World*, the fitness landscapes which such sparsely connected metabolisms inhabit and explore are highly correlated. Instead of the skyscrapers being scattered randomly around its fitness landscape, potential adaptive solutions to similar environmental problems will often be found clustered together in a single downtown Manhattan region. Geneless metabolisms may in principle skip from one skyscraper to another, finding adaptive solutions with varying degrees of fitness in their immediate search vicinity. The dynamics of networks which are under the influence of small attractors should, in addition, be relatively unaffected by random changes in the concentration of their components.

Sparsely connected networks thus demonstrate exactly the type of robustness, homeostatic stability and evolvability that is necessary, both to maintain their core informational structure across time, and to allow it to be systematically and incrementally modified so that it is able to update its representation of the outside world. Highly connected, chaotic networks, on the other hand are hugely sensitive to damage, mutations and perturbations and inhabit landscapes that are not easily navigable by evolutionary processes. The deletion of a single component from the metabolism would vastly reduce the size of the *Concentration Landscape World* that the network inhabited and have profound effects on both the attractors and state cycles of the system.

The change from orderly to chaotic behaviour in these networks occurs as a phase transition of the type we saw earlier in the button and thread analogy. Christopher Langton has suggested that the best compromise between ordered stability and flexibility, and thus the most complex mode of information processing, occurs when networks are poised within the ordered region at the edge of chaos. Natural selection may actually take systems to the point at which they teeter within the ordered region of the transitional zone, that just precedes and is consequently 'on the edge' of the phase transition to chaos. At this point they are primed to be both stable and maximally evolvable and are thus able to respond most efficiently to environmental challenges.

Networks can be brought to this point artificially, either by finely tuning the number of interconnections between their components, or by changing the nature of the Boolean logic gates that are deployed. As we know, the bias of each gate can be adjusted. Increasing the bias reduces the potential for feedback and tends to make the dynamics more ordered. A chaotic network in which each component receives more than two inputs may thus be coaxed into a more orderly mode of dynamics by tuning the bias of the logic gates that compute and control the output response of each component.

Although it is possible to model the self-assembly and self-organization of disembodied abstract chemistries without making any explicit reference to the material nature of the chemical units from which they might self-assemble, the biological meaning and utility of these models can only be assessed with reference to particular chemistries. Despite the broad generalities that can be derived from simulations of abstract Boolean network architectures, the dynamics of real chemical networks are likely to be exquisitely sensitive to details. These include the precise topology of the network and the kinetic parameters associated with the chemical reactions from which it is constructed. The assumption that each component can only be ON or OFF is furthermore a chemically unrealistic idealisation, as real enzymes give graded responses. It is also inaccurate to model the connections between the network components as if they were spatially fixed or 'hard wired', as they are, in fact, formed by the diffusion of chemical components within the reaction space of the network. Thus, despite the model's apparent robustness and insensitivity to details, and the fact that the broad behaviour of any real chemical network should be computable, the construction of accurate simulations is likely to require a comprehensive catalogue of all of the relevant physical and chemical parameters. These include, amongst many other things, the half-life of each component and, in the case of enzymes, the rates and equilibrium constants of the reactions in which they participate.

The question of the extent to which such theoretical models correspond to the real world and real chemistries may be subdivided into two components. One of these relates to whether historically plausible chemistries might have been able to self-assemble into sparsely connected, self-organizing metabolic networks. The other relates to the question of whether it might in principle be possible to construct these types of networks and their associated dynamics from artificial chemical components that do not necessarily resemble the natural historical candidates. One might also, as mentioned before, question the extent to which computer models of artificial chemistries and the idealizations they incorporate constitute plausible representations of real chemistries.

The answer to these questions lies principally in: (1) the extent to which the components of the network can be generated by non-informed chemical processes, (2) the extent to which repertoires of components contain suitable profiles of mutually compatible cross-catalytic activities, (3) the frequency with which such components can be found within combinatorial libraries, (4) the extent of the connectivity between the components from which the self-assembling chemical networks are constructed and (5) the extent to which the components are controlled by logic gates that generate orderly dynamics.

At the end of the day, the questions about plausibility can only be determined empirically using appropriate candidate chemistries. Thus, although the cross-catalytic and autocatalytic landscapes that are generated by different types of chemical polymers can, in principle, be modelled on the basis of their chemical structures, they are best explored using the types of combinatorial libraries that we examined in our discussion of the *RNA World*. Once the frequency and distribution of such activities have been mapped for different classes of natural chemical systems, it should be possible to predict whether any particular collections of polymers of a defined size range would be able to self-assemble and spontaneously generate a profile of logic gates, connections and biases capable of generating rudimentary static and dynamical order. There is no intrinsic reason why such networks should be constructed from only one class of polymer, and it might be the case that only certain mixtures of different types of polymers are able to produce orderly, self-sustaining networks.

One important question is whether polymers of the limited size range likely to have been generated by abiological prebiotic chemical processes would have harboured a range and frequency of catalytic activities sufficient to generate a collectively autocatalytic set. The reaction specificity of these components is also a critical issue, perhaps even unsurmountable, in which case the chances of generating a collectively autocatalytic network from natural chemistries would be negligible. The model in which Boolean logic gates are assigned entirely randomly to each component is, however, likely to represent a worst case scenario. Naturally occurring metabolisms are likely to utilize a more limited set of Boolean gates, that favour the creation of orderly dynamics. We should perhaps be encouraged by the fact that almost all known genetic circuits are regulated by very few inputs and are under the controlling influence of a subset of Boolean logic gates which favour the creation of dynamical order.

Despite these potential difficulties, experimental studies have provided the tangential, tentative, but nevertheless suggestive beginnings of an empirical foundation for the geneless metabolism hypothesis. In 1996 David Lee and Reza Ghadiri demonstrated that a peptide of length thirty-six amino acids can autocatalytically increase the rate of its own production from the ligation of two peptide reactants of lengths fifteen and seventeen amino acids. The reaction demonstrated high amino acid sequence and structural specificity, with less than 15% of the reactants being converted into side products. This provided the first evidence for the chemical feasibility of autocatalytic peptide self-replication. A year later, Lee and Ghadiri were able to demonstrate that two different autocatalytically self-replicating peptides could be cross-catalytically linked into a mini-network. Although this miniature network is not itself collectively autocatalytic as each of the components is individually autocatalytic, it nevertheless incorporates the type of cross-catalytic connections which could form the basis of an exclusively collectively autocatalytic metabolism. Shao Yao and Joan Chmielewski were able to take this one step further by generating a self-organizing peptide network consisting of four autocatalytically self-replicating peptide ligation reactions. Each individual autocatalytic reaction was interconnected to the others by a web of superimposed cross-catalytic connections. The concentrations of the four products of this reaction were, furthermore, shown to be sensitive to various aspects of the surrounding environment including its pH and salt concentration. This suggests that the concentration profiles generated by mini-networks of simultaneously autocatalytic and cross-catalytic chemical reactions are able to adapt to environmental conditions and evolve. It also provides an inkling of how the state cycles generated by small attractors within orderly collectively autocatalytic networks might translate into a unique set of concentration phenotypes, each of which may have distinct adaptive consequences.

The extent to which these results can be generalized to more complex networks that rely exclusively on cross-catalysis and which are collectively but not individually autocatalytic, remains to be seen. They do, however, constitute preliminary evidence for the chemical plausibility of collectively autocatalytic chemical networks constructed from peptide molecules of a length that could, in principle, have been synthesized under prebiotic conditions. The *Geneless World* may thus have been inhabited by analog organisms that utilized peptide technologies for the spontaneous generation of their metabolisms. Although the naive peptide components themselves would have been composed from digital sequences of amino acids, they would, at least in the very first instance, have been utilized in an exclusively analog fashion.

We will shortly be leaving the hypothetical *Geneless World*. However, before doing so we will briefly address the question of whether these hypothetical geneless organisms were compartmentalized and thus able to act as discreet units of selection. We will also discuss whether they could reproduce themselves, mutate and evolve, and examine how they might have generated the flux of matter or energy necessary to maintain their far-from-equilibrium state. We will then attempt to draw a limit to the degree of complexity that such potentially quasi-living networks might in principle have attained and end by discussing how the geneless, self-organizing, ahistorical order of these networks might eventually have been enslaved, historically processed and subsequently overwritten by the digitally encoded logic of genes.

The earliest flora and fauna of the *Geneless World* is likely to have consisted of ephemeral naked, non-compartmentalized or 'liquid' geneless creatures that quasi-lived on mineral or other types of two-dimensional surfaces, or perhaps freely within concentrated or dehydrated solutions. It is unclear whether such poorly defined, non-compartmentalized geneless metabolisms could have reproduced themselves, evolved, or produced extended lineages. In the case in which two such non-compartmentalized collectively autocatalytic metabolisms co-inhabited the same space, there would have been no clear distinction between them. Although non-compartmentalized creatures would have been incapable of the discreet, high-fidelity replication processes that underwrite the unlimited heredity of contemporary living things, they may have been able to reproduce their concentration profiles in a low-fidelity and more continuous manner. They may consequently have been self-sustaining and capable of exerting a strong focusing influence on the concentration landscape of the micro-environment they inhabited. In this sense, they may be considered as having possessed the rudiments of a system of heredity.

Richard Bagley, Doyne Farmer and Walter Fontana have provided computer modelling evidence which suggests that collectively autocatalytic metabolisms might have been able to function as 'seeds', that, even at very low concentrations, would have been able to take over a medium with appropriate properties and block the formation of other metabolisms. Even a subset of the collectively autocatalytic components could, in principle, have recreated the entire network. In the case where a non-compartmentalized metabolism was the sole inhabitant of a micro-environment, there would be no population of alternative metabolisms and thus no potential for competition and evolution. This does not mean to say that such isolated self-sustaining metabolisms were not able to accumulate incremental changes that influenced their fitness and continued survival. It is perhaps most likely that they were continuously created and destroyed, until eventually conditions favoured their encasement within some type of geometrically well-defined boundary. This may have occurred as the result of the parasitism of pre-existing compartments such as the autocatalytic self-replicating micelles described by Pascale Bachmann and Jacques Lang, or as a consequence of reactions within the network itself, that resulted in the *de novo* synthesis of a compartmentalized structure. The transition from non-compartmentalized to compartmentalized proto-life was not, however, trivial and would have introduced several new problems associated, for example, with the transport of food components across the surface of the compartment.

Once rudimentary proto-cells had self-assembled, the concentration of each of the network components could be much increased. Each geneless proto-cell might also have been able to reproduce itself by a low-fidelity process that involved fissioning into two when its volume exceeded a threshold limit. Any mutations that occurred within the parent compartment might, in this way, have been passed on to daughter compartments. Unequal transfer of network material to daughter compartments would have introduced an additional possibility for analog mutations and assisted in the generation of a repertoire of analog mutants. Each compartment would have been able to act as a discreet unit of selection, and the unique concentration landscape generated by the attractors within any particular network would have acquired the potential to be translated into a phenotype that could influence the survival and reproductive efficiency of the proto-cell.

Core proto-genetic information could then, for the first time, be directly linked to a selectable phenotype. This would have allowed metabolisms to homeostatically define the nature of their encapsulated micro-environments and to be differentially selected from an ecosystem of heterogeneous compartments. Complex communities of compartmentalized geneless proto-organisms may eventually have been established, where different proto-organisms interacted with one another in a host of different ways that included: co-operation, competition, symbiosis and parasitism. At least one of these communities might have spawned the historical lineage that resulted in the first cellular forms of genetically encoded life, and in so doing established both the possibilities and constraints for all natural and artificially generated forms of life.

Although individual metabolisms may be modified as a result of the accumulation of analog mutations, in order for a true process of Darwinian evolution to occur, a population of variants must be generated. Only then can the relative correspondence of the phenotypes of these alternative copies of the founder metabolism with the structure of their environment be iteratively tested and retested. Some variants would presumably be better adapted and thus reproduce more efficiently than others. These mutants would eventually take over the micro-environment they inhabited. But how might a repertoire of mutant proto-organisms have been established in the first place?

Richard Bagley, Doyne Farmer and Walter Fontana have suggested that mutations could be introduced into collectively autocatalytic metabolisms by the addition and deletion of the cross-catalytic loops from which they are composed.

The molecules from which collectively autocatalytic metabolisms are constructed may be divided into those that form part of the collectively autocatalytic set and the 'background' molecules produced by non-catalyzed, spontaneous reactions, which do not. Some of the background molecules, however, are produced by reactions that involve members of the collectively autocatalytic set, and may be imagined as forming a 'shadow' around it. Members of the shadow consequently exist at higher concentrations than the rest of the background and their concentrations closely track those of the autocatalytic set.

A perturbation or spontaneous fluctuation in the concentration of one of the shadow molecules may, directly or indirectly, result in the catalysis of a spontaneous background reaction within the shadow. If one of the products of such a reaction was able autocatalytically to increase its own production, the fluctuation would be greatly amplified. Furthermore, if one of the new products participated in a reaction that could be catalyzed by a member of the collectively autocatalytic set, a new chemical reaction loop might be generated and cross-catalytically connected into the metabolism. Deletion of a metabolic loop might similarly occur if, for example, one of the new products inhibited the activity of a network component. The shadow set thus constitutes an important breeding ground for molecules that have the potential to introduce spontaneous analog mutations into the metabolism. A single fluctuation may initiate a complex cascade of network modifications which in some cases will be irreversible. Fixation of the ramifying consequences of spontaneous fluctuations is, furthermore, likely to alter the future behaviour of the metabolism by, for example, influencing the probability that a new fluctuation will occur and the likelihood that such a fluctuation will introduce a new modification. The manner in which a metabolism might be modified by a given fluctuation is also likely to influenced by any permanent change to the composition and architecture of the metabolism.

Another potential mechanism for introducing mutations into collectively autocatalytic metabolisms is the violation of metabolisms by an influx of invading components that are chemically unrelated to those of the parent metabolism. Invader components could in principle mutate a network in much the same way as the spontaneous fluctuations of the shadow set. The deletion or emigration of members of the core set would produce similar results. One might also imagine a situation in which a metabolism could be violated by an influx of chemically related components. The influx of a new mutant peptide might, for example, function like a prion protein and induce a conformational change in related network peptides.

In addition to structural mutations, identical or very similar networks could converge onto different small attractors and suffer dynamical mutations. Thus despite having a similar or identical wiring or topology, populations of compartmentalized proto-cell metabolisms might be able to generate very different chemical concentration landscapes within their boundaries. If particular dynamical patterns translated into differential abilities to reproduce the proto-cell and its contents, they would tend to out-compete any proto-cells that were under the influence of different attractors. Once a complex ecosystem of geneless proto-organism compartments had been established, the repertoire of variants might include sets of stable cooperative, parasitic or symbiotic interactions between different proto-organisms.

The material composition of a network might also change as the result of the substitution of one type of binding site or catalytic activity for another similar or identical one realized in a different chemical technology. Alternatively, two components with different binding site geometries might catalyze the same reaction using different mechanisms. The kinetics of these mechanisms may differ, however, in that the substitution of one component for another might, for example, influence the dynamics or evolvability of the network and thus its overall fitness. On the other hand, one might imagine a situation in which it were possible to replace a metabolism composed exclusively from ribozymes with single-stranded DNA, peptides or other components without altering the fundamental phenotype of the network.

It appears then, that even individual non-compartmentalized

metabolisms are able to explore small regions of the *Space of All Possible Collectively Autocatalytic Metabolisms* and their associated self-organizing dynamics. Whilst not constituting a true process of Darwinian evolution, as there is no differential selection of mutants from a repertoire of alternatives, the metabolism can nevertheless incrementally modify its structure and dynamics across time. The inability of such isolated metabolisms to generate a population of mutants would, however, make them extremely susceptible to changes in the micro-environments they inhabited. It is therefore unlikely that any particular non-compartmentalized metabolism would have survived longer than seconds, minutes, hours, days, or perhaps weeks. Compartmentalized metabolisms, on the other hand, would be able to generate a repertoire of mutants that could be selected on the basis of their differential ability to reproduce and survive. This would allow them to evolve by classic Darwinian processes and to establish informational lineages that may have had the potential for unlimited heredity. These hypothetical quasi-living creatures, which may have populated the geneless plains of the *Geneless World*, would thus have been able to evolve across a fitness landscape of alternative geneless proto-creatures. In the same way that genetic organisms which encode their essential information using DNA explore DNA sequence spaces, geneless metabolisms may be imagined as exploring the set of alternative mathematical structures housed within the *Geneless Zoo*.

As we enter the *Geneless Zoo*, the shelves are packed with the analog geneless kits of potential geneless proto-organisms. As was the case with the gene kits housed in *Gene Space*, most have been untouched by the hand of history and are dusty and unopened. The kits in the *Geneless Zoo* hypermarket, however, represent only a fraction of the kits housed in the far larger *Collectively Autocatalytic Network Space* hypermarket. This contains the complete set of all possible collectively autocatalytic metabolisms, the majority of which are chaotic or rigidly frozen into simple, repetitive dynamics. The black life-lines within this hypermarket represent the thin seam of geneless kits that are able to generate proto-life and which populate the *Geneless Zoo*. The vast *Sea of Whiteness* that separates them contains the kits that are unable to generate the types of dynamical patterns compatible with proto-life. The *Geneless Zoo* is itself separated into multiple sub-spaces, as the components from which its kits are constructed may be fashioned from one or several different chemical technologies. One of these sub-spaces, for example, is the *Peptide Molecules Only Geneless Zoo*, which contains the complete collection of peptide metabolisms able to generate quasi-living things. Another is the *RNA Molecules Only Geneless Zoo* and yet another the *Peptide and RNA Molecules Only Geneless Zoo*. Still others may contain kits constructed from chemical technologies that are quite unexpected and unfamiliar. Who would have imagined that some of these mathematical spaces may have contained the connections and conduits, the very portals into a new world, which would eventually lead, via a circuitous route through a host of different mathematical spaces, into the vast combinatorial expanses of *Gene Space* and the immense potential for complexity that it houses?

For such collectively autocatalytic metabolisms to sustain their non-equilibrium state, they would have required a flux of matter or energy. This capacity to dissipate matter and energy might initially have been attained in a number of different ways. Chemical reactions may be subdivided into exothermic, or heat-giving, reactions, and endothermic reactions, which consume energy. In modern encoded metabolisms, the excess energy released from exothermic chemical reactions can be captured in the bonds of energy transporter molecules. The most important of these is known as adenosine triphosphate (ATP) and stores energy in its phosphate bonds. If energy is required to drive an endothermic reaction, the phosphate bonds of ATP can be broken down to release energy. This injection of chemical energy into a system is analogous to the heat energy that generates the geometrical honeycomb patterns in Rayleigh-Bénard systems. However, these types of complex energy-transporting systems have evolved over huge time scales and are unlikely to have been available to proto-organisms. The collectively autocatalytic metabolisms of geneless proto-organisms are consequently likely to have used simpler types of energy-shuttling systems, or to have coupled exothermic and endothermic reactions into mutually compatible modes of co-operation. However, the simplest way of driving such systems away from equilibrium is by maintaining an influx of molecules that help regenerate the individual components of the network.

Now that we have completed our discussion of the minimal

requirements for life, it seems reasonable to conclude that compartmentalized, collectively autocatalytic metabolisms and perhaps even their non-compartmentalized precursors, were at the very least quasi-living. Although they were members of the broad set of systems that we might reasonably define as 'living', they belonged more specifically to the subset of living things that we have called 'quasi-life'. These quasi-living creatures may be imagined as having an organization lying somewhere between that of an inanimate dissipative system, such as that described by Rayleigh and Bénard and a genetically encoded living thing, in which chemistry is transcended. Although it is unlikely that one of these hypothetical geneless proto-creatures would have been able to write *The Marriage of Figaro* or to perform even the most rudimentary of the tasks carried out by modern unicellular creatures, we should not let expectations derived from our familiarity with complex contemporary genetically encoded creatures, colour the way in which we perceive and define the ancient proto-creatures that preceded them. It is unreasonable to make the same demands of creatures whose lineages were in existence for perhaps only 3.6 days, as those that benefit from around 3.6 billion years of evolutionary experience programmed into both their components and regulatory elements.

It is interesting, at this point, to make some attempt at drawing a limit to the extent of the complexity that the hypothetical geneless proto-creatures that may have inhabited the *Geneless World* might have been able to attain. Although a lack of detail prevents us from making specific quantitative predictions, we can examine some of the broad limitations that are likely to have constrained their complexity.. Perhaps the most fundamental of these is the nature of the mathematical landscapes that these creatures would have been free to explore. Unlike the infinite mathematical landscapes available to genetically encoded organisms, the mathematical landscapes available to proto-organisms were finite. Whereas each component and regulatory element of an encoded 'smart' metabolism can be historically processed and programmed so that an invariant input food set can be channelled, via encoded synthetic programs, into a limitless number of alternative molecules, in non-encoded metabolisms the boundaries of the available component landscape are clearly delineated and fixed by the nature of the available food set. These naive, unprogrammed components consequently lack the evolutionary intelligence of encoded components and the historically defined means for their coordination.

The number of possible components may occasionally change as a result of the immigration and subsequent incorporation of new chemical species into the metabolism, but the components themselves cannot be historically processed and programmed. Thus, although new components may be grafted on to or deleted from the metabolism, the components themselves remain historically naive and thus invariable. This constraint severely limits the extent of the computational power that geneless proto-organisms might have attained. Furthermore, the addition or deletion of any individual loop to or from a metabolism is unlikely to have a profound effect on the selectable behaviour it is able to generate. The extent to which new loops can be added to a metabolism and the degree to which such analog mutations can generate new phenotypes will themselves be constrained by the size of the compartment in which the metabolism is housed. Several other constraints might also be expected to limit the complexity of geneless organisms. These include the fact that their behaviour is limited by the dictates of only one or a few inflexible, small attractors; that their fidelity of reproduction is likely to be low; and that they would have no mechanism for regulating their dynamics across time. The amount of information per unit component that such metabolisms would be able to hold would also be very low. How, then, might such geneless organisms have overcome the information crisis that threatened to undermine their existence?

Perhaps the most compelling possibility is that compartmentalized geneless proto-organisms were infected by digital material which, in the first instance, functioned as what we have called pre-genes. These digital molecules, whilst not directly encoding information, may as we have suggested earlier, have been able to influence either the dynamics of non-encoded metabolisms, or the nature of their individual components. Pre-genes are seen as being the precursors of proto-genes, which are themselves envisaged as being the first molecules that were able to use their digital template-based structures to symbolically encode biological information. This was achieved by coupling the information encoded in their digital sequences to biochemical machinery that specialized in template-based polymer synthesis. Although rudimentary, the coding abilities of proto-genes would have provided the basis for their eventual evolution into true genes. Unlike the limited complexity of proto-genes and their associated 'proto-code', genes utilize a complex combinatorial code to synthesize encoded metabolisms according to well-defined historical mandates. Although they would, at least initially, have been constrained by the intrinsic geneless order of the system that they were in the process of enslaving, proto-genes would eventually have been able both to rewrite and in some cases overwrite both the self-assembling structure and the self-organizing dynamics of non-encoded metabolisms. Furthermore, by fixing tried and tested clusters of genes together into genomes, mutually compatible collections of components could be conserved.

Until the time when genes became sufficiently complex and powerful to dictate and imprint their own unique patterns and dictates onto self-assembling and self-organizing systems, pre-genes and proto-genes may have functioned to ensnare, tune, trap and manipulate the natural geneless order of self-programming, non-encoded metabolisms. Not, in the first instance, generating the complexity, but limiting and refining it instead; forcing and tweaking it with the injection of their digital inoculum; not specifying the structure of the network, but rather modifying the properties of its components, constraining its degrees of freedom and adjusting the outcome of naturally occurring, self-organizational processes; parasitically exploiting and subverting the unstructured edifices of analog proto-living creatures and constraining the way in which they explored their state spaces. Digital, non-coding pre-genes might, for example, have influenced a network's dynamics by functioning as analog computational devices that served to integrate information, in much the same way as the non-coding DNA regulatory sequences of modern genomes. They might, alternatively, have influenced the folding and stability of network components, or their ability to form oligomeric and multimeric complexes.

The eventual emergence of true genes and fully encoded metabolisms placed life squarely onto an infinite combinatorial mathematical space. By establishing a reliable communication channel between the past and the present and thus a capacity for unlimited heredity, genes provided a 'time tunnel' that enabled aspects of the information of the past to be communicated and reliably connected with that of the future. The individual components of biological machines could now be historically processed and programmed by Darwinian evolution by natural selection. The essential aspects of a system's structure and dynamics could also be captured symbolically within sets of abstract digital instructions that could be systematically interpreted and executed from a central command centre.

Before finally leaving the *Geneless Zoo* and the potential geneless creatures it contains, we will briefly address the question of where the first geneless organisms might actually have originated from. As there is no direct experimental evidence to substantiate any of the possibilities, it would be unreasonable to rigidly fixate on one alternative technology as opposed to another. What we can do, however, is review some of the candidates, any one of which might have provided a plausible material basis for the types of processes we have outlined. Having generated a hypothesis, we are obliged to sit back and wait for the appropriate experiments to be conducted, in order to determine whether any one of them contains even a grain of truth.

The basis of the connectivity in collectively autocatalytic metabolisms is cross-catalysis, and the binding site and active site of an enzyme can be mimicked by several different chemical technologies which include RNA, peptides, proteins and single-stranded DNA molecules. It thus seems reasonable to consider the possibility that the first metabolisms were constructed from a complex mixture of different technologies. The other possibilities, however, are either that each metabolism was constructed from only one class of chemicals, or that life had dual or multiple origins, each of which utilized a different technology. If this was the case, then the flora and fauna of the *Geneless World* may have included a mixture of proto-organisms, some constructed exclusively from RNA or peptides and others whose metabolisms were constructed from very different classes of molecules, or indeed from a combination of several different technologies.

It is worth emphasizing that if the first metabolisms did utilize RNA or single-stranded DNA technology, then these would almost certainly have been employed in a non-genetic manner. That is to say that they would have functioned as ribozymes or DNAzymes, but not as templates for the self-replication, digital codification and translation of genetic information. According to the definition of genes outlined earlier, such RNA and single-stranded DNA molecules could not, in the absence of even a rudimentary codification competence, be referred to as genes or even proto-genes. It is instead, as we have already said more reasonable to call them pre-genes. This acknowledges that they may have been the precursors of proto-genes and genes, but were not themselves at that point utilizing the processes of symbolic representation that characterize genes.

When attempting to construct a coherent hypothesis to explain the origin of living things, it is important to remember that one is building sandcastles upon a fragmentary and often incomplete collection of indirect pieces of experimental evidence and theoretical speculations. Perhaps one should not speculate at all. The value of falsifiable speculation, however, is that it may help establish the basis for an experimental agenda. We will now pause to review the argument and then proceed to formulate a sketch for a number of alternative hypotheses.

With the exception of RNA viruses, the core information of modern life is encoded in genes, genetic control elements and historically assembled collections of genes (genomes) that are made from DNA. Double-stranded DNA and the ancillary components necessary for it to function as a genetic material constitute a highly complex piece of machinery. Double-stranded DNA is consequently unlikely to have been the first genetic material. This suggests that the first living things utilized a simpler genetic technology and that one or several different genetic takeover events have occurred during the history of life. Life may thus over large durations, have switched successively from one template-based genetic technology to another. Eventually, however, virtually all living things came to utilize digital DNA databases for archiving their essential information.

The principal structural components of contemporary creatures on the other hand are made from proteins. Although the material technology of life may have persisted unchanged despite possible changes in the way in which its genetic information was represented, it is also conceivable that genetic takeover events were accompanied by corresponding material takeover events. If this were the case, then the very first organisms may not have used proteins or peptides as their structural components. So both the informational and material aspects of living things may have undergone technological changes. Each genetic takeover event presupposes a more rudimentary template-based precursor technology. This raises the question of how the very first template-based self-replicating genetic systems came into existence. How, indeed, was the first genetic program programmed? And how were these systems of heredity able to form cooperative interactions with proteins and peptides and eventually come to encode the information of these molecules?

The following appears reasonable. The first living creatures may not have utilized genes, and the origin of life may have pre-dated template-based chemistry and codification. The rudimentary non-encoded metabolisms of the very first organisms may indeed have been generated by the spontaneous self-assembly and self-organization of components that were incapable of template-based self-replication. These analog quasi-creatures which lacked the capacity for residue-by-residue copying of informational polymers, would have had only a very limited capacity for heredity. Furthermore, as they were unable to encode their core information digitally, they would have lacked mechanisms for error-correction. Their complexity and ability to sustain and transmit their information across time would thus have been severely constrained. Such hypothetical quasi-organisms may, nevertheless in a rudimentary and attenuated manner, have displayed many of the properties that we associate with living things.

The collectively autocatalytic metabolisms of the first organisms might have been constructed from peptides, RNA, a combination of RNA and peptides, or from other chemical technologies such as pyranosyl RNA (p-RNA), peptide nucleic acid (PNA) or informational organic polymers that are very different to nucleic acids. A hypothetical geneless proto-organism constructed from RNA for example, however, is imagined as replicating its individual RNA components in a non-template-based manner. RNA molecules should be capable of the type of non-template-based autocatalytic ligation reactions described by David Lee and Reza Ghadiri. A large RNA product might thus catalyze its own production by ligating the smaller fragments from which it was composed.

Let us now consider what we will call the 'peptides first' hypothesis, which suggests that the first proto-creatures were made from collectively autocatalytic peptide networks. If this was the case, the first rudimentary peptide-based creatures may eventually have been infected and parasitized by either: (1) non-self-replicating RNA molecules, (2) RNA molecules which self-replicated by means of autocatalytic ligation reactions, or (3) autocatalytically self-replicating RNA molecules which utilized a template-based mechanism of replication. But would random libraries of peptides have been able to furnish components of a complexity sufficient for the generation of a self-assembling metabolism? Would template-based self-replicating RNA molecules have been able to arise *de novo* in the absence of a pre-existing metabolism, and if not, how might peptide metabolisms have assisted the formation of the first template-based self-replicating RNA molecules? In what way, furthermore, might the incorporation of RNA molecules have conferred a selective advantage on peptide metabolisms?

James Ferris and Leslie Orgel have shown that peptides of up to fifty-five amino acids long can be synthesized abiologically on mineral surfaces. In the light of this result it seems reasonable to speculate that under more optimal conditions, it should be possible to synthesize even longer peptides. But can peptides of this order of magnitude fold to produce the types of stable structures necessary for the generation of catalytic activities? Alan Davidson and Robert Sauer have shown that folded peptides occur extremely frequently in artificially constructed, random combinatorial libraries which are eighty to one hundred amino acids long and which are composed principally of only three different amino acids. This demonstrates how even a highly restricted amino acid alphabet can generate a huge repertoire of folded structures, and indicates that a significant number of random sequences should fold into unique structures. Structural polypeptides are thus extremely common among random, low-complexity sequences, suggesting that the rapid folding of natural peptide sequences is not a consequence of extensive evolutionary optimization, but an intrinsic and robust property of the 'uneducated' peptides themselves. The fact that folded structures can be generated using a very reduced amino acid alphabet also indicates that repertoires of folded peptide structures could, in principle, have been generated abiologically using a minimal set of amino acid starting materials. Different restricted sets of amino acids are likely to generate their own unique set of folds and physico-chemical properties. Peptides composed mainly of random combinations of the amino acids glutamine, leucine and arginine, for example, are exceptionally stable at high temperatures. Mary Struthers and Barbara Imperiali have furthermore demonstrated that peptides of less than thirty amino acids are able to fold into discreet structures. The discovery by Kay Severin and Reza Ghadiri of a peptide ligase activity in a peptide that is only thirty-three amino acids long, indicates that these abiologically constructed libraries might have provided a repertoire of catalytic activities sufficient for the generation of a collectively autocatalytic set.

We have suggested that the first step towards template-based self-replication may have been the infection of pre-existing self-assembling peptide metabolisms by abiologically synthesized RNA molecules which were, at that point, incapable of template-based self-replication. Although not initially encoding symbolic information within their digital sequences, these RNA pre-genes may have been able to force and subvert the spontaneous self-organizing dynamics of geneless peptide-based proto-organisms. They may also have been able to modify the properties of the network components by, for example, chaperoning their folding, increasing their stability, or regulating their activity. The discovery by Biliang Zhang and Thomas Cech of ribozymes able to catalyze the formation of peptide bonds suggests that RNA molecules may have helped synthesize peptides *de novo* from monomer precursors. In some instances, however, RNA molecules might simply have provided new catalytic activities that had not previously been present within the peptide network and thus filled some of the holes in its catalytic repertoire. A peptide library constrained both by size and amino acid complexity could in this way supplement the repertoire of folds and associated catalytic activities available to its analog machinery. As with peptides, we know that it is possible to synthesize RNA molecules of up to fifty-five ribonucleotides long abiologically, on mineral surfaces. But this assumes that the ribonucleotide monomers were themselves present in the prebiotic broth.

Most importantly, however, digitally infected peptide organisms would have provided a stable context for the evolution of template-based self-replication and codification. They may thus have formed a breeding ground for proto-genes and perhaps eventually genes. With their fully digitalized, template-based self-replicating metabolisms and rudimentary genetic code, these more informationally complex proto-organisms would eventually have been able to use their powerful and historically informed genetic programs to overwrite the simple dynamical behaviour of self-organizing chemical systems, and to generate the type of complexity, error-correcting machinery and capacity for unlimited heredity characteristic of modern life. Instead of utilizing naive components which lacked the experience of history, the individual components of metabolisms and the logic that controlled their behaviour could be successively optimized by natural selection and eventually assembled into the first genome. If this is what happened, then the basic outline of living things and the rudiments of heredity necessary for processes of life were established without the agency of genes. Life with imperfect and very rudimentary heredity, but heredity nevertheless. The RNA molecules which parasitized the first geneless creatures, however, provided proto-life with a conduit into the infinite digital expanses of *RNA Sequence Space* and thence into *DNA Sequence Space* and the complexity and unlimited heredity it had on offer.

It is unlikely that the first RNA molecules that became incorporated into proto-creatures were capable of either complementary or self complementary template-based self-replication. Indeed, all the experimental evidence to date indicates that the completion of a full cycle of replication and the initiation of subsequent cycles in an unlimited manner requires a host of organized catalytic activities and thus the presence of a pre-existing metabolism. Any hypothesis that excludes the possibility of rudimentary life from pre-template chemistry and which is based exclusively upon couplings between template-based, self-replicating RNA molecules such as that suggested by Manfred Eigen and Peter Schuster must, consequently offer some explanation of how template-based, self-replicating RNA molecules came into existence in the first place. Thus although such 'hypercyclic' couplings between RNA molecules may have been important in the generation of complex informational structures in the later phases of the origin of life, the apparent complexity of such structures makes it unlikely that they could have been important right from the very beginning of life. Hypercycles have, furthermore, been shown to be prone to various constraints and catastrophes which include short-circuiting, collapse and susceptibility to parasitic RNA sequences that may out-replicate the other molecules in the hypercycle.

An 'RNA first' scenario in which the first proto-organisms were made from collectively autocatalytic RNA networks in which no individual RNA molecule was capable of template-based self-replication is equally plausible. RNA metabolisms are imagined to have become infected by autocatalytically self-replicating peptide components. Some of these may have functioned as RNA chaperones that helped fold RNA molecules, thereby allowing them to realize their secondary and tertiary structures. In the absence of proteins, both simple and complex RNA molecules often misfold, or fail to fold at all. Peptide molecules may also have assisted in the evolution of template-based self-replication, by providing missing catalytic activities that were needed to establish a full replication cycle, but which were not present in the catalytic repertoire of ribozymes. They might also have assisted in the mechanics of the replication cycle by, for example, stabilizing single-stranded forms of RNA molecules and thus helping a newly synthesized complementary or self-complementary strand to be released from its template. One may also imagine a situation in which a quasi-organism made entirely from RNA was able to fuse with one made exclusively from peptides. Following such an event, the two originally distinct organisms may have remained symbiotically entwined.

We have now completed our discussion of the origins of life and heredity. And as we leave the analog amnesia of the *Geneless World*, a world of quasi-organisms and pre-template chemistry barely capable of history and inheritance, we linger for a moment to watch the apparent futility of these continual cycles of creation and destruction. To behold life stripped of its most essential machinery, and which is unable to be much guided by the past and is too readily imprinted by the present. A world in which the cogs of evolution were barely able to turn the wheels of existence, and in which information disappeared like rainwater seeping into the ground. But a world primed to be torn apart by an informational renaissance and explosion. A world populated by the ghosts of our most ancient past; the tentative formless ancestors which we have long since forgotten. As we stand squarely on this now solid turf, their screams are louder and more coherent. The tiger paces restlessly in the cupboard, and the impossible giraffes with their thick black stripes and gleaming matted hair graze peacefully on the plains of this new Atlantis, whilst the cries of monkeys, birds and aeroplanes fill the space of the previously quiet air.

13

The Future of Life

The screams are still louder now and the twisted mathematical prisoners seek redemption. And you become *my* Lazarus. We are led into the desert to be tempted. And the voices haunt us like the hollow, leaden ring of a church bell. But this metal is quite ugly and we are left with nothing but the laughter of Portuguese sailors. A strange kind of justice you might say. And I remember lying under the piano, silently watching as the hammers struck the vibrating strings. The harp that almost didn't fit through the door and falling asleep on the sofa.

Your wing span is, I estimate, something between five and seven feet. But I confess that I am only guessing. Something like a parrot or a penguin, stretched taut like a piece of canvas. If only we had taken a moment to realize that *this* was what it was about. The timeless palms creating shade for the woman who brings the mangoes. Still quivering one supposes, oblivious to our absence. And Timothy patient as ever, where are you now? Was the bridge ever built? And what of the ferryman's daughter? The dragons finally did make their appearance, but it was too late. We were already imprisoned on that island.

The distillations of the broken lineages appear as the words that we cannot speak. Perhaps we have no place for this strange logic? And we were so confident that things would never change. The bluebells in the wood, the fleeting glimpse of a deer. But we will never be as small as the insects, or like the pivots of a butterfly valve. Remember that tall trees should not bear large fruit on slender branches. And you were serene and majestic, dark-faced so they say. In colour maybe grey and quite defenceless against your enemy, the feral pig. The rustic foreigners looked on with disbelief and there was much merry-making. For it had been a long voyage and they were much in need of a treat. The clumsy pelicans with their orange beaks search for silver fish, which they are convinced might occasionally fly. And we walk hand in hand, the ocean and the salt sea spray, with its memories of walking barefoot on the baking hot paving stones, the rows of coloured light bulbs, passion fruit ice lollies and cracking open peanuts on the beach. The Cape baboon that ran off with the camera lens and the smell of the sunshine on your car seat. Playing with plastic animals held somehow behind green fences, and vanilla milkshakes which we drank with red and white paper straws.

The racing car is bright and shiny red. And now we are hurtling along narrow country roads. The roof is down and we are organically connected with the ground. Each movement tense, like the horses by the motorway. And we are talking under the arches. I still remember that conversation, our discussion of the cages. And the ark is still emptying, spilling out its complex load to the song of the mystic's pipes. The shamans have had their day, and the vacuum, where things might have been, spins like a Catherine wheel into the darkness. Faster and faster and faster we go. The signposts have long since disappeared. Still faster, and as we look back we see *Gene Space* somewhere far in the distance. And the spinning tops and giant springs jostle with one another for our attention, like the pirate beads washed up on the shore. The light is quite brilliant and soon we find ourselves on a new and unfamiliar landscape. The crystal mountains tower high above the now frozen road that snakes its careful route through the alpine villages. And the fractal patterns that assist our navigation lead us into the infinite tunnels from which we may never escape. Our memories of the gold mine are not so distant you know. They touch us at every moment, as surely as the pick axe strikes the face. But I am still playing in the rock pools. And in my dream, the people are leaving the villages. Exchanging their colourful robes for linen. And the lost languages fade slowly into obscurity, dancing dispassionately to the now listless beat of their ancestors.

Forward into the *Sea of Holes*. The creatures are getting restless. And we fall once again for what seems like days. Our destination this time is a place far more fearful as we are *absolutely* forbidden from being here. The place where information must be passed over in silence. A place devoid of heroes. And I do not recognize the countries on this atlas, for this continent has no limits. We have said our goodbyes and, like eighteenth-century explorers stumbling across some new and unknown terrain, we approach this place of darkness with trepidation. Our rudder cuts a fine line in this *Forbidden Sea*. And even as we leave the town of stones, our firmly closed eyes are unable to dispel the vision of the strange beast. For this firmament is filled with things unknown. Indeed, we cannot touch them. But we will continue nevertheless towards this, the horizon of your new world.

We have now entered the *Secret Sea* and are aware that our descent is being driven by something far more powerful than gravity. The object, for that is what it is, defines a curvature in space that appears to extend indefinitely. We land gently on its surface and find that it is sticky to the touch and emits a sweet odour reminiscent of liquorice. Our curiosity is such that we are led to explore the surface on foot. And, as our own dimensions change, the true horror strikes us as we realize that we have been traversing an insect's mandible. But this is a sleeping beauty that should never be woken. For it belongs to a fairy tale that cannot be written and which will never be realized. Indeed, we defy you to be born.

The tunnels that lead further underground now offer glimpses of other creatures. And like visitors to the Tower, we view the woman with seven arms and the *Quismachrunchion*. I turn my head in disgust, I would rather not have been here. The *Jagayag* and the *Pomperion* elegant in their apparent serenity. But we cannot help you. You must remain in your own prison. And there are more, so many more. There is hardly space to house them.

The toys are alive and now they are leaving their boxes. The spotless leopard and the five-horned rhino. We run together hand in hand. Opening the cages, as the creatures rain down from the turbulent skies. The mallard flaps its cotton wings and floats almost imperceptibly by. And *Punch and Judy* merge effortlessly with the *Jack-in-the-Box*, who is apparently quite angry to have been overlooked. And as they line up on the hillside, defining and revising the rules for their mortal combat, they hold their banners high. The crazed beetle formations are nothing to the disembodied scorpion. Indeed we know of no shapes which should be like this.

And you embrace these things with open arms. The cowboy greets the Indian. Together we kiss the ether, our saddle is every tomorrow. But we have no space for this conjuring deception, for these tricks, this *hocus-pocus*. I have no wish to fall inside your microscope. You may keep the pieces of your foreign machines. And I feel you drawing us with your disguised magnets to a place that we know all too well, but will never see. And now you are my limbs and, as we return, the voices around us are strange and unintelligible. But I am afraid that I am unable to recognize this language. And there is thunder in the sky as the giant wings beat menacingly. The predator that is the future.

The engine that enables natural processes of life to explore the complex network of interconnected, multidimensional mathematical spaces on which the machinations of life unfold is unlimited heredity. The guiding forces are evolution by natural selection, historical contingency and the deeply-rooted constraints that define the limits of possibility. Each space has its own unique properties, constraints and pattern of interconnections. The limited proto-heredity of the *Geneless World* was the magician's cloak which delivered the keys to unlock the gateway to these infinite mathematical surfaces. Each individual space is as littered with 'wormholes' that lead into and interconnect with multiple other mathematical spaces. These invisible mathematical tunnels enable living things to explore new domains of possibility and to search for novel ways of circumventing the barriers to existence that the material world relentlessly invents.

History suggests that all known forms of life have a minimal level of complexity, and that although there is no inherent drive to do so, the complexity of some lineages tends to increase beyond the minimal requirement necessary for unlimited heredity. Some creatures, on the other hand, such as bacteria, have maintained a relatively constant level of complexity for over three billion years. We should not, however, confuse a lack of morphological sophistication with the informational complexity of the individual historically programmed coding components and non-coding computational control elements from which the genomes of these organisms are constructed. But if some living things have sustained though not greatly augmented the basic nature of their anatomical sophistication, why should others have increased their informational complexity in a more dramatic manner?

Existence is characterized by uncertainty. It is the ability to predict and rationalize the vagaries of an uncertain future that enables living things to perpetuate their informational structures across successive generations. Living things are in this sense machines for representing the informational texture of the environment in which they operate. They are, in fact, a type of imperfect time machine, which, although unable to recreate and revisit any particular event, can nevertheless reach deep into the past and retrieve some of its features.

Any particular habitat may be thought of as being sub-divided into different frequency bands of informational flux. In low-frequency bands, aspects of the environment change only very slowly and the flux of information across the space per unit time is negligible. In high-frequency bands, however, the flux of information is both more prodigious and complex. An example of an extremely low-frequency change is the coming and going of an ice age. A low-to-medium-frequency band may include seasonal changes and an ability to respond to them: examples of this are migration in birds and hibernation in hedgehogs. A higher frequency band may, on the other hand, incorporate such things as fluctuations in the availability of essential nutrients, environmental toxins and parasites, extremes of temperature and pH, and day/night cycles. A still higher band may include the ability to avoid an apple falling from a tree, catch a fly (if you are a frog), escape the jaws of a hungry lion (if you are an impala), remain airborne at night (if you are a bat), or pass a test of your knowledge of the streets of London (if you are a taxi-driver).

The flux of information within any particular environment is generated both by the creatures which inhabit it and by physico-chemical factors. The more complex a living thing, the higher the frequency of the informational flux it generates. The challenge presented to living things is to exploit new opportunities that emerge in previously unexplored environments or different frequency bands within the same micro-environment, and to track change inherent to the frequency band they occupy and which directly effects their survival potential. This in turn involves generating a nested set of hierarchical strategies for information acquisition, representation and perpetuation, each of which is tuned into and sensitive to changes occurring within the appropriate level.

A principle of heredity enables living things to benefit from features of the experience of their ancestors by generating a database of historically acquired information. This can then be used to construct biological computing machines that can pre-empt and thus provide appropriate responses to the types of changes likely to occur within the living space or 'informational niche' they inhabit. It also confers lineages of creatures with the ability to encode any new changes they encounter, and to use aspects of this information to update their representation of the outside world. In this way, living things come to mirror features of the environment that are relevant to their continued survival. It should thus be possible, at least in principle, to infer aspects of the environment in which an organism has evolved by examining its informational structures. If there is competition between different examples of living things for limited space and resources, those whose genetic notation enables them to generate a more accurate representation of the outside world are more likely to survive than those with less proficient representations.

In a clockwork world devoid of change, living things would be able to survive with only a rudimentary representation of the causal texture of their environment. They would, furthermore, have no need for mechanisms that enabled them to update their representations, as there would be no informational flux for them to track. But in the real world the environmental landscapes that creatures inhabit are constantly twisting, stretching and buckling as a consequence of the informational flux that cuts their surface. The greater the flux they are subjected to, the more exaggerated the deformations of the landscape are likely to be and the harder it becomes to sustain an existence on its surface. A good representation, however, will enable a creature to predict the general nature of these deformations, to track incumbent change and to respond to it appropriately.

The different frequency bands of informational flux may be imagined as subdividing each environmental niche into multiple sub-niches. By occupying different, but spatially overlapping, frequency bands, different species are able to occupy the same general environment without having to directly compete with one another for available space and resources. This situation is somewhat analogous to the way in which air traffic controllers enable several different aeroplanes to fly over the same region of space simultaneously without their colliding, simply by allocating them flight paths of different altitudes. Whereas low-technology aeroplanes fly successfully at low altitudes, increasingly complex aeronautical engineering designs and mechanisms are needed to operate machines at higher altitudes. The very first geneless creatures, that possessed only an extremely limited capacity for heredity, would have occupied niches in which the information flux was minimal. Any chance events that allowed an individual to increase the complexity of its representation would have enhanced its capacity to explore environments that were informationally more complex. However, it was only with the evolution of digital genetic notations and the associated unlimited heredity which they confer that existence could be maintained indefinitely in the face of a sustained and non-uniform informational flux.

We may think of niches as having filled in from the bottom up, with low-frequency niches being occupied first. Once all the available space within low-frequency slots had been filled by the simple precursors of modern unicellular organisms, any changes resulting in the construction of representational apparatus able to track higher frequencies or qualitatively different types of change, would have conferred mutant creatures with a competitive advantage. Instead of being limited to exploring the crowded lower frequency niches and living life in the slow lane, mutants could exploit the novel resources and untapped opportunities scattered across the empty plains of higher frequency niches. The sustained minimal complexity of unicellular creatures such as bacteria can be contrasted with the profound complexity of multicellular creatures and taken as reflecting the fact that organisms are only as complex as they need be. The informational asymmetry between a plankton and a porcupine and indeed between all anatomically simple unicellular organisms and more complex multicellular organisms, may be rationalized in a similar way.

Bacteria may consequently have maintained an approximately constant level of anatomical complexity for 3.6 or so billion years, simply because these types of machines operate perfectly adequately in the niches they inhabit. For if something works, in terms of both its general logic, design, and the resources available for its continued operation, there is no reason for it to become more complex. We should, though, be aware that despite the fact that they have remained unicellular, lack the structural sophistication of eukaryotic cells and are consequently subject to all of the constraints associated with this mode of existence, the programs, genomes and protein components of these creatures have experienced a considerable degree of historical processing. Each protein component in a modern bacterium is more precisely optimized, designed and complex than those of its antecedents. Bacteria are also able to benefit from their very short inter-generational times and huge population sizes, that enable their genes to track change far more efficiently than multicellular organisms, which usually have small populations and in extreme cases can have inter-generational times more than a hundred years long. The short inter-generational times of unicellular creatures such as bacteria, however, introduce a constraint that prevents them from obtaining access to the complex anatomical machineries generated by processes of differentiated multicellular development.

Bacteria have, in addition, evolved other genetic strategies for tracking change. These include the production of a finite number of hypermutating individuals that produce genetic variability at a rate which far exceeds that of the general population, and mechanisms for the horizontal transfer of genetic material between one unrelated bacterium and another. Some bacteria also appear to have evolved SOS programs, which encode survival mechanisms that enable them to increase the rate at which they mutate their genes. They do this by inhibiting specific components of their DNA repair machinery. When they find themselves in stressful situations, SOS programs are activated that allow bacteria to 'reinvent' themselves by increasing the rate at which their genomes are mutated. This increases the likelihood of them producing a mutation that will rescue them from their adversity. The use of biochemical mechanisms to alter their DNA sequences in response to environmental conditions enables bacteria to manipulate the usually random processes that introduce mutations into genes.

It is clear, then, that bacteria are able to compensate for their structural simplicity by using a variety of strategies that redeploy and reprogram the relatively limited set of genes they possess. By making imaginative use of their finite genetic means, they are able to live life in the evolutionary fast lane. Some bacteria, however, have not maintained their basic level of genetic complexity, and have instead down-sized their genomes. The gene kit of the internal parasite *Mycoplasma genitalium*, for example, contains a mere 470 genes, as compared with the more substantial kit of *Haemophilus influenza*, that has a total of 1,703. But there is a sense in which these minimalistic bacteria have merely substituted one kind of (genetic) complexity for another type of (parasitic) complexity. This parasitic strategy involves harnessing and tapping into the powerful genetic and extragenetic informational databases of another organism. In so doing, *Mycoplasma genitalium* gains access to many of the benefits of complex multicellular existence without having to carry any of the information from this impossibly large database within its own small genome.

The fact that it took over three billion years for creatures confined to exploring the mathematical plains which specify unicellular forms of life to discover an entrance into the mathematical space specifying differentiated multicellular creatures, indicates that this transition was by no means trivial to accomplish. However, once the key that unlocked the door into the mathematical *Space of All Possible Differentiated Multicellular Creatures* had been discovered, it was to take a mere 600 million or so years of historical perambulations to retrieve creatures with the potential to discard the genetic legacy they had inherited and to redesign themselves artificially from first principles. The discovery of such complex creatures was by no means inevitable. Indeed, a key to this particular mathematical surface may never have been found. The probability of finding such a key depends both on the nature and frequency of their distribution within the mathematical expanses of the *Space of All Possible Unicellular Creatures*. Nevertheless, the discovery of the *Space of All Possible Differentiated Multicellular Creatures* and the mathematical surface that we may call the *Space of All Possible Developmental Mechanisms*, to which it is interconnected, enabled life to discover a host of new technological innovations at a number of different scales. These included eyes, ears, antennae, arms, legs, leaves, wings, teeth, fur, horns, fins, gills, livers, kidneys, lungs, hearts and brains. This treasure chest of structures and functions enabled life to explore the world of complex form, and in so doing enabled its protean manifestations to track change of a nature and frequency that had hitherto been impossible.

The five great evolutionary transitions that underpinned and paved the way for the historical discovery of the diverse and complex forms of life with which we are currently familiar are: (1) the transition from an analog non-encoded *Geneless World* of limited heredity into a *Gene World* inhabited by populations of ancestral self-replicating unicellular creatures, whose informational structures were encoded using digital DNA genetic technology that conferred them with a capacity for unlimited heredity; (2) the evolution from the universal unicellular ancestor of all existing life of the three different types of cells, *Bacteria*, *Eukarya* (the complex cells whose genetic material is partitioned into a discreet nucleus and from which all modern animals and plants are constructed) and *Archaea* (that are somewhat intermediate between *Bacteria* and *Eukarya* and are often found to inhabit extreme, high-temperature environments). These define the three different genealogical domains of life from which all creatures on Earth have arisen; (3) the evolution of multicellular animals (Metozoa) and plants, that is the discovery of developmental programs for patterning cells in three-dimensional space and mechanisms for differentially regulating gene expression in time and space; (4) the evolution of brains (and the capacity for learning, complex behaviour, representation, awareness and, in the case of man, consciousness); and (5) the evolution of culture, that, in its most complex manifestations, includes the taming of fire, and the development of language, agriculture, written notation, societies, art and science, but which also includes all the simpler examples of extragenetic information acquisition, storage and transmission that are utilized by a small number of species.

The first great transition from an analog, non-encoded *Geneless World* of proto-heredity into a digital, encoded world of unlimited heredity, populated by the universal unicellular ancestor that was the genealogical precursor of *Bacteria*, *Eukarya*, *Archaea* and most likely an accompanying collection of dead end beasts that contributed little if any information to the lineages of modern life, itself involved a long string of contingent complexification events. The order and manner in which these major transitions occurred and the mechanisms that underpinned them are, however, by no means clear. This is, amongst other things, because our genealogical tree cannot as yet be traced beyond the types of creatures represented by modern day *Bacteria*, *Eukarya* and *Archaea*. Perhaps, when the complete genomic sequences of all or most of the known unicellular creatures have been determined, it will be possible to infer and reconstruct some of the more ancient events in a historically meaningful manner. But it is doubtful that it will ever be possible to trace the precise chronology of life beyond the point at which DNA was first employed as the genetic material industry standard. Despite these uncertainties, it is briefly worth surveying some of the most significant types of informational changes that these major evolutionary transitions which define the first great transition, may have entailed.

One of the first major transitions by which the first great transition was defined is likely to have involved the discovery of a mathematical space that we may call the *Space of All Possible Compartmentalized Geneless Metabolisms*. As mentioned earlier, once compartmentalization had been discovered, any advantages that were idiosyncratic to a particular geneless metabolism could be differentially selected from alternative metabolisms which lacked these characteristics. The components of metabolisms could also become more concentrated. Furthermore, once it had been partitioned off from the outside world, the composition of the chemical reaction space enclosed by the compartment could be regulated so as to make it more favourable for the chemical reactions it accommodated.

The next set of major transitions probably involved technological changes to the catalytic components from which the metabolisms were constructed. We have examined the possibility that the first metabolisms were composed either from peptide components, RNA components, or a mixture of both. Indeed each type of possible metabolism might have existed independently, or co-existed with one or several different types of metabolisms within the same micro-environment. But one can also envisage several other chemically plausible scenarios in which RNA and peptide technologies constituted a later addition to an already flourishing community of quasi-living metabolisms that used a different set of chemical components to represent their proto-genetic information. Whatever the precise nature, method and order in which different chemical technologies were employed by proto-living creatures, it is clear that each technological takeover or innovation event involved movement from one type of interconnected mathematical space to another.

Perhaps the most profound transition of all, however, involved the discovery of the wormhole that led from the analog mathematical plains of the *Geneless World* into the digital expanses of the *Proto-Gene World* and the proto-genetic specifications housed within it. This major transition no doubt occurred incrementally, and one can imagine proto-genetic systems in which only a single component of the network was digitally encoded, whilst the majority remained analog and non-encoded. The discovery of a rudimentary proto-genetic principle nevertheless laid the foundations for the development of more complex genetic systems able to sustain unlimited heredity. The origins of such partially-encoded protein metabolisms depended on the specific chemical recognition that can occur between nucleic acids and amino acids. We can imagine, then, that the first proto-genes were molecules of RNA that either self-replicated autocatalytically, or were replicated with a variable degree of fidelity by encoded or non-encoded molecules within the metabolism, and were able to instruct the synthesis of low complexity amino acid chains according to the information encoded within their ribonucleotide base sequences.

The evolution of genetic encoding principles could not, however, occur without the co-evolution of a corresponding decoding principle and appropriate translation machinery. Because of its inherent complexity, it seems reasonable to speculate that the near-universal genetic code used by all contemporary forms of life was preceded by several different and far simpler proto-codes. It has been suggested that the first genetic proto-code utilized only two RNA bases (**A** and **U**) which formed eight possible triplet codons (**AAA**, **UUU**, **AAU**, **AUU**, **UAA**, **UUA**, **UAU** and **AUA**). A simple coding system of this sort would have been able to specify a maximum of seven different amino acids and a single STOP signal. The fact that many different types of protein folds can be generated using very restricted sets of amino acids suggests that this hypothesis is plausible and that the very first genetic systems may well have encoded proteins of only a relatively minimal amino acid sequence complexity. Such simple RNA-based genetic systems are imagined to have eventually increased their complexity by the sequential introduction of the **G** and **C** ribonucleotide bases utilized by contemporary RNA molecules. The use of only one additional base would have provided a coding capacity sufficient to encrypt all twenty of the naturally occurring amino acids utilized in modern proteins.

The first proto-genes were unlikely to have been collected together into a genome and probably existed as free-standing molecules. The universal unicellular ancestor might, however, have had a proto-genome that included multiple free-standing proto-genomic fragments of RNA and/or DNA that contained a small number of contiguous genes, but which were nevertheless smaller and less organized than the genomes of modern unicellular organisms. This major transition thus involved the discovery of the wormhole that led into *RNA Proto-Genome Space* and from there into *RNA Genome Space* and then *DNA Genome Space*. By grouping genes together into genomes, successful combinations of genes could be linked together and maintained across time.

One of the most important and enigmatic changes underpinning the first great transition involved the hypothetical genetic takeover event that enabled proto-genes constructed from single-stranded RNA molecules to switch technologies and be refashioned from double-stranded DNA. It is unclear whether this, and the accompanying major transition, that involved the evolution of a distinction between the genetic material and messenger molecules that communicated the information of genes to the protein synthesis machinery, would have occurred before or after the discovery of RNA proto-genomes or genomes. Nevertheless, once life had discovered the gateway that led out from the darkness of the *RNA World* into the previously unknown *DNA Gene Space* world and the mathematical structures of the double-stranded DNA genes and universal genetic code that it contained, the stage was set for the quantum boost in informational complexity that was to create the possibility for all of the other great transitions. There is not as yet any clear model of how such an RNA to DNA genetic takeover event might have occurred. However, it seems likely that during this period life passed through a transitionary *RNA/DNA Gene Space* world in which some proto-genes were upgraded to DNA technology, whilst others continued to be synthesized from the informationally inferior and by that time largely out-dated RNA technology. The evolution of genes constructed from DNA and double-stranded DNA genomes provided the structural basis for the highly advanced genetic replication, error-correcting and proofreading mechanisms that played a crucial role in securing the foundations for informationally more complex processes such as development, learning and the extra-genetic mechanisms of culture. These new high-tech, DNA-based genetic machines would eventually enable the replication and expression of genetic information to reach previously unobtainable levels of fidelity and complexity. In so doing they would change the face of life forever.

The universal unicellular ancestor and mother of all modern cells was the principal ambassador for life as it cruised towards the second great transition. Although, as a consequence of the major transitions endured during the first great transition, it was significantly more technologically advanced than its precursors, it was nevertheless technologically and thus informationally far inferior to the *Bacteria*, *Eukarya* and *Archaea* kingdoms of life that it would eventually give rise to. Had things been slightly different, this universal ancestor may have suffered extinction, in which case one of the other forms of proto-life that had doubtless persisted up until that time could have carried the torch of life into the future in some quite unique and unpredictable manner. Had this happened, it is most unlikely that life's impetuous dance of mathematical exploration would have led it onto the mathematical planes of the *Space Of All Possible Humans*. Indeed even a minor inclemency might have changed the balance of events, perhaps generating a fourth domain of life, or omitting the complex *Eukarya* lineage of cells from which all known animals and plants have been derived.

So what were the major transitions that underpinned the second great evolutionary transition and how did the three fundamentally different forms in which modern life is packaged arise from the universal unicellular ancestor? Unlike *Bacteria* and *Archaea*, which are small and have a simple anatomical structure and internal organization, *Eukarya* have a complex structural anatomy and organization. The hallmark of eukaryotic cells is the way in which their genetic material is packaged into a set of compact, highly structured chromosomes that are enclosed within a discreet centralized compartment, known as the nucleus. This contrasts with the rudimentary organization of *Bacteria* and *Archaea*, whose genetic archives consist of only a single, relatively unstructured circular chromosome that lacks a distinct nuclear compartment and is in direct contact with the rest of the cell. Furthermore, in addition to being significantly larger than the cells of *Bacteria* and *Archaea* – the volume of a eukaryotic cell being typically around 10,000 times greater than that of an average *Bacterial* or *Archaean* cell – the cytoplasm of eukaryotic cells is also far more complex. Unlike *Bacteria* and *Archaea*, the cytoplasm of *Eukarya* is partitioned into an elaborate series of membranes, sub-cellular compartments and organelles which include: (1) the endoplasmic reticulum (which plays a critical role in the production and export of newly synthesized proteins), (2) mitochondria (which supply energy to the cell in the form of ATP), (3) peroxisomes (which perform a variety of metabolic functions), (4) the golgi apparatus (which functions to chemically modify newly synthesized proteins by, for example, attaching sugars to them), and, in the case of algae and plants, (5) plastids (which are the site of photosynthesis).

In addition to this morphological complexity, eukaryotic cells are characterized by considerable behavioural complexity that is not seen in *Bacteria* and *Archaea*. One hugely important example of this is the existence of proteins that control the rate and timing of cell division and that are consequently able to define a cell cycle. The cell cycle is divided into four stages which include: a resting phase, a period of DNA synthesis, an intermediate period and a phase in which the newly synthesized genetic material is separated from its precursors and the cell divides in two. The cytoplasm of eukaryotic cells also contains a cytoskeleton, that ramifies throughout the cell and provides a scaffolding that helps structure the cell and, with the assistance of molecular motors, enables it to change its shape and reposition its contents. The emergence of skeletal elements within eukaryotic cells enabled the evolution of both motility and gross movements of the cell membrane which allowed cells to engulf materials from the outside world (endocytosis) and to secrete products that they had synthesized (exocytosis).

The presence of microfossils resembling modern bacteria, and the putative fossilized remains of strange mat-forming micro-organisms called stromatolites in rocks that are up to 3.5 billion years old, suggests that the universal common ancestor was well entrenched at a time which followed the origin of solid rocks on the Earth by only 0.9 billion years. The first fossilized eukaryotic cells only appear in the geological record around two billion years ago. It thus appears to have taken around 1.5 billion or so years for evolution to accomplish the second great transition. The relative simplicity of *Bacteria* and *Archaea* suggests that they are likely to have been the direct precursors of the *Eukarya*. However, kinship relations between the three domains of life, derived from molecular genetic considerations, suggest that the genealogical history of early unicellular life may not be as simple as initially supposed.

On the 23rd of April 1996, the genome sequence of the Archaean organism *Methanococcus jannaschii* was published in its entirety. The completion of this sequence allowed for a direct comparison between genomes that stood as ambassadors for each of life's three major domains. With the genome sequences of the representative Bacteria *Haemophilus influenzae* and Eukarya *Saccharomyces cerevisiae* (baker's yeast) in hand, the full inventory of genes from the three domains could be contrasted. Although the sequences of the 'operational' genes of *Archaea* (which include those involved with metabolism, cell division and energy production) had similarities to their bacterial counterparts, suggesting that *Bacteria* and *Archaea* share a genealogical connectivity that does not extend to the *Eukarya*, when 'informational' systems (such as those involved with transcription, translation and DNA replication) were compared, it turned out that, *Archaea* shared genes with *Eukarya* that had nothing in common with *Bacteria*. This paradoxical finding demonstrates that, although anatomically similar to *Bacteria*, *Archaea* are at the core molecular informational level far more like *Eukarya*. This suggests that *Archaea* and *Eukarya* last shared a common ancestor more recently with one another than either of them do with *Bacteria*.

Sequence analysis of the small complement of extra-nuclear genes found within the mitochondria of eukaryotic cells provides evidence for an endosymbiotic hypothesis of eukaryotic origins. Mitochondrial genes are in fact clearly related to bacterial genes, suggesting that at some point in history a primitive bacterial symbiont was engulfed by an ancestral proto-eukaryotic cell and subsequently took up residence inside its pre-mitochondrial host. Over time, some of the bacterial genes are thought to have been transferred to the proto-eukaryotic host genome, while others were expunged. The tiny mitochondrial genome of around five or more genes that remains inside all eukaryotic cells may thus be viewed as a highly compressed and edited bacterial genome; the informational carrion of a once fully functional bacterial genome. It is possible, however, as William Martin and Miklós Müller have suggested, that eukaryotic cells arose as a result of the symbiotic association between a bacterial symbiont and an Archaean host, rather than from an ancestral proto-eukaryotic host.

Reconstructed genealogical lineages based on select groups of DNA sequences and protein molecules, however, have a tendency to yield distinctly different family trees, depending on which genes and proteins are selected to construct the molecular chronometer and the particular species and geological events used to calibrate it. The incompleteness of the sequence database does not help matters. However, as larger numbers of representative *Bacteria*, *Archaea* and *Eukarya* genome sequences become available, the resolution and reliability of the kinship relationships between the three domains of life should increase considerably. The fact that the greater majority of living species are unicellular, and that the majority have not yet been characterized, suggests that these teeming microscopic 'rainforests' of unicellular diversity may even throw up a new domain of life, which, along with the sequence information derived from the other genome sequences, would help root the tree of life more firmly in historical fact.

The transition from the mathematical space containing the universal unicellular ancestor to the mathematical *Space of All Possible Eukarya*, paved the way for the discovery of the wormhole that led into the mathematical *Space of All Possible Differentiated Multicellular Creatures* and defined the beginning of the third great transition. The several critical structural features of eukaryotic cells which provided the basis for the cellular complexity that enabled the evolution of differentiated multicellularity and the accompanying panoply of complex plant and animal forms encompassing dinosaurs, beetles, butterflies and beech trees, included, amongst others: the evolution of the nucleus, golgi apparatus, endoplasmic reticulum, microtubules and modern chromosomes organized into periodic structures known as nucleosomes. These consist of a stretch of around 200 base pairs of chromosomal DNA, bound, packaged and organized by an octamer sheath consisting of two copies of four different histone proteins.

These major microstructural innovations were complemented by a host of nanostructural innovations which included the discovery of new functional programs and molecules that regulated pre-existing programs in novel ways. Important examples of these included: the evolution of intrinsic apoptotic 'self-destruction' programs, the evolution of new families of genes, such as the homeotic transcriptional regulators which play a central role in the generation of three-dimensional form, the evolution of complex regulatory programs within the sequences of mRNA transcripts that influence their stability, editing and in some cases the rules by which they are decoded; and the evolution of molecular mechanisms for the reversible modification of gene activity and protein structure. These included the process of genomic imprinting in which genes can be 'shut down' either by acquiring an imprint of methylation on selected cytosine residues, or by the acetylation of histone proteins within the modular nucleosome elements from which higher level chromatin structures self-assemble.

In addition to innovations involving the coding regions of genes, the origin of development was also likely to have been dependent on the evolution and reshuffling of highly programmed and increasingly 'smart' non-coding modular genetic components. The non-coding regulatory module of the *Endo 16* gene of the sea urchin *Strongylocentrotus purpuratus*, for example, functions as a miniature analog microprocessor device that is sensitive to the presence of up to thirty different protein 'data bits'. Each of these bind to specific control sites and the overall pattern of binding is integrated to produce an appropriate ON/OFF, or modulation of gene expression response. The modular nature of these types of non-coding regulatory units implies that they could easily be duplicated and occasionally mutated, reshuffled and relocated so as to influence the activity of previously 'dumb' genes. A gene that was in this sense relatively computationally 'naive' could thus, in a stroke, be supplied with a device that made it sensitive to a plethora of environmental cues of which it had previously been unaware. This information could then be integrated by the genetic micro-processor and by influencing the timing and extent to which pre-existing programs were activated, could help compute the developmental fate of individual cells. Other examples of non-coding regulatory elements are the control sequences that influence the way in which the modular components of some genes are differentially spliced to generate a host of different mRNA molecules and their corresponding protein products.

The evolution of the nucleus allowed gene transcription and translation to become uncoupled and provided a basis for the evolution of complex programs of differential gene expression. It also enabled the symmetry of cells to be broken by allowing mRNA molecules to be selectively concentrated within discreet regions of the cytoplasm. The evolution of protein modification organelles such as the golgi apparatus, on the other hand, allowed a new dimension of technological complexity to be superimposed onto protein structure and dynamics. This involved the discovery of a wormhole that connected the mathematical *Space of All Possible Proteins* to the mathematical *Space of All Possible Glycoproteins*. The addition of sugar molecules to newly synthesized proteins can, amongst other things, alter their function, stability and ability to fold correctly and over biologically realistic timescales.

The evolution of the protein-sorting mechanisms of the endoplasmic reticulum enabled newly synthesized proteins to be directed to specific addresses within the cell and thus provided a finely tuned mechanism for the micro-patterning of protein components at the sub-cellular level. It also provided a pathway for the secretion of proteins, which could, amongst other things, form the basis of morphogenetic (form-giving) gradients. These help define the chemical coordinates that allocate a spatial address to some developing cells and in so doing enable their genomes to be differentially programmed.

The cytoskeletal structures of eukaryotic cells and their ability to migrate during the process of developmental patterning rely on the existence of microtubules. Microtubules also perform many of the other critical functions essential for generating the complexity of multicellular creatures. These include assisting with the process of messenger mRNA polarization within egg cells, which, as we have already seen, provides a basis for the intrinsic asymmetry of the development process in some organisms. By mediating the generation of chemical inhomogeneities in the form of morphogenetic gradients, microtubules help establish the fields of chemical instructions that program the experientially naive genomes of the developing cells that encounter them.

The evolution of chromosomes that are highly organized into nucleosomes and their associated higher order chromatin structures allowed larger volumes of DNA to be packed into a smaller volume of space. They also provided a mechanism for protecting DNA from chemical damage and facilitated the development of the complex and powerful mechanisms of transcriptional regulation that enable eukaryotic cells to generate an enormous repertoire of tightly regulated patterns of differential gene expression. The evolution of the cell cycle allowed differential cell growth and asymmetric cell division to occur. It also provided the opportunity for the success and fidelity of DNA replication events to be closely monitored at a series of checkpoints and thus for damage and replication errors to be efficiently repaired.

The third great transition which resulted in the evolution of multicellular creatures composed from non-identical cellular constituents that are patterned in space to generate complex three-dimensional architectures was itself defined and dependent on several major transitions. These included: the evolution of mechanisms for forming symmetrical collections of interconnected cells, the evolution of mechanisms for breaking this symmetry and generating the asymmetry that 'kick-starts' and in some cases sustains developmental processes, the evolution of mechanisms for generating the intercellular chemical gradients that encode developmental information and the evolution of robust mechanisms that enable individual cells to read, interpret and translate the positional information contained within these three-dimensional chemical gradients into differential patterns of gene expression. These, in turn, translate at the microscopic level into the differentiated cellular architectures and functions that define the core modular units from which multicellular creatures are assembled, and at the macroscopic level into their three-dimensional form and higher order functions.

As with all evolutionary innovations, each major transition event was itself underpinned by the discovery of both new genes and new patterns of gene expression. The protein products of these additions to the core unicellular eukaryotic genetic tool kit were able to tune and modify the self-assembly and self-organizational properties of pre-existing proteins, and the networks and complexes in which they participated. They were also able to provide new structures whose intrinsic self-assembling properties could themselves be organized, patterned and modified to generate new processes and functions. The redeployment of the products of old genes to new spatial locations and time windows of expression also made a significant contribution to the programs and molecular patterns that underwrote the process of developmental complexification. The completion on 11th December 1998 of the 97 million base pair genome of the worm *Caenorhabditis elegans*, which is the first animal genome to be fully sequenced, is helping to shed some light on the ways in which genomes have changed and expanded to adapt to the challenges of multicellular life.

The first step in the evolution of developmental complexification is likely to have been the evolution of undifferentiated or poorly differentiated multicellularity. This entailed the evolution of mechanisms which enabled the daughter cells produced by cell division to adhere to one another. If a band of interconnected cells had a selective advantage over their isolated companions, then evolution would have favoured the perpetuation of the multicellular state. An insight into both the manner and ease with which such a process might occur can be obtained from the soil-dwelling amoeba *Dictyostelium*. When food is in short supply, individual amoebae band together to form a multicellular slug. This then migrates to greener pastures, where it disperses by building a stalk that releases spores. Bin Wang and Adam Kuspa have demonstrated that this whole series of events is mediated by a single protein known as *PKA*. This indicates how the evolution of a single new protein may have been sufficient for the emergence of multicellularity. But development requires more than just multicellularity. A mechanism for breaking the symmetry of a band of cells is also needed, so that cells can be differentially programmed and made non-equivalent.

The breaking of the symmetry of a single cell as it divides into two daughter cells may occur as a result of the intervention of either extrinsic factors relating to the interactions of the daughter cells with one another and the outside world, or intrinsic factors originating from within the cell itself. The end result of an asymmetric process of cell division is two daughter cells that have different developmental potentials and are able to acquire different fates. During the development of the peripheral nervous system of the fly, for example, a single sensory organ precursor (SOP) cell undergoes two rounds of asymmetric cell division to produce four sensory bristle cells, two outer hair and socket cells and two inner neuron and sheath cells. In a mutant fly that lacks a protein called *Numb*, however, the SOP divides symmetrically. The amount of *Numb* in SOP daughter cells thus appears to determine their fate.

The evolution of developmental complexification may consequently have begun with a major transition event that involved the discovery of a wormhole that led from the mathematical *Space of All Possible Symmetrical Cell Divisions* into the *Space of All Possible Asymmetrical Cell Divisions*. But this is unlikely to have occurred directly. Indeed Lewis Wolpert has argued that intrinsic mechanisms for generating asymmetry were preceded by far simpler, environmentally generated extrinsic mechanisms. The contact of a cell with a surface might, for example, have induced a localized change in the cell membrane that influenced the symmetry of cell division. But how might this have occurred? When microtubules assemble they polymerize out in largely random directions. But this type of self-organizing exploratory behaviour has the potential to be trapped and stabilized into an asymmetrical distribution. The changes induced focally at the membrane of a cell when it makes contact with a surface may thus have provided exactly the type of signal necessary for asymmetrical polymerization to occur. Once established, an asymmetrical distribution of microtubules could have provided the structural basis of a mechanism for defining an axis within cells that had previously been symmetrical.

There are several different types of intrinsic mechanisms by which the symmetry of cell division can be broken. All of these involve processes which asymmetrically distribute key regulatory proteins such as *Numb* that function either as transcription factors which modify the expression of key genes, or as agents that influence cell-to-cell interactions between daughters. This asymmetry is invariably cell cycle related and, when the cell is not dividing, the distribution of these proteins is usually symmetrical.

Asymmetry can be achieved by processes that selectively transport or degrade the master regulatory protein itself by asymmetrically localizing the mRNA encoding the protein, or by a combination of both. Modern eukaryotic cells employ a complex set of protein machines to localize the intrinsic determinants of asymmetric cell division. The asymmetric localization of some proteins and mRNAs is, for example, dependent on microfilament or microtubular components of the cytoskeleton. This suggests that they may be bound and transported by a carrier molecule that moves along the microfilament or microtubule like a cable car. In the case of asymmetrically distributed mRNA molecules such as *Bicoid*, part of the sequence itself marks it for transportation so that it can become sharply localized within the cell. Lower-tech methods for inducing the asymmetric distribution of proteins may, however, involve a simple reduction in the local dimensionality of the system, by, for example, using anchors to trap the proteins onto the surfaces of membranes so that their diffusion is restricted to two dimensions instead of three.

Although some multicellular creatures, such as worms, rely extensively on sequential rounds of intrinsically controlled programs of asymmetric cell division to determine their major developmental decisions, others, such as vertebrates and insects, do not. Instead, once the initial symmetry of the parent cell has been broken, the extrinsic signals encoded within the complex three-dimensional chemical gradients that surround them function as the principal motors that pattern their development. The exact position of a cell within these gradients specifies both the manner in which it is programmed and its destiny. Each cell reads the local information in the gradient and translates it via the activation or inhibition of an array of transcription factors into a differential pattern of gene expression. The concentration gradient of a single morphogen signal is thus able to subdivide sheets of cells into zones of digital ON/OFF states, depending on whether or not the concentration threshold for a response has been exceeded. If more than one morphogen contributes to the information gradient, then the morphogenic code may be combinatorial, in which case the response of a given cell may depend on the relative concentration values of both morphogens. All of these genetically controlled morphogenetic processes, however, rely ultimately on both the fundamental self-assembling properties of the protein components and on the self-organizational properties of the higher order structures that they directly or indirectly generate.

The bestiary of multicellular plants and animals that have graced the surface of the Earth since the origin of the ancestral multicellular creature from which all differentiated multicellular life has presumably arisen, have, as discussed earlier, been clustered, on the basis of shared fundamental characteristics of form, into three broad kingdoms known as: *Animalia*, *Plantae* and *Fungi*. Each of these may, in turn, be more finely subdivided on the basis of shared anatomical features into groups known as phyla. The phylum *Chordata*, for example, that includes vertebrates, contains animal species which have a spinal cord. The invertebrate phylum *Arthropoda*, on the other hand, which includes insects, spiders, crabs and the now extinct trilobites, contains animal species with segmented bodies, jointed limbs and which lack a spinal cord. But what is the genetic basis for these different morphologies and how did the fundamental body plans that define all existing and extinct phyla originate? Do different innovations of form, such as the evolution of limbs or the chassis of a dragonfly, involve the utilization of distinctive and non-overlapping sets of genetic components and processes? How is historically encoded patterning information represented and delivered to the developing organism and how, furthermore, does a collection of undifferentiated cells differentiate into a baboon rather than a python?

One of the most extraordinary findings of modern genetics is that almost all animals share an ancient family of around thirty-one to thirty-nine genes known as homeotic or *Hox* genes, that are critically important in establishing body patterns. There is thus a fundamental unity of process underlying the development of every animal on Earth and the same basic set of genes can apparently be used repeatedly to generate an immense diversity of forms. *Hox* genes are able to translate linear DNA sequences into the three-dimensional architecture of complex living things and have enabled life to explore the mathematical expanses of what we might call *MorphoSpace*.

Hox genes are expressed in discreet regions of the body along the head-to-tail (anterior-posterior) axis, and encode transcription factors with a conserved DNA-binding motif, known as the Homeobox, that controls the expression of a large and diverse collection of target genes involved in development. The fly *Hox* gene *Ultrabithorax*, for example, has been estimated to regulate the expression of between 85 and 170 genes, each of which contains a Homeobox binding site. Some of these genes are themselves transcription factors which initiate a host of downstream events and amplify the effects of *Hox* genes. The genes controlled by *Hox* proteins initiate the cellular processes that define morphogenesis. Many of these appear to influence the rate of cellular proliferation, the initiation of programmed cell death and the processes that generate morphogen gradients. The study of *Hox* genes has demonstrated that the conundrum of the generation of complex form is a bit like that of a ship a the bottle in the sense that, although the end product appears quite impossible, it can in fact to be achieved by the implementation of relatively simple engineering principles. All insects, for example, from a centipede to a butterfly and a beetle to a fly, appear to have evolved from an ancestral body plan constructed by the same set of *Hox* genes. So if creatures as different as a fly and a ladybird share the same set of *Hox* genes, how is it that they have such different forms?

The evolutionary modification of the ancestral body plans of the thirty-five or so modern phyla could have been achieved in several different ways. The number of *Hox* genes of a given type could have been increased by gene duplication, or decreased by selective deletion; new types of *Hox* genes could have been introduced into developmental circuits by *Hox* gene duplication coupled with functional divergence; the spatial location, timing or level of *Hox* gene expression could have been differentially modified; and the logic and manner in which *Hox* proteins influence their targets could have been modified. In fact all of these mechanisms appear to have been utilized in some form or other.

The evolutionary importance of changes in the level and regional localization of *Hox* gene expression, for example, has been demonstrated by studies in which *Hox* genes have been over-expressed, or targeted to new spatial locations. Robert Pollock and Charles

Bieberich have demonstrated that artificially induced alterations in the spatial location and amount of *Hoxb-8* and *Hoxc-8* gene expression in mice can 'atavistically' run the clock of evolution backwards and transform modern skeletal structures into earlier evolutionary forms which have not been seen in vertebrates for millions of years. This illustrates how the level and spatial location of *Hox* gene expression is able to generate a complex combinatorial code with the potential to translate into a multiplicity of different developmental fates. It also demonstrates how evolution has, at least in some cases, generated modern architectures by laying additional regulatory layers of control onto ancient morphogenetic programs: economically adjusting and adding to what already exists, rather than redesigning new functions from first principles.

We have seen that in the case of insects, new animals can apparently be created without a need for the evolution of a new *Hox* gene tool kit. This is because *Hox* genes do not directly instruct the formation of anatomical structures such as claws, wings, legs or antennae, but rather modify the activity of the genetic programs involved in their formation. The potential for generating these appendages appears to be present within every cell in a spatially defined region. *Hox* genes may consequently be imagined as selecting one particular outcome (for example the development of a leg) from a repertoire of possible outcomes that might, for example, include the development of antennae. They constitute an apparently universal regulatory tool kit of master 'designer' genes that can, in conjunction with a host of effecter circuits and ancillary control genes, be used to divert and redirect developmental programs in a host of different ways to construct a vast array of different tissue patterns.

The diversity of insect species within the *Arthropoda* phylum appears to have been generated principally from the evolution of new regulatory interactions between *Hox* proteins and the cluster of morphogenetic programs they regulate. But what about the diversity of form between phyla? Are the vertebrate members of the phylum *Chordata* such as giraffes, chimpanzees and humans constructed by the same set of *Hox* genes as members of the *Arthropoda* phylum such as insects, lobsters and spiders, or did the evolution of vertebrates require the agency of a new and highly specialized set of *Hox* genes?

The fact that the cluster of *Hox* genes present in *Arthropoda* appears to have been duplicated several times within different members of the *Chordata* phylum suggests that the generation of major innovations in body plan requires the evolution of new *Hox* genes. Interestingly, however, as is the case in insects, the morphological diversity within vertebrates appears to have been generated by essentially the same set of *Hox* genes. One might conclude then, that major differences in body plan, such as the differences between *Arthropoda* and *Chordata*, are principally achieved by increasing the number of *Hox* genes, whereas relatively minor variations upon an anatomical theme, such as the anatomical differences seen between modern man, neanderthal man, elephants, meerkats and hyenas, may be achieved by reprogramming the regulatory circuits of a fixed complement of *Hox* genes. The ability of a relatively discreet set of differentially regulated *Hox* genes to generate forms as dissimilar as those of a rhinoceros and a kangaroo appears to stem from: the multiple levels of hierarchically organized control circuits in which they play a central orchestrating role, the combinatorial manner in which different *Hox* gene regulatory circuits can be deployed at different levels of expression both in body plan time and space, and the fact that all of the *Hox* regulatory systems appear to have a modular organization, such that one *Hox* gene subsystem can be selectively modified or redeployed without affecting others.

But things are not quite this simple. Vertebrate species do in fact vary quite considerably in their *Hox* gene complements. Mice, for example, have thirty-nine *Hox* genes, whereas pufferfish *Fugu* have only thirty-one. It appears, then, that vertebrates do not utilize a universal *Hox* gene kit for the construction of their novelties of form and that some species may have achieved particular morphological ends by adding or deleting individual *Hox* genes to or from their genomes.

Around 530 million or so years ago, in the big bang of animal evolution known as the Cambrian Explosion, effectively all of the major animal phyla (including a tiny worm-like creature called *Pikaia* and another *Yunnanozoon lividum* the first known chordates and one of which may have been the ancestor from which we have descended) were created in over a geologically sudden period of creative frenzy that lasted only six to ten million years. It may not, furthermore, have been exclusively dependent on changes in *Hox* gene numbers. Indeed William Jeffery and Billie J. Swalla have shown that a single gene called *Manx*, which encodes a transcription factor that does not contain a Homeobox domain, may be responsible for switching on the cascade of genes that generate the chordate-like body plans and characteristics of larval tunicates, which include a tail and rudimentary spinal cord. A single gene, that is apparently unrelated to the *Hox* family of transcription factors, may thus have played a critical role in the evolution of the *Chordata* phylum.

Many other genes are involved in developmental patterning at various levels of hierarchically organized control systems. The *lefty* and *nodal* genes, for example, appear to be involved in determining left/right asymmetry, whereas the *T-box* family of transcriptional activators are, amongst other things, involved in heart and limb morphogenesis. The non-homeotic *eyeless* transcriptional factor gene, on the other hand, appears to be one of the master controllers of eye development by means of recognizing regulatory elements on the set of genes involved in eye construction. Georg Halder, Patrick Callaerts and Walter Gehring have shown that if *eyeless* is artificially expressed in parts of flies where it wouldn't normally be active, they may grow as many as fourteen extra eyes in the targeted region, for example their wings, legs or other tissues. The ship in the bottle once again. That is to say, a single *eyeless* gene is able to take a tissue that would normally make a wing or an antenna and turn it into a highly complex and therefore *a priori* improbable structure like an eye. But even a gene as powerful as *eyeless* appears to be nested hierarchically beneath, and therefore under the ultimate control of an apparently omnipotent *Toy* gene, which contains a Homeobox domain and is consequently a member of the *Hox* gene family of transcription factors. It may well turn out that *Hox* genes sit at the apex of all of the multiple-layered hierarchies of control processes that influence the construction of animal body plans.

However, despite the extraordinary power of these types of master developmental control genes, it is unlikely that evolution has routinely proceeded through *MorphoSpace* by means of quantal *Hox*-gene-mediated leaps. This is largely because a change to a master control gene would produce so many downstream effects that the resulting body plan is unlikely to constitute a plausible candidate for successful existence. It nevertheless remains the case that all of the thirty-five or so anatomical ground plans of modern animals appear to have been formed within a breathtaking geological moment lasting only six to ten million years. It may consequently have been the case that at the dawn of animal form, when life discovered the route that led into the *Hox Gene Space* subregion of *Gene Space*, evolution was able to experiment relatively freely with different body forms. However, once a set of thirty-five body plans had been selected from this playful repertoire of creativity as the result of systematic decimation by historically contingent events and natural selection, and life locked in on these designs, major constraints set in and precluded the possibility of redesign or fundamental alteration to the basic logic of body plans.

Indeed no new phyla have arisen since the onset of the Cambrian Explosion. Working within the framework of the deeply ingrained constraints of the past, all subsequent evolutionary innovations have involved a mere tuning and tweaking of these ancient designs. As a result, natural processes of evolutionary discovery have most likely only dipped into and explored a tiny volume of the *Space of All Possible Body Plans* that might in principle have been realized on Earth. This is not to say that wildly new forms cannot be retrieved from the *Hypermarket of All Possible Body Plans* by artificial processes of exploration and discovery within the corresponding *Hox Gene Space*. It is not, however, by any means clear how many different potentially realizable forms this mathematical space contains, or how many different body plans history actually experimented with during the Cambrian Explosion. We can in addition, observe once again the contingent nature of mankind's existence; as if the chordate body plan for which *Pikaia* and *Yunnanozoon lividum* were the first ambassadors, had for one reason or another been contingently decimated, there would most likely be no human life on Earth and no temples, paintings, steamships or computers.

Suzanne Rutherford and Susan Lindquist demonstrated in 1998 that many of the mutations in the genes that form key components of developmental patterning programs, are phenotypically silent. The heat shock protein known as *Hsp90* was shown to mediate this phenomenon by buffering the functional effects of mutant proteins. *Hsp90* thus allows clusters of mutations in several different genes to accumulate without inducing any obvious phenotypic changes. However, when this buffering is compromised, for example by high temperatures, the phenotypes of this cluster of silent mutants may be expressed and maintained, even when *Hsp90* function is restored. This generates 'escape routes' that form the basis of a powerful means by which rapid and coordinated developmental change can be accomplished in response to environmental contingencies and a way of elaborating, but not radically redesigning, developmental processes that might otherwise remain firmly entrenched.

The fourth great evolutionary transition in the history of life was the evolution of the brain. Like the rest of animal morphology, the generation of this complex edifice depended on the evolution and coordinated activity of a unique subset of master regulatory genes, that contributed the intrinsic and extrinsic instructions necessary for patterning the forebrain, midbrain and hindbrain structures from which brains are composed. Philip Crossley, Salvador Martinez and Gail Martin have demonstrated that ectopic midbrain structures can be induced in regions of tissue that normally gives rise to forebrain structures, simply by artificially over-expressing a single gene known as *Fgf8*. A similar set of genes appears to regulate the patterning in the brains of both insects and vertebrates, which suggests that there may be a universally conserved program for establishing the ground plan of every known type of brain.

The morphological evolution of brains and their component neurones and glial cells was accompanied by the concurrent evolution of an ancestral set of chemical neurotransmitter molecules which communicate information between neurones by means of a frequency-based and perhaps also a temporal code. The fine-tuning of the brain's neuronal circuitry is experience dependent. In the developing cerebral cortex, for example, the loss of activity that results from the occlusion of an eye early on in development produces a permanent change in cortical micro-architecture and connectivity. The mature micro-anatomical structure of the brain is thus not exclusively hard-wired within genes, but plastic and dependent on activity within immature neuronal circuits during development.

With the evolution of brains came the capacity for complex information processing, individual learning, sophisticated behaviour, cognition and, in the case of mankind, consciousness. The evolution of a capacity for learning enabled informational databases relating to features of the outside world to be compiled within the lifetime of an individual. These databases allowed behavioural responses to contingent conditions to be modified on the basis of the past experience of the individual, rather than on the genetically encoded historical experience of the organism's ancestors. The strength of the neuronal interconnections within brains thus provided an additional source of experientially acquired information, that could be used to supplement the genetically represented databases encoded within the DNA sequences of genes. Unlike genetic archives, however, information acquired by learning could not, in the first instance, be communicated between generations.

One example of this type of information is associative learning, which enables organisms to link environmental cues such as odours and colours with a particular outcome, such as the presence of food or sickness. Although information acquired in this way is not directly dependent on genes, it can, in some instances, be constrained by genes, in that associations between some types of events and cues are easier to establish than others. The evolution of brains also generated a host of new niches for parasites, which were able to manipulate and co-opt the behaviour of other organisms. The malaria-parasite *Plasmodium falciparum*, for example, is able to alter the behaviour of its *Anopheles* mosquito host so as to increase the rate at which it moves about and bites its prey. Although this is clearly useful for the *Plasmodium falciparum* as it increases its chance of infecting new hosts, it is most disadvantageous for the mosquito which usually dies when it bites someone, as a result of being swatted.

Brains also constituted a generalized computational resource that could be used as the basis for the development of advanced sensory systems such as those involved with vision, smell, hearing, taste and touch and the associated higher-level processing systems involved, for example, with pattern recognition. They also provided the computational basis for the evolution of advanced motor systems, the coordinated behaviour of groups of different motor elements and the types of perceptual systems that enable, for example, a monkey or squirrel to jump long distances from one tree to another without falling to the ground.

The evolution of brains and processes of learning also paved the way for the discovery of the key that led into a very different type of mathematical space known as awareness or in the case of mankind, consciousness. This involved the discovery of the *Space of All Possible Mental Representations* of which consciousness occupies a small, but highly refined region. A capacity for mental representations enabled the generation of mental models of worldly scenarios. This enabled the consequences of alternative courses of potential action to be tested in mental simulations before any commitment to a particular alternative was made. The probability of an unfavourable outcome could in this way be much reduced. Unlike *Gene Space*, which could only be navigated slowly and sequentially, the *Thought Space* generated by conscious human brains has no such constraints. Vast volumes of this information space can be navigated instantaneously and appropriate solutions retrieved effortlessly in a relatively unconstrained manner from its most distant corners. The evolution of brains thus enabled living things to track frequencies of change that were infinitely higher and more complex than anything that their brainless ancestors could ever have achieved. Adaptive solutions to environmental problems such as thermoregulation could also, for the first time, be furnished by extra-genetic mechanisms. Instead of evolving thick fur, for example, man was able to invent clothing and to learn how to tame fire.

The fifth great transition involved the evolution of a mechanism by which the extra-genetic information acquired through individual learning could be communicated between individuals by an extra-genetic process of inter-individual information transmission, known as culture. The evolution of cultural mechanisms of information acquisition, representation and transmission accelerated processes of life into the informational fast lane and constituted a quantum leap in the size, nature and extent of the available informational databases where solutions to environmental problems could be systematically encoded and searched for.

In the more advanced cases of cultural information transmission, in addition to being communicated from one brain to another, information can be down-loaded on to non-biological representations in the form of such things as crafted tools, paintings, primitive symbolic notations and mature systems of writing. In the case of mankind, who first emerged in Africa around 4.5 million years ago and gave rise to anatomically modern man *Homo sapiens* around 150,000 years ago, these activities were taken to their extremes. This began with the manufacture of stone tools at least 2.5 million years ago and culminated in the origin of spoken languages more than 40,000 years ago and symbolic notation around 13,000 years ago. It was these major cultural innovations that helped establish the foundations for the beginning of agriculture in southwest Asia's Fertile Crescent, stretching from southeast Turkey to western Iran, around 11,000 years ago. Archaeological records show that only a few centuries at the most were then required to transform hunter-gatherer villages harvesting wild plants into farming villages that planted and harvested fully domesticated crops and held domesticated animals.

The development and subsequent expansion of a system of agriculture enabled these peoples to eventually out-compete rival hunter-gatherer economies and provided the basis for the establishment of large, sedentary and densely populated human populations. The subsequent generation of complex stratified societies in which food-producing activities could be restricted to only some members of the population freed others to become professional soldiers, scribes, bureaucrats, tradesmen, artists, metalworkers and craftsmen. This in turn paved the way for the emergence of the great civilizations of human history, the first cities around 10,000 years ago, and modern industrialized societies.

Man is not, however, the only species to have evolved cultural mechanisms of information acquisition, representation and perpetuation. The use of tools has for a long time been taken as the hallmark of human culture, as the manufacture of a tool requires that a new shape be artificially imposed upon an object that might originally have looked very different and had a different purpose. This implies the existence of some type of mental representation of the end product in the 'mind' of the manufacturer. The New Caledonian crow *Corvus momeduloides* has been shown to manufacture and utilize two different types of hook tool to assist with the capture of small invertebrate prey. It thus provides an excellent example of tool use and hence rudimentary 'culture' amongst animals. The tools share features with those that first appeared in the stone and bone-using cultures of early humans and indicate that tool manufacture in crows has attained a considerable degree of technological competence.

Chimpanzees in Guinea have been shown to use long, hooked sticks to help them reach the branches of fig trees that they would otherwise be unable to reach. They also produce a range of other task-specific tools which vary distinctly in their weight, size, material that they are made from, hardness and length. There is also evidence to suggest that the evolution of sperm whales can be influenced by culturally transmitted traits which involve learned behaviours that are passed down from mothers to their offspring. The songs of birds such as parrots and hummingbirds which have important functions both in mate attraction and territorial defence provide yet another example of culturally transmitted information in animals.

We are now nearing the end of our journey. And having completed our exploration of the history and structure of living things, we will shortly commence a brief examination of the future of life. Before doing so, however, we will pause for a moment to take stock of some of the principles that have emerged on our way. Most importantly, it should by now be clear that the information of living things may be envisaged as being distributed across multiple sets of interconnected mathematical spaces, each with its own unique properties and constraints. If we consequently wish to describe an organism rather than merely detailing its genetic specification, the information contained within each of these levels has to be taken into consideration. These include genetic, developmental, learning and cultural levels of information acquisition, representation and transmission. Higher levels are nested hierarchically above lower levels, such that lower levels constrain higher levels. The extent, however, to which this occurs may be imagined as being attenuated at each successive level, so that although human thought may ultimately be constrained by the genes that construct brains, and culture is in a broad sense deeply rooted in genetic history, these effects are unlikely to be as powerful as phenotypes such as eye colour, which are directly controlled by genes.

Living creatures are best thought of as information kits which are housed within an *Information Zoo*. Instead of talking about DNA sequence distances between different creatures, it is more appropriate to think in terms of information distances. Although some units of information are clearly more fundamental than others, information at every relevant level is necessary to create an exact facsimile of a living thing. If, for example, we wanted to recreate an historically accurate dodo, a gene kit specification would be quite inadequate. This arises from the fact that genetic information is itself associated with multiple tiers of complexity and that aspects of the information of a dodo are likely to be distributed across several information spaces, including development, learning and culture.

What use, after all, is genomic DNA sequence information if we do not know which genes have been functionally silenced by inheritable 'epigenetic' imprinting processes such as DNA methylation and histone protein acetylation? These epigenetic processes allow identical DNA sequences to be maintained in completely different functional states. What we need then is to detail not just genetic information, but also any associated epigenetic information. Indeed 'epimutations' can have just as much informational impact on a creature as a mutation to a nucleotide base itself. Cancer cells, for example, often have altered patterns of DNA methylation, and imprinted genes are known to have an important role in the development of many plants and animals.

Genes are not, furthermore, the only source of inheritable information within an organism. Fungi and yeast appear to utilize protein-based (prion-like) systems of inheritance to an extremely limited extent. The process of development also contains many tiers of information that are not represented within the sequences of DNA. The asymmetry of the egg that kick-starts the developmental process in many organisms and the follicular signals delivered by the mother, may not, for example, necessarily be computable from a free-standing genome. The development of mature structures is also influenced by the environmental context in which they are constructed. The architecture of cortical circuits in the brain is, as we have already seen, critically dependent on appropriate patterns of neuronal activation during development. Changes in temperature can also have profound effects on developmental processes. Some cellular structures like mitochondria are, furthermore, directly inherited from the mother and cannot consequently be inferred from a free-standing genomic sequence. The physical continuity of membrane structures between a parent cell and its daughters constitutes another source of information that may not have a direct genetic representation. The information stored within brains and peripheral nervous systems also lacks a genetic representation, as does the extra-genetic cultural information communicated from one individual to another and stored in non-biological representational machinery.

Information and the asymmetric distribution of information between different living things are thus the unifying concepts in the study of living things. Different species of organisms may be thought of as forming private 'information clubs', as their information is jealously guarded and partitioned from that of other species. Life itself may thus be viewed as a mathematical pattern in a multidimensional and substrate-free information space. Epigenetic processes, development, learning and culture may consequently be viewed as the higher dimensions of *Gene Space*, or what we might call *Genetic Hyperspace*. The historical process of evolution by natural selection has involved the discovery of a host of successively higher-dimensional mathematical solution spaces, each with its own unique topology, search rules, constraints, ease of navigation and 'solution' distribution. A logical problem presented by the environment may be imagined as ramifying through the qualitatively different set of solution spaces accessible to a given organism. If a solution to a problem cannot be found in one information space, it may be searched for in another. If reasonable solutions exist in more than one space, the level at which the solution is furnished will be determined by a host of different optimization considerations.

Each solution space extracts its own price, in terms, for example, of the constraints that it imposes, the ease with which the space can be searched, the cost of maintaining the representation, the stability of the solution, and its validity. A solution furnished at the genetic level, for example, although slow to acquire, is based on the tried and tested experience of history and thus reliable. It is also relatively stable thanks to a number of information-preserving mechanisms which we have discussed. These include: the existence of DNA repair and proofreading enzymes, checkpoint proteins, the design of the genetic code, the redundancy and robustness of biochemical pathways, and the presence of molecular chaperones such as *Hsp90* which can mask the phenotypic effects of some mutant proteins by helping them to fold. However as a result of the hard-wired nature of genetically represented information, the utilization of DNA-based solutions introduce profound constraints into an organism.

A solution furnished at the cultural level, on the other hand, can be acquired very rapidly, may be expensive to maintain (as brains are prodigious consumers of glucose) and may suffer from sampling error if the solution has not been tried and tested by history. Extra-genetic databases may, in addition, be less stable than genetic archives, as a computer can for example crash and books may be burnt or lost. Unlike information stored at the genetic level, however, cultural information is less likely to introduce irreversible constraints into a system. Indeed one of the great assets of cultural information is that it may easily be reprogrammed and its effects are therefore reversible. The strategy of sketching out the broad outline of a program genetically and then allowing the details to be filled in by experience furthermore conveys organisms with a capacity for behavioural plasticity which enables them to adapt to new situations in a rapid and highly efficient manner.

In addition to the general notion that living things are informational structures, some of the other central concepts we have encountered include both the fact that living things are deeply historically contingent entities, and that it is possible to make a distinction between the formal and material aspects of living things. History itself may be viewed as a sort of animal that consumes the irreversible medium of time as it meanders indifferently through the interconnected mathematical spaces that specify the complete range of all possible creatures and the relationships and events by which they might be interlinked.

Like fashion designers exhibiting small samples of their collections on a Paris catwalk, the history of living things has been a mere 'showcase' of life's possibilities. If any single event had been only slightly modified, things could have been quite different. Indeed if the tables had been turned it may have been our ancestors that became extinct and Neanderthal man who survived to dominate the Earth. On the other hand, mankind may not have evolved at all if the chordate body plan had not been discovered during the Cambrian Explosion and, if one of the very early lineages of proto-organisms had by chance been decimated as opposed to another, modern life might still be RNA-based, or utilize a completely different informational polymer. Indeed life might not have got going at all. Furthermore when we encounter the word 'nature', there is a sense in which we might just as well substitute the word 'history'.

But what of the future? In the future living things will be designed from first principles, much like motor cars. The invention of powerful digital computers and perhaps eventually massively parallel or quantum computing devices coupled with advanced mechanisms for reading and engineering DNA sequences, has enabled (and will to an ever increasing extent continue to enable) the formal logical and material historical aspects of life to be understood in the finest detail. Having shaped the trajectory of life for over several billion years, DNA and the natural and historically contingent processes that shaped its evolution will eventually be subjugated, supplemented, or replaced by artificial and profoundly ahistorical strategies for the construction of living things.

It is at this point worth attempting to sketch out the basis of a paradigm for the future of life that might provide a framework for structuring and charting future technological innovations. The aim here is not to be prescriptive, but simply to state what is likely to be the case. The six-point plan that I will tentatively suggest is as follows: (1) the complete documentation of the formal and material aspects of all existing forms of life; (2) the design of experimental methods and algorithms for inferring the molecular structures and morphologies of extinct forms of life and perhaps proto-life; (3) the engineering of existing forms of life; (4) the reconstruction and engineering of extinct forms of life; (5) the design of new life from first principles using 'natural', historical protein, DNA and RNA construction materials and natural information-encoding polymer systems and genetic logics; (6) the design of new life from first principles using 'artificial', ahistorical construction materials and artificial information-encoding chemical systems and genetic logics.

The complete documentation of all existing forms of life entails the compilation of several different kinds of information. This includes amongst many other things: (a) the determination of the complete genomic sequence of every known living thing and its associated 'proteome' (expressed protein sequences); (b) the derivation of the inferred evolutionary connectivities between all known living things and between the different proteins from which they are constructed; (c) the identification of the core sets of essential genes shared by related organisms; (d) the compilation of an inventory of all the classes of genes and associated protein folds used by every type of organism; (e) the annotation of the functions of every known gene and the identification of any redundancy of function; (f) identifying the patterns of tissue-specific gene expression in time and space throughout development and across a representative range of environmental circumstances in both health and disease; (g) the identification and clarification of the regulatory hierarchies that control all known biochemical pathways and processes; (h) the acquisition of a complete set of epigenetic data that documents the methylation status of every cytosine residue in every imprinted genome; (i) the documentation and annotation of all the non-coding sequences in every genome; (j) documenting any non-universal genetic codes and recoding principles; (k) documenting every form of post-translational modification such as the glycosylation, phosphorylation, acetylation and cholesterol modification of proteins and processes such as protein splicing; (l) identifying all mRNA regulatory elements and the characterization the physico-chemical properties of mRNA molecules such as their half-life; (m) documenting all the representative viral and other parasitic sequences found within genomes; (n) documenting all genetic polymorphisms; (o) documenting the complete set of alternative splicing patterns of genes and their patterns of expression both in time and space and (p) documenting the relative positions of species with respect to one another in *DNA Sequence Space*.

It also includes: (q) documenting all higher-order protein interactions including the formation of every type of oligomeric and multimeric protein/protein complex as well as interactions with regulatory DNA elements; (r) documenting the spatial location of every protein within every cell; (s) documenting the three-dimensional structure and dynamic properties of the complete set of protein folds utilized in nature; (t) documenting the physical and chemical profiles of every known protein including their kinetic and thermodynamic profile of properties and the relative properties and positions of their nearest neighbours in *Protein Sequence Space*; (u) documenting the dynamic properties of cellular machines and processes such as the noise involved in transcription and translation and the nature, efficiency and dynamics of DNA repair systems; (v) documenting the development of every organism in full detail; (w) documenting both the information learned by every species and the core cultural information utilized by any particular species; (x) documenting the individual behaviour and in some cases social behaviour of every species; (y) documenting the ecosystem and physico-chemical environment in which every organism is embedded; and (z) the production of a complete set of abstract mathematical algorithms and representations that fully describe the processes which formally define every known type of living thing.

Once all known existing forms of life have been documented and rationalized to form a single evolutionary family tree, it should be possible with a greater or lesser degree of success, to infer the molecular nature of their ancestral precursors. In some cases these hypotheses will be testable by rescuing ancient DNA or protein material that can be sequenced and compared with the predicted sequences. The first demonstration that ancient DNA molecules could be successfully recovered from extinct creatures came from *Equus quagga*, a member of the horse family that looks like a cross between a horse and a zebra and became extinct in 1883. Since then it has been claimed that small fragments of DNA can be extracted from samples that are as much as eighty million years old.

At all of the other epigenetic, developmental, learning and cultural levels, however, the information from ancient organisms appears to have been lost irretrievably. Thus although it might, for example, be possible to infer the cry of a dodo from eyewitness reports, or from morphological, physiological and perhaps biochemical considerations, there does not as yet appear to be any way in which the historical validity of such hypotheses could be falsified. It might, however, be possible to infer epigenetic patterns of imprinting and developmental processes from the evolutionary tree and from any DNA and protein sequence information that can be retrieved from fossils. Without a time machine we will probably never be able to trace the true genealogy of the extinct lineages of molecules from which our most ancient ancestors were constructed, but it may one day be possible to rationally infer the sequences of individual ancestral molecules using computational analysis. We will also, one presumes, eventually be able to create proto-living organisms in a test tube, from non-living chemical precursors.

Thomas Jerman and Steven Benner have attempted to reconstruct the protein sequences of ancient organisms using a rational parsimony procedure which assumes that descendent proteins arose from extinct ancestors by the smallest number of independent evolutionary events. The properties of such ancient molecules can subsequently be tested and compared with those of their more highly programmed descendants. The reconstruction of the poorly programmed and non-encoded molecular skeletons of our most distant ancestors is likely to remain one of the biggest challenges of the future.

The profound engineering of existing forms of life either before or after the completion of the initial phase of documentation should not represent a great challenge. Engineering may be performed either at the level of the germ line in which case the changes will be inherited, or somatically, in which case they will not. Genes can be inserted, deleted, or modified either globally, or in a tissue-specific manner. Cells of a given type might also be reprogrammed so as to change their pattern of gene expression, behaviour and developmental fate. Once the molecular basis of aging, cancer and all other such processes are understood, genomes could in principle be modified so that many of the diseases that afflict mankind may become a thing of the past. Prophylactic engineering to protect against infectious diseases may also be an option. The demonstration that a reduction in the expression of the *daf-2* gene in the worm *Caenorhabditis elegans* doubles its lifespan and that a mutation in the *methuselah* gene of fruit flies increases their lifespan by up to 35% suggests that human lifespans might be altered in a similar way. Once a function has been assigned to each of the 100,000 or so genes in the human genome and their regulatory circuits are properly understood, there will doubtless be pressures to modify many other genetically controlled human characteristics.

The design of completely new life forms from first principles using natural (historical) protein, DNA and RNA technologies presents an interesting and entirely feasible challenge to molecular biologists. Such an enterprise might be approached in two different ways. New living things could be designed using knowledge gained from the study of existing forms of life, or computer algorithms devised to scan and search *DNA Sequence Space* for regions that contain sequence features which are characteristic of living things. These could then be selected, constructed and tested in the real world to see if the creature they specify is viable and resembles the predicted morphology and behaviour in a meaningful way. The construction of such 'design by selection' algorithms should, in principle, constitute a relatively trivial task once the database of all possible genome sequences has been compiled. However given the immense size of this space, an extremely high degree of computational power would be necessary to complete even a partially exhaustive search of a defined region.

The discovery of the mathematical structures of the creatures that inhabit the *DNA Zoo* regions of *DNA Sequence Space* is exactly analogous to the discovery of territories such as the New World or the Belgian Congo. These creatures are indeed 'out there' in a very real and tangible mathematical sense, and may one day be realized. One can imagine a time when it will be possible to produce a complete map of *DNA Sequence Space* that charts all of the regions of interest and the routes that history might in principle have taken to navigate this *DNA Sequence Sea*. It may turn out that some islands of form are surrounded by large expanses of sequences that encode shapeless and lifeless nonsense and could never be realized by natural, historical processes of navigation. Although we will invariably have neither the time nor space to realize the greater majority of the potential creatures that our computers will discover, their presence will be known to us, in much the same way as we are aware of the icy expanses of the North Pole, the sunny beaches of Cannes and the tropical rain forests of northeastern Australia. But we must not forget that some creatures may not be computable from their DNA sequences alone and that the ultimate test of the validity of a potential creature is to run its genetic programs in the real world. Furthermore, because of the importance of epigenetic phenomena, it might be more appropriate to begin our search within *Imprinted DNA Sequence Space*.

The designer strategy gives the genetic architect the opportunity to recombine the modular elements from which living things are constructed in a host of different ways. The option of seconding genes and modular functions from different species and transferring them to new species must also be considered. One need not, furthermore, restrict oneself to using molecular components that have a historical precedent. New protein, DNA and RNA components can be designed, modified and mapped onto functions that have not been discovered by history. Although the introduction of any particular design feature may constrain several others, it nevertheless makes sense to draw a distinction between the chassis or gross external morphology of a creature, its internal morphology in terms of its organ structures and its complement of unique function-generating circuits.

Hox genes provide a logical starting point for the construction of computer programs that can model and explore the potentialities for form inherent within a given set of *Hox* genes. These programs would be based upon *Hox* gene expression databases, which have themselves been assembled by monitoring *Hox* gene expression during development using either DNA micro-arrays, or genes tagged with fluorescent or luminescent markers. Manuel Calleja and Ginés Morata have, for example, used this type of approach to monitor the expression of genes during the development of a living *Drosophila* fruit fly. Other computer programs could be written to explore the whole universe of gene expression patterns, in order to define new programs of expression that generate different cell types, or functions. This may include ones which have not been realized by history.

One might also play with gene kits in order, for example, to build the smallest or largest possible creature. Children may even have competitions at school to see who can design a protein-based organism able to survive at a temperature that is 1°C hotter than the hottest environment known to be compatible with life. The existence of unicellular 'extremophile' organisms that inhabit a range of unfavourable habitats and include barophiles that grow at high pressures, acidophiles such as *Thermoplasma acidophilum* which grow at pH values as low as zero, halophiles such as *Halobacterium cutirubrum* that only grow at high salt concentrations, hyperthermophiles which grow at temperatures greater than 75°C and 'hell's bacterium' *Bacillus infernus* that lives 2.7 kilometres under the ground in the absence of oxygen and at a temperature of 60°C, provides some notion of the diversity of conditions that living things are able to tolerate and the types of alien biochemistries that might in principle be engineered into the circuits of artificially constructed living things.

One thing, however, is clear. And this is that following sufficient experimentation using both computer simulations and accompanying construction processes, a moment will come when the constraints inherent to particular types of designs and functions will begin to reveal themselves. We might at this point, decide to compile a list that describes which phenotypes can and cannot be achieved and how the introduction of one type of modification into a particular structure or module will invariably clip or profoundly modify the wings of another. But what types of constraints might we encounter and do some of these arise as a consequence of the natural protein, DNA and RNA technologies that we utilize?

One of the most fundamental constraints associated with restricting our processes of design to the 100,000 or so genes utilized by modern man is that the large majority of the proteins encoded by these genes appear to map to little more than around 1,000 different protein families and their associated folds. The strategy of evolution appears then to have involved the discovery and elaboration of a relatively small set of ancestral folds that were recombined and decorated in a host of different and subtle ways to generate new functions. Some, such as the immunoglobulin fold are so versatile that they have been duplicated and co-opted to new functions repeatedly to generate a large superfamily of ancestrally related protein folds. Once the relatively minimal set of ancestral folds had been discovered, evolution locked in on them and adopted a strategy that principally involved the tuning and tweaking of pre-existing folds by the combinatorial reshuffling of their modular structural elements to create a host of chimeric molecules, rather than the discovery of entirely new folds. The diversity of folds in the rest of the animal kingdom is, however, uncertain. It is unclear whether the utilization of only around 1,000 different folds in man represents an intrinsic stereochemical constraint of protein architecture, or just a reflection of historical contingency. Either way, the end result is the same, and the repertoire of three-dimensional shapes that historical protein technologies are able to offer, is constrained.

Numerous other constraints, that are independent of any particular construction technology, also appear to limit the nature of living things. Peng Chai and Robert Dudley have, for example, shown that the ruby-throated hummingbird *Archilochus colubris* operates at the very limits of vertebrate locomotor energetics in terms of its oxygen consumption and muscle power output. So energetic constraints may also set limits to possibility. The metabolic demands of flight also place a constraint on genome and cell size. Birds with high metabolic rates require a correspondingly higher rate of gas exchange across their cell surfaces. As the surface-to-volume ratio of the cell is inversely related to its volume, small cell volumes permit a greater rate of gas exchange per unit volume. Large, flightless birds thus tend to have relatively large genomes whereas small, highly metabolically active birds tend to have smaller genomes and cell sizes. The evolution of flight thus appears to have set a constraint on the genome size of birds.

Developmental programs also presumably impose their own constraints. Is it possible, for example, to produce a winged horse, a centaur, or a flying giraffe? The dung beetle *Onthophagus* uses its horns to repel competing suitors. One might have expected evolution to select for increased horn size, but this does not happen, as the developmental program that produces big horns also generates small eyes and body parts. These trade offs between different structures provide a clear example of a developmental constraint in action. Thus, although the specification for a long-horned dung beetle with large eyes and body parts exists within the *Information Zoo*, there may be no natural developmental process by which it can be realized. This hypothetical long-horned, large-eyed dung beetle is thus apparently confined to dwell forever in an informational shadowland of impossibility, that we have called the *Secret Sea*. Other constraints include the complex physico-chemical laws of self-assembly and self-organization and the need for a naturally evolving creature to be optimized for evolvability.

The redesign of life from first principles using artificial, ahistorical building materials and logic, introduces a number of interesting possibilities into the processes of life which incorporate all of the advantages that new technologies and logics are able to bring. It would, one presumes, have been impossible to build the Eiffel Tower from bricks, or the Taj Mahal from cardboard. Although protein construction technology has been the industry standard for several billion years, the very first living things may, as we have noted, have utilized very different technologies to make their catalysts and structural building blocks. Proteins do not have a monopoly over the construction of folds business. In fact a fairly large number of natural and synthetic molecules appear to be able to fold into complex structures. Single-stranded RNA and DNA molecules can, for example, generate a prodigious repertoire of binding sites and catalytic activities. However, synthetic molecules such as polymers constructed from the hydrocarbon phenylacetylene, are also able to spontaneously adopt folded structures. Phenylacetylene polymers can in fact generate alpha helices and cavities, that could be utilized as a binding site. The properties of such synthetic molecules might introduce interesting modifications into living things, if indeed an artificial biochemistry could be found to mediate their synthesis.

There may be nothing special about the protein building block materials that have been used to construct nature's profusion of complex living forms. Their use in modern biological machines may indeed have been highly contingent. The other possibility, however, is that proteins and the *Protein Sequence Space* they inhabit have unique properties that make them ideally suited for natural processes of life-construction. This might, for example, result from the topography of stable *Protein Sequence Space* and thus the relative distribution and intrinsic evolvability of the folds scattered across it. The intrinsic flexibility of proteins that enables them to function as the well-oiled molecular components of the tiny nano-machines that perform the essential tasks within cells, their capacity for self-assembly, and complex formation, or their ability to undergo conformational changes as a result of covalent modification are also likely to be important factors.

We need not, however, restrict ourselves to utilizing natural proteins, or indeed the set of twenty naturally occurring amino acids. Artificial proteins and indeed any other natural or synthetic polymer molecules that are able to fold, can be modified and patterned by mixing in other technologies that take advantage of self-assembly and self-organizational processes. Indeed the set of folds and activities that repertoires of such molecules contain may well be associated with their very own zoos, that have not yet been explored. The informational database of life might also be replaced with alternative structures and technologies. Joseph Piccirilli and Steven Beener have, for example, demonstrated how the genetic alphabet could be expanded from four to six letters by the introduction of a new base pair into DNA and RNA. Synthetic nucleic acid analogues such as those containing pyranosyl could also be used to generate artificial genetic materials and the logic of the genetic code and ribosomal machinery modified so that proteins were constructed from twenty-one or more types of amino acid, instead of twenty.

If the formal and material aspects of living things can be precisely defined and delineated, then there is no reason why living things could not eventually be constructed from entirely synthetic components. Furthermore, if their logic is able to transfer directly to machines, it is possible that natural historical forms of life might eventually become yet another example of evolution's predilection for decimation. Indeed, it is by no means certain that living things constructed from natural biological materials would be able to out-compete their synthetic and ahistorically designed, machine-based rivals.

Bibliography

Abouheif, E., Akam, M., Dickinson, W. J., Holland, P. W. H., Meyer, A., Patel, N. H., Raff, R. A., Roth, V. L., and Wray, G. A. Homology and developmental genes. *TIG* **13**, 432–433 (1997).

Adleman, L. M. Computing with DNA. *Scientific American*, 34–41 (August 1998).

Adleman, L. M. Molecular computations of solutions to combinatorial problems. *Science* **266**, 1021–1024 (1994).

Ahari, H., Bedard, R. L., Bowes, C. L., Coombs, N., Dag, O., Jiang, T., Ozin, G. A., Petrov, S., Sokolov, I., Verma, A., Vovk, G., and Young, D. Effect of microgravity on the crystallization of a self-assembling layered material. *Nature* **388**, 857–860 (1997).

Aiello, L. C. Thumbs up for our early ancestors. *Science* **265**, 1540–1541 (1994).

Alberts, B. The cell as a collection of protein machines: Preparing the next generation of molecular biologists. *Cell* **92**, 291–294 (1998).

Alexander, R. M. How dinosaurs ran. *Scientific American*, 62–68 (1991).

Alexander, R. M. Tyrannosaurus on the run. *Nature* **379**, 121 (1996).

Alivisatos, A. P., Johnsson, K. P., Peng, X., Wilson, T. E., Loweth, C. J., Bruchez, M. P., and Schultz, P. G. Organization of 'nanocrystal molecules' using DNA. *Nature* **382**, 609–611 (1996).

Allain, F. H. T., Gubser, C. C., Howe, P. W. A., Nagai, K., Neuhaus, D., and Varani, G. Specificity of ribonucleoprotein interaction determined by RNA folding during complex formation. *Nature* **380**, 646–650 (1996).

Alper, J. Assembling the world's biggest library on your desktop. *Science* **281**, 1784–1786 (1998).

Altamirano, M. M., Golbik, R., Zahn, R., Buckle, A. M., and Fersht, A. R. Refolding chromatography with immobilized mini-chaperones. *Proc. Natl. Acad. Sci. U.S.A.* **94**, 3576–3578 (1997).

Amato I. Fomenting a revolution, in miniature. *Science* **282**, 402–405 (1998).

Ambler, R. P. The distance between bacterial species in sequence space. *J. Mol. Evol.* **42**, 617–630 (1996).

Amlani, I., Orlov, A. O., Toth, G., Bernstein, G. H., Lent, C. S., and Snider,

G. L. Digital logic gate using quantum-dot cellular automata. *Science* **284**, 289–291 (1999).

Anderson, K. V. New views of life. *Cell* **90**, 593–594.

Andersson, S. G. E., Zomorodipour, A., Andersson, J. O., Sicheritz-Pontén, T., Alsmark, U. C. M., Podowski, R. M., Näslund, A. K., Eriksson, A-S., Winkler, H. H., and Kurland, C. G. *Nature* **396**, 133–140 (1998).

Arnheim, N., and Shibata, D. DNA mismatch repair in mammals: role in disease and meiosis. *Current Opinion in Genetics and Development* **7**, 364–370 (1997).

Arnold, F. H. When blind is better: protein design by evolution. *Nature Biotechnology* **16**, 617–619 (1998).

Atkins, J. F., and Gesteland, R. F. A case for *trans* translation. *Nature* **379**, 769–771 (1996).

Atkins, P. *Creation Revisited: The Origin of Space, Time and the Universe* (Penguin Books, London, 1994).

Atwell, S., Ultsch, M., Abraham, M., De Vos, A. M., and Wells, J. A. Structural plasticity in a remodeled protein-protein interface. *Science* **278**, 1125–1128 (1997).

Audic, S., and Béraud-Colomb, E. Ancient DNA is thirteen years old. *Nature Biotechnology* **15**, 855–858 (1997).

Bachmann, P. A., Luisi, P. L., and Lang, J. Autocatalytic self-replicating micelles as models for prebiotic structures. *Nature*, **357**, 57–59.

Bairoch, A., and Murzin, A. G. Sequences and topology predicting evolution. *Current Opinion in Structural Biology* **7**, 367–368 (1997).

Baker, D., and Agard, D. A. Kinetics versus thermodynamics in protein folding. *Biochemistry* **33**, 7505–7509 (1994).

Baker, D., Sohl, J. L., and Agard, D. A. A protein-folding reaction under kinetic control. *Nature* **356**, 263–265 (1992).

Baldwin, R. L., and Rose, G. D. Is protein folding hierarchic? I. Local structure and peptide folding. *TIBS* **24**, 26–33 (1999).

Balter, M. In Toulouse, the weather – and the science – are hot. *Science* **269**, 480–481 (1995).

Bard, A. J. Electron transfer branches out. *Nature* **374**, 13 (1995).

Bargmann, C. I. Neurobiology of the *Caenorhabditis elegans* genome. *Science* **282**, 2028–2033 (1998).

Barinaga, M. New clues to how proteins link up to run the cell. *Science* **283**, 1247–1249 (1999).

Baringa, M. New clue to brain wiring mystery. *Science* **270**, 581 (1995).

Barkai, N., and Leibler, S. Robustness in simple biochemical networks. *Nature* **387**, 913–917 (1997).

Barlow, D. P. Gametic imprinting in mammals. *Science* **270**, 1610–1613 (1995).

Bass, B. L. RNA editing and hypermutation by adenosine deamination. *TIBS* **22**, 157–162 (1997).

Bassett, D. E., Basrai, M. A., Connelly, C., Hyland, K. M., Kitagawa, K.,

Mayer, M. L., Morrow, D. M., Page, A. M., Resto, V. A., Skibbens, R. V., and Hieter, P. Exploiting the complete yeast genome sequence. *Current Opinion in Genetics and Development* **6**, 763–766 (1996).

Batey, R. T., and Doudna, J. A. The parallel universe of RNA folding. *Nature Structural Biology* **5**, 337–340 (1998).

Bauchmann, P. A., Luisi, P. L., and Lang, J. Autocatalytic self-replicating micelles as models for prebiotic structures. *Nature* **357**, 57–59 (1992).

Bauer, C., and Jelkmann, W. Carbon dioxide governs the oxygen affinity of crocodile blood. *Nature* **269**, 825–827 (1977).

Baum, E. B. Building an associative memory vastly larger than the brain. *Science* **268**, 583–585 (1995).

Baylin, S. B. Tying it all together: epigenetics, genetics, cell cycle, and cancer. *Science* **277**, 1948–1949 (1997).

Beardsley, T. Smart Genes. *Scientific American*, 73–81 (August 1991).

Beaudry, A. A., and Joyce, G. F. Directed evolution of an RNA enzyme. *Science* **257**, 635–641 (1992).

Beddington, R. Left, right, left . . . turn. *Nature* **381**, 116–117 (1996).

Bell, T. W. Molecular trees: A new branch of chemistry. *Science* **271**, 1077–1078 (1996).

Bender, M. Metamorphosis in *Drosophila*: from molecular biology to mutants. *TIG* **11**, 335–336 (1995).

Benkovic, S. J., and Ballesteros, A. Biocatalysts – the next generation. *TIBTECH* **15**, 385–386 (1997).

Benowitz, L. I., and Routtenberg, A. GAP-43: an intrinsic determinant of neuronal development and plasticity. *TINS* **20**, 84–91 (1997).

Benton, D. Bioinformatics – principles and potential of a new multidisciplinary tool. *TIBTECH* **14**, 261–272 (1996).

Berek, C., and Milstein, C. The dynamic nature of the antibody repertoire. *Immunological Reviews* **105**, 5–26 (1988).

Berendsen, H. J. C. A glimpse of the holy grail? *Science* **282**, 642–643 (1998).

Bernal, J. D. *The Origin of Life* (Weidenfeld and Nicolson, London, 1967).

Bestor, T. H. Methylation meets acetylation. *Nature* **393**, 311–312 (1998).

Bhattacharya, S. K., Ramchandani, S., Cervoni, N., and Szyf, M. A mammalian protein with specific demthylase activity for mCpG DNA. *Nature* **397**, 579–583 (1999).

Bickerton, D. Chattering classes. *Nature* **382**, 592–593 (1996).

Binzel, R. P. Eros's extended family. *Nature* **388**, 516–517 (1997).

Bird, A. Gene number, methylation and biological complexity. *TIG* **11** 383–384 (1995).

Bird, A. P. Gene number, noise reduction and biological complexity. *TIG* **11**, 94–100 (1995).

Birge, R. R. Protein-based computers. *Scientific American*, 66–71 (March 1995).

Blackstock, W. P., and Weir, M. P. Proteomics: quantitative and physical mapping of cellular proteins. *TIBTECH* **17**, 121–127 (1999).

Blattner, F. R. *et al.* The complete genome sequence of *Escherichia coli* K-12. *Science* **277**, 1453–1462 (1997).

Blaxter, M. *Caenorhabditis elegans* is a nematode. *Science* **282**, 2041–2046 (1998).

Block, S. M. Real engines of creation. *Nature* **386**, 217–219 (1997).

Blondelle, S. E., and Houghten, R. A. Novel antimicrobial compounds identified using synthetic combinatorial library technology. *TIBTECH* **14**, 60–65 (1996).

Bloom, B. R. A microbial minimalist. *Nature* **378**, 236 (1995).

Bock, L. C., Griffin, L. C., Latham, J. A., Vermaas, E. H., and Toole, J. J. Selection of single-stranded DNA molecules that bind and inhibit human thrombin. *Nature* **355**, 564–566 (1992).

Boesch, C. The question of culture. *Nature* **379**, 207–208 (1996).

Bogarad, L. D., and Deem, M. W. A hierarchical approach to protein molecular evolution. *Proc. Natl. Acad. Sci. USA* **96**, 2591–2595 (1999).

Bogdanov, A., and Weissleder, R. The development of in vivo imaging systems to study gene expression. *TIBTECH* **16**, 5–10 (1998).

Bonner, J. T. The evolution of life's complexity. *Nature* **374**, 508–509 (1995).

Bordignon, C., Notarangelo, L. D., Nobili, N., Ferrari, G., Casorati, G., Panina, P., Mazzolari, E., Maggioni, D., Rossi, C., Servida, P., Ugazio, A. G., and Mavilio, F. Gene therapy in peripheral blood lymphocytes and bone marrow for ADA – immunodeficient patients. *Science* **270**, 470–475 (1995).

Borges, J. L. *Labyrinths: Selected Stories and Other Writings* (Penguin Books, London, 1970).

Bork, P., and Koonin, E. V. Protein sequence motifs. *Current Opinion in Structural Biology* **6**, 366–376 (1996).

Botstein, D., Chervitz, S. A., and Cherry, J. M. Yeast as a model organism. *Science* **277**, 1259–1260 (1997).

Bowerman, B. The worm keeps turning. *Nature* **390**, 228–229 (1997).

Brack, A. Life on Mars: a clue to life on Earth? *Chemistry & Biology* **4**, 9–12 (1997).

Bradley, J. P., Harvey, R. P., and McSween, H. Y. No 'nanofossils' in martian meteorite. *Nature* **390**, 454 (1997).

Bradshaw, A. D. Genostasis and the limits to evolution. The Croonian Lecture 1991. *Phil. Trans. R. Soc. Lond. B.* **333**, 289–305 (1991).

Bray, D. Protein molecules as computational elements in living cells. *Nature* **376**, 307–312 (1995).

Bray, D. Signaling complexes: biophysical constraints on intracellular communication. *Annu. Rev. Biophys. Biomol. Struct.* **27**, 59–75 (1998).

Breaker, R. R. DNA enzymes. *Nature Biotechnology* **15**, 427–431 (1997).

Breaker, R. R., and Joyce, G. F. A DNA enzyme that cleaves RNA. *Chemistry & Biology* **1**, 223–229 (1994).

Brenner, S. Biological computation. *The limits of reductionism in biology, Wiley, Chichester (Novartis Foundation Symposium 213)*, 106–116 (1998).

Brenner, S. E., Chothia, C., and Hubbard, T. J. P. Population statistics of protein structures: lessons from structural classifications. *Current Opinion in Structural Biology* **7**, 369–376 (1997).

Brenner, S. Errors in genome annotation. *TIG* **15**, 132–133 (1999).

Brenner, S., Elgar, G., Sandford, R., Macrae, A., Venkatesh, B., and Aparicio, S. Characterization of the pufferfish *(Fugu)* genome as a compact model vertebrate genome. *Nature* **366**, 265–268 (1993).

Breslow, R., Huang, Y., Zhang, X., and Yang, J. An artificial cytochrome P450 that hydroxylates unactivated carbons and regio- and stereoselectivity and useful catalytic turnovers. *Proc. Natl. Acad. Sci. U.S.A.* **94**, 11156–11158 (1997).

Bridges, B. A. Hypermutation under stress. *Nature* **387**, 557–558 (1997).

Bridges, B.A. DNA repair: Getting past a lesion – at a cost. *Current Biology* **8**, R886-R888 (1998).

Brookfield, J. F. Y. Evolving Darwinism. *Nature* **376**, 551–552 (1995).

Bryson, J. W., Betz, S. F., Lu, H. S., Suich, D. J., Zhou, H. X., O'Neil, K. T., and DeGrado, W. F. Protein design: A hierarchic approach. *Science* **270**, 935–941 (1995).

Buchanan, M. Travels on rugged landscapes. *Nature* **382**, 302–303 (1996).

Bult, C. J., *et al.* Complete genome sequences in the methanogenic Archaeon, *Methanococcus jannaschii . . . Science* **273**, 1058–1073 (1996).

Byrne, J. H. Plastic plasticity. *Nature* **389**, 791–792 (1997).

Cairns-Smith, A. G. *Seven Clues to the Origin of Life: a scientific detective story* (Cambridge University Press, Cambridge, 1993).

Callahan, H. S., Pigliucci, M., and Schlichting, C. D. Developmental phenotypic plasticity: where ecology and evolution meet molecular biology. *BioEssays* **19**, 519–525 (1997).

Callebaut, W. *Taking the Naturalistic Turn or How Real Philosophy of Science is Done* (The University of Chicago Press, Chicago, 1993).

Calleja, M., Moreno, E., Pelaz, S., and Morata, G. Visualization of gene expression in living adult *Drosophila*. *Science* **274**, 252–255 (1996).

Calude, C. S. Parallel thinking. *Nature* **392**, 549–551.

Cane, D. E., Walsh, C. T., and Khosla, C. Harnessing the biosynthetic code: combinations, permutations, and mutations. *Science* **282**, 63–68 (1998).

Capra, F. *The Web of Life: A New Synthesis of Mind and Matter* (HarperCollins Publishers, London, 1996).

Carroll, R. L. Revealing the patterns of macroevolution. *Nature* **381**, 19–20 (1996).

Carroll, S. B. Homeotic genes and the evolution of arthropods and chordates. *Nature* **376**, 479–485 (1995).

Casari, G. *et al.* Challenging times for bioinformatics. *Nature* **376**, 647–648 (1994).

Casti, J. L. Creation in silicon. *Nature* **385**, 399 (1997).

Cate, J. H. Gooding, A. R. Podell, E. Zhou, K., Golden, B. L., Kundrot,

C. E., Cech, T. R., Doudna, J. A. Crystal structure of a group 1 ribozyme domain: principles of RNA packing. *Science* **273**, 1678–1679 (1996).

Cavalier-Smith, T. How selfish is DNA? *Nature* **285**, 617–618 (1980).

Caves, C. M. A tale of two cities. *Science* **282**, 637–638 (1998).

Chamberlain, A. K., and Marqusee, S. Touring the landscapes: partially folded proteins examined by hydrogen exchange. *Structure* **5**, 859–863 (1997).

Chan, H. S. Matching speed and locality. *Nature* **392**, 761–763 (1998).

Chaudhuri, J. B. Biochemical engineering – past, present and future. *TIBTECH* **15**, 383–384 (1997).

Chen, R. Z., Pettersson, U., Beard, C., Jackson-Grusby, L., and Jaenisch, R. DNA hypomethylation leads to elevated mutation rates. *Nature* **395**, 89–93 (1998).

Chervitz, S. A., Aravind, L., Sherlock, G., Ball, C. A., Koonin, E. V., Dwight, S. S., Harris, M. A., Dolinski, K., Mohr, S., Smith, T., Weng, S., Cherry, J. M., and Botstein, D. Comparison of the complete protein sets of worm and yeast: Orthology and divergence. *Science* **282**, 2022–2028 (1998).

Cho, R. J., Fromont-Racine, M., Wodicka, L., Feierbach, B., Stearns, T., Legrain, P., Lockhart, D. J., and Davis, R. W. Parallel analysis of genetic selections using whole genome oligonucleotide arrays. *Proc. Natl. Acad. Sci. USA* **95**, 3752–3757 (1998).

Chothia, C. One thousand families for the molecular biologist. *Nature* **357**, 543–544 (1992).

Chothia, C. Protein families in the metazoan genome. *Development Supplement*, 27–33 (1994).

Chothia, C., and Gerstein, M. How far can sequences diverge? *Nature* **385**, 579–581 (1997).

Chothia, C., Hubbard, T., Brenner, S., Barns, H., and Murzin, A. Protein folds in the ALL-and ALL-classes. *Annu. Rev. Biophys. Biomol. Struct.* **26**, 597–627 (1997).

Chown, M. Seeds, soup and the meaning of life. *New Scientist*, 6 (August 1996).

Churchland, P. M., and Churchland, P. S. Could a machine think? *Scientific American*, 26–31 (January 1990).

Chyba, C. Buried beginnings. *Nature* **395**, 329–330 (1998).

Chyba, C. F. Life beyond Mars. *Nature* **382**, 576–577 (1996).

Chyba, C.F. Life on other moons. *Nature* **385**, 201 (1997).

Clapham, D. E. The G-protein nanomachine. *Nature* **379**, 297–299 (1996).

Clarke, N. D. Sequence 'minimization': exploring the sequence landscape with simplified sequences. *Current Opinion in Biotechnology* **6**, 467–472 (1995).

Clarke, N. D., and Berg, J. M. Zinc fingers in *Caenorhabditis elegans*: Finding families and probing pathways. *Science* **282**, 2018–2022 (1998).

Claude, C. S., and Casti, J. L. Parallel thinking. *Nature* **392**, 549–551 (1998).

Clayton, R. A., White, O., Ketchum, K. A., and Venter, J. C. The first genome from the third domain of life. *Nature* **387**, 459–462 (1997).

Cohen, P. T. W. Novel protein serine/threonine phosphatases: variety is the spice of life. *TIBS* **22**, 245–251 (1997).

Cohn, M. J., and Tickle, C. Limbs: a model for pattern formation within the vertebrate body plan. *TIG* **12**, 253–257 (1996).

Cole, S. T., *et al.* Deciphering the biology of *Mycobacterium tuberculosis* from the complete genome sequence. *Nature* **393**, 537–544 (1998).

Cole-Turner, R. Religion and gene patenting. *Science* **270**, 52 (1995).

Conway-Morris, S. *The Crucible of Creation: The Burgess Shale and the Rise of Animals* (Oxford University Press, Oxford, 1998).

Conway-Morris, S. The fossil record and the early evolution of the Metazoa. *Nature* **361**, 219–225 (1993).

Conway-Morris, S. Why molecular biology needs palaeontology. *Development* **Supplement**, 1–13 (1994).

Connell, C. Are partial gene sequences patentable? *TIBTECH* **16**, 197–198 (1998).

Cordes, M. H. J., Walsh, N. P., McKnight, C. J., and Sauer, R. T. Evolution of a protein fold in vitro. *Science* **284**, 325–327 (1999).

Cossins, A. Cryptic clues revealed. *Nature* **396**, 309–310 (1998).

Couture, L. A., and Stinchcomb, D. T. Anti-gene therapy: the use of ribozymes to inhibit gene function. *TIG* **12**, 510–515 (1996).

Coveney, P., and Highfield, R. *Frontiers of Complexity: The Search for Order in a Chaotic World* (Faber and Faber, London, 1996).

Cox, J. C., Cohen, D. S., and Ellington, A. D. The complexities of DNA computation. *TIBTECH* **17**, 151–154 (1999).

Cox, R., Mirkin, S. M. Characteristic enrichment of DNA repeats in different genomes. *Proc. Natl. Acad. Sci. U.S.A.* **94**, 5237–5242 (1997).

Coyne, J. A., Crittenden, A. P., and Mah, K. Genetics of a pheromonal difference contributing to reproductive isolation in *Drosophila*. *Science* **265**, 1461–1464 (1994).

Crow, J. F. Molecular evolution – who is in the driver's seat? *Nature Genetics* **17**, 129–130 (1997).

Crozier, R. H., and Pamilo, P. One into two will go. *Nature* **383**, 574–575 (1996).

Crutchfield, J. P. and Mitchell, M. The evolution of emergent computation. *Proc. Natl. Acad. Sci. U.S.A.* **92**, 10742–10746 (1995).

Csermely, P. Proteins, RNAs and chaperones in enzyme evolution: a folding perspective. *TIBS* **22**, 147–149 (1997).

Csink, A. K., and Henikoff, S. Something from nothing: the evolution and utility of satellite repeats. *TIG* **14**, 200–204 (1998).

Culligan, K. M., and Hays, J. B. DNA mismatch repair in plants. *Plant Physiol.* **115**, 833–839 (1997).

Culotta, E. A boost for "adaptive" mutation. *Science* **265**, 318–319 (1994).

Cunningham, R. P. DNA repair: Caretakers of the genome? *Current Biology* **7**, R576-R579 (1997).

Czerny, T., Halder, G., Kloter, U. Souabni, A., Gehring, W. J., and Busslinger, M. *twin* of *eyeless*, a second *Pax-6* gene of *Drosophila*, acts upstream of *eyeless* in the control of eye development. *Molecular Cell* **3**, 297–307 (1999).

Dahiyat, B. I., and Mayo, S. L. De novo protein design: fully automated sequence selection. *Science* **278**, 82–87 (1997).
Dalal, S., Balasubramanian, S., and Regan, L. Protein alchemy: Changing β-sheet into α-helix. *Nature structural biology* **4**, 548–552 (1997).
Daltry, J. C., Wüster, W., and Thorpe, R. S. Diet and snake venom evolution. *Nature* **379**, 537–540 (1996).
Davidson, A. R., and Sauer, R. T. Folded proteins occur frequently in libraries of random amino acid sequences. *Proc. Natl. Acad. Sci. U.S.A.* **91**, 2146–2150 (1994).
Davidson, A. R., Lumb, K. J., and Sauer, R. T. Cooperatively folded proteins in random sequence libraries. *Nature Structural Biology* **2**, 856–863 (1995).
Davis, A. P., and Capecchi, M. R. Axial homeosis and appendicular skeleton defects in mice with a targeted disruption of *hoxd-11*. *Development* **120**, 2187–2198 (1994).
Davis, M. C. Making a living at the extremes. *TIBTECH* **16**, 102–104 (1998).
Dawkins, R. *Climbing Mount Improbable* (Viking, London, 1996).
Dawkins, R. *River out of Eden: A Darwinian View of Life* (Weidenfeld & Nicolson, London, 1995).
Dawkins, R. *The Blind Watchmaker* (Penguin Books, London, 1988).
De Duve, C. The birth of complex cells. *Scientific American*, 38–45 (1996).
De Robertis, E. M. Dismantling the organizer. *Nature* **374**, 407–409 (1995).
De Robertis, E. M., Oliver, G., and Wright, C. V. E. Homeobox genes and the vertebrate body plan. *Scientific American*, 26–32 (July 1990).
Dean, J. Animats and what they can tell us. *Trends in Cognitive Sciences* **2**, 60–67 (1998).
Deckert, G., Warren, P. V., Gaasterland, T., Young, W. G., Lenox, A. L., Graham, D. E., Overbeek, R., Snead, M. A., Keller, M., Aujay, M., Huber, R., Feldman, R. A., Short, J. M., Olsen, G. J., and Swanson, R. V. The complete genome of the hyperthermophilic bacterium *Aquifex aeolicus*. *Nature* **392**, 353–358 (1998).
Degrado, W. F. Proteins from scratch. *Science* **278**, 80–81 (1997).
Deming, T. J. Facile synthesis of block copolypeptides of defined architecture. *Nature* **390**, 386–389 (1997).
Dennett, D. C. *Darwin's Dangerous Idea: Evolution and the Meanings of Life* (Simon & Schuster, New York, 1995).
Dennis, P. P. Ancient ciphers: Translation in archaea. *Cell* **89**, 1007–1010 (1997).
Denu, J. M., Stuckey, J. A., Saper, M. A., and Dixon, J. E. Form and function in protein dephosphorylation. *Cell* **87**, 361–364 (1996).
DeRisi, J. L., Iyer, V. R., and Brown, P. O. Exploring the metabolic and genetic control of gene expression on a genomic scale. *Science* **278**, 680–686 (1997).
Desmond, A., and Moore, J. *Darwin* (Penguin Books, London, 1992).
Devlin, J., Panganiban, L. C., and Devlin, P. E. Random peptide libraries: A source of specific protein binding molecules. *Science* **249**, 404–405 (1990).

Diamond, J. Location, location, location: The first farmers. *Science* **278**, 1243–1244 (1997).
Dickinson, W. J. Molecules and morphology: where's the homology? *TIG* **11**, 119–121 (1995).
Dickman, S. Pääbo S: Pushing ancient DNA to the limit. *Current Biology* **8**, R329-R330 (1998).
Dill, K. A., and Chan, H. S. From Levinthal to pathways to funnels. *Nature Structural Biology* **4**, 10–19 (1997).
DiVincenzo, D. P. Quantum computation. *Science* **270**, 255–261 (1995).
Dixon, R. A., and Arntzen, C. J. Transgenic plant technology is entering the era of metabolic engineering. *TIBTECH* **15**, 441–444 (1997).
Doll, J. J. The patenting of DNA. *Science* **280**, 689–690 (1998).
Doolittle, R. F. A bug with excess gastric avidity. *Nature* **388**, 515–516 (1997).
Doolittle, R. F. The multiplicity of domains in proteins. *Annu. Rev. Biochem.* **64**, 287–314 (1995).
Doolittle, R. F. The origins and evolution of eukaryotic proteins. *Phil. Trans. R. Soc. Lond. B* **349**, 235–240 (1995).
Doolittle, R. F., Feng, D-F., Tsang, S., Cho, G., and Little, E. Determining divergence times of the major kingdoms of living organisms with a protein clock. *Science* **271**, 470–477 (1996).
Doolittle, W. F. A paradigm gets shifty. *Nature* **392**, 15–16 (1998).
Dorit, R. L., and Gilbert, W. The limited universe of exons. *Current Opinion in Genetics and Development* **1**, 464–469 (1991).
Doudna, J. A. A molecular contortionist. *Nature* **388**, 830–831 (1997).
Dover, G. A. On the edge. *Nature* **365**, 704–706 (1993).
Dover, G. Ignorant DNA? *Nature* **285**, 618–620 (1980).
Drexler, K. E. Building molecular machine systems. *TIBTECH* **17**, 5–7 (1999).
Drickamer, K., and Taylor, M. E. Evolving views of protein glycosylation. *TIBS* **23**, 321–324 (1998).
Dronamraju, K. R. *Haldane's Daedalus Revisited* (Oxford University Press, Oxford, 1995).
Duan, Y., and Kollman, P. A. Pathways to a protein folding intermediate observed in a 1-microsecond simulation in aqueous solution. *Science* **282**, 740–744 (1998).
Duboule, D. Temporal colinearity and the phylotypic progression: a basis for the stability of a vertebrate Bauplan and the evolution of morphologies through heterochrony. *Development*, Supplement, 135–142 (1994).
Dujon, B. The yeast genome project: what did we learn? *TIG* **12**, 263–269 (1996).
Dyer, B. D., and Obar, R. A. *Tracing the History of Eukaryotic Cells: The Enigmatic Smile* (Columbia University Press, New York, 1994).
Dyson, F. J. A model for the origin of life. *J. Mol. Evol.* **18**, 344–350 (1982).
Dyson, F. J. *Infinite in all Directions: Gifford Lecture Given at Aberdeen, Scotland, April-November 1985* (Penguin Books, London, 1990).

Eaton, W. A., Henry, E. R., Hofrichter, J., and Mozzarelli, A. *Nature Structural Biology* **6**, 351–357 (1999).

Eckland, E. H., and Bartel, D. P. RNA-catalysed RNA polymerization using nucleoside triphosphates. *Nature* **382**, 373–376 (1996).

Edgell, D. R., and Doolittle, W. F. Archaebacterial genomics: the complete genome sequence of *Methanococcus jannaschii*. *BioEssays* **19**, 1–4 (1997).

Editorial. Entering the total-genomic era. *Nature Genetics* **15**, 111–112 (1997).

Editorial. To affinity . . . and beyond! *Nature Genetics* **14**, 367–370 (1996).

Eigen, M. *Steps towards Life: A Perspective on Evolution* (Oxford University Press, Oxford, 1992).

Eisenberg, D. Into the black of night. *Nature structural biology* **4**, 95–97 (1997).

Ekland, E. H., Szostak, J. W., and Bartel, D. P. Structually complex and highly active RNA ligases derived from random RNA sequences. *Science* **269**, 364–370 (1995).

Ekland, E. H., and Bartel, D. P. RNA-catalysed RNA polymerization using nucleoside triphosphates. *Nature* **382**, 373–376.

Elena, S. F., Cooper, V. S., and Lenski, R. E. Punctuated evolution caused by selection of rare beneficial mutations. *Science* **272**, 1802–1804 (1996).

Elgar, G., Sandford, R., Aparicio, S., Macrae, A., Venkatesh, B., and Brenner, S. Small is beautiful: comparative genomics with the pufferfish. (*Fugu rubripes*) *TIG* **12**, 145–150 (1996).

Ellegren, H., and Fridolfsson, A-K. Male-driven evolution of DNA sequences in birds. *Nature Genetics* **17**, 182–184 (1997).

Ellington, A. D., and Szostak, J. W. Selection *in vitro* of single-stranded DNA molecules that fold into specific ligand-binding structures. *Nature* **355**, 850–852 (1992).

Ellis, R. J. Molecular chaperones: Avoiding the crowd. *Current Biology* **7**, R531-R533 (1997).

Ellis, R. J. Roles of molecular chaperones in protein folding. *Current Opinion in Structural Biology* **4**, 117–122 (1994).

Emmeche, C. *The Garden in the Machine: The Emerging Science of Artificial Life* (Princeton University Press, Princeton, New Jersey, 1994).

Emmons, S. W. Simple worms, complex genes. *Nature* **382**, 301–302 (1996).

Emmons, S. W. Worms as an evolutionary model. *TIG* **13**, 131–134 (1997).

Epstein, A. H., and Senturia, S. D. Macro power from micro machinery. *Science* **276**, 1211 (1997).

Farabaugh, P. J. Alternative readings of the genetic code. *Cell* **74**, 591–596 (1993).

Feng, Q., Park, T. K., and Rebek, J. Crossover reactions between synthetic replicators yield active and inactive recombinants. *Science* **256**, 1179–1180 (1992).

Ferré-D'Amaré, A. R., Zhou, K., and Doudna, J. A. Crystal structure of a hepatitis delta virus ribozyme. *Nature* **395**, 567–573 (1998).

Ferris, J. P., Hill, A. R., Liu, R., and Orgel, L. E. Synthesis of long prebiotic oligomers on mineral surfaces. *Nature* **381**, 59–61 (1996).

Ferster, D., and Spruston, N. Cracking the neuronal code. *Science* **270**, 756–757 (1995).

Finkelstein, A. V. Implications of the random characteristics of protein sequences for their three-dimensional structure. *Current Opinion in Structural Biology* **4**, 422–428 (1994).

Finkelstein, A. V., and Ptitsyn, O. B. Why do globular proteins fit the limited set of folding patterns? *Prog. Biophys. Molec. Biol.* **50**, 171–190 (1987).

Firestine, S. M., Nixon, A. E., and Benkovic, S. J. Threading your way to protein function. *Chemistry & Biology* **3**, 779–783 (1996).

Fischer, T. M., Blazis, D. E. J., Priver, N. A., and Carew, T. J. Metaplasticity at identified inhibitory synapses in *Aplysia*. *Nature* **389**, 860–865 (1997).

Fischmann, J. Have 25-million-year-old bacteria returned to life? *Science* **268**, 977 (1995).

Fishman, M. C., and Olson, E. N. Parsing the heart: Genetic modules for organ assembly. *Cell* **91**, 153–156 (1997).

Flam, F. The chemistry of life at the margins. *Science* **265**, 471–472 (1994).

Fleischmann, R. D. *et al.* Whole-genome random sequencing and assembly of *Haemophilus influenzae* Rd. *Science* **269**, 496–512(1995).

Fontana, W., and Schuster, P. Continuity in evolution: On the nature of transitions. *Science* **280**, 1451–1455 (1998).

Forrest, S. Genetic algorithms: principles of natural selection applied to computation. *Science* **261**, 872–878 (1993).

Fraser, C. M. *et al.* Complete genome sequence of *Treponema pallidum*, the syphilis spirochete. *Science* **281**, 375–388 (1998).

Fraser, C. M., *et al.* The minimal gene complement of *Mycoplasma genitalium*. *Science* **270**, 397–403 (1995).

Fredrickson, J. K., and Onstott, T. C. Microbes deep inside the earth. *Scientific American*, 42–47 (1996).

Fried, M., Nosten, F., Brockman, A., Brabin, D. J., and Duffy, P. E. Maternal antibodies block malaria. *Nature* **395**, 851–852 (1998).

Fullerton, M. Phylogeny and molecular biology: reconstructing the tree of life. *TIG* **12**, 533 (1996).

Fussenegger, M., Schlatter, S., Dätwyler, D., Mazur, X., and Bailey, J. E. Controlled proliferation by multigene metabolic engineering enhances the productivity of Chinese hamster ovary cells. *Nature Biotechnology* **16**, 468–471 (1998).

Fuster, J. M. Network memory. *TINS* **20**, 451–459 (1997).

Gaasterland, T. Structural genomics taking shape. *TIG* **14**, 135 (1998).

Ganfornina, M. D., and Sánchez, D. Generation of evolutionary novelty by functional shift. *BioEssays* **21**, 432–439 (1999).

Garcia-Fernàndez, J., and Holland, P. W. H. Archetypal organization of the amphioxus *Hox* gene cluster. *Nature* **370**, 563–566 (1994).

Gee, H. It's a mole-rat, Jim, but not as we know it. *Nature* **380**, 584 (1996).

Gelbart, W. M. Databases in genomic research. *Science* **282**, 659–661 (1998).

Gellon, G., and McGinnis, W. Shaping animal body plans in development and evolution by modulation of *Hox* expression patterns. *BioEssays* **20**, 116–125 (1998).
Gerstein, M. Patterns of protein-fold usage in eight microbial genomes: a comprehensive structural census. *Proteins* **33**, 518–534 (1998).
Gesteland, R. F., and Atkins, J. F. (Eds) *The RNA World: The Nature of Modern RNA Suggests a Prebiotic RNA World* (Cold Spring Harbor Laboratory Press, USA, 1993).
Gesteland, R. F., Atkins, J. F. Recoding: dynamic reprogramming of translation. *Annu. Rev. Biochem.* **65**, 741–768 (1996).
Ghadiri, M. R., Granja, J. R., Milligan, R. A., McRee, D. E., and Khazanovich, N. Self-assembling organic nanotubes based on a cyclic peptide architecture. *Nature* **366**, 324–327 (1993).
Gibbons, A. Biologists trace the evolution of molecules. *Science* **257**, 30–31 (1992).
Gibbons, A. On the many origins of species. *Science* **273**, 1496–1498 (1996).
Gibbons, A. The species problem. *Science* **273**, 1501 (1996).
Gibbons, A. When it comes to evolution, humans are in the slow class. *Science* **267**, 1907–1908 (1995).
Gibbs, A. Tracing pedigrees of genes. *Science* **267**, 35–36 (1995).
Gibson, E. K., McKay, D. S., Thomas-Keprta, K., and Romanek, C. S. The case for relic life on mars. *Scientific American*, 36–41 (1997).
Gilbert, S. F., Opitz, J. M., and Raff, R. A. Resynthesizing evolutionary and developmental biology. *Developmental Biology* **173**, 357–372 (1996).
Gilbert, W. The RNA world. *Nature* **319**, 618 (1986).
Gimble, F. S. Putting protein splicing to work. *Chemistry & Biology* **5**, R251-R256 (1998).
Gleick, J. *Chaos: Making a New Science* (Abacus, London, 1995).
Godzik, A. Counting and classifying possible protein folds. *TIBTECH* **15**, 147–151 (1997).
Goffeau, A. Life with 482 genes. *Science* **270**, 445–446 (1995).
Golden, B.L., Gooding, A.R., Podell, E. R., and Cech, T. R. A preorganized active site in the crystal structure of the *Tetrahymena* ribozyme. *Science* **282**, 259–263 (1998).
Golding, B. Evolution: When was life's first branch point? *Current Biology* **6**, 679–682 (1996).
Goldstein, D. B., and Harvey, P. H. Evolutionary inference from genomic data. *BioEssays* **21**, 148–156 (1999).
Goldstein, D. J. An unacknowledged problem for structural genomics? *Nature Biotechnology* **16**, 696 (1998).
Gonzalez, L., Brown, R. A., Richardson, D., and Alber, T. Crystal structures of a single coiled-coil peptide in two oligomeric states reveal the basis for structural polymorphism. *Nature structural biology* **3**, 1002–1010 (1996).
Goodfellow, P. Complementary endeavours. *Nature* **377**, 285–286 (1995).
Goodman, M. F. Mutations caught in the act. *Nature* **378**, 237–238 (1995).

Gould, S. J. Of it, not above it. *Nature* **377**, 681–683 (1995).
Gould, S. J. The evolution of life on the Earth. *Scientific American*, 63–69 (October 1994).
Gould, S. J., and Lewontin, R. C. The spandrels of San Marco and the Panglossian paradigm: A critique of the adaptionist programme. *Proc. R. Soc. Lond. B.* **205**, 581–598 (1979).
Goyenechea, B., and Milstein, C. Modifying the sequence of an immunoglobulin V-gene alters the resulting pattern of hypermutation. *Proc. Natl. Acad. Sci. USA* **93**, 13979–13984 (1996).
Grady, D. Quick-change pathogens gain an evolutionary edge. *Science* **274**, 1081 (1996).
Grady, M., Wright, I., and Pillinger, C. Opening a martian can of worms? *Nature* **382**, 575–576 (1996).
Graham-Rowe, D. March of the biobots. *New Scientist*, 26–30 (December 1998).
Graves, D. J. Powerful tools for genetic analysis come of age. *TIBTECH* **17**, 127–134 (1999).
Gray, M. D., Shen, J-C., Kamath-Loeb, A. S., Blank, A., Sopher, B. L., Martin, G. M., Oshima, J., and Loeb, L. A. The Werner syndrome protein is a DNA helicase. *Nature Genetics* **17**, 100–103 (1997).
Gray, M. W. *Rickettsia*, typhus and the mitochondrial connection. *Nature* **396**, 109–110 (1998).
Gray, M. W. The third form of life. *Nature* **383**, 299 (1996).
Gray, N. S., Wodicka, L., Thunnissen, A-M. W. H., Norman, T. C., Kwon, S., Hernan Espinoza, F., Morgan, D. O., Barnes, G., LeClerc, S., Meijer, L., Kim, S-H., Lockhart, D. J., and Schultz, P. G. Exploiting chemical libraries, structure, and genomics in the search for kinase inhibitors. *Science* **281**, 533–538 (1998).
Gribbin, J. *In the Beginning: The Birth of the Living Universe* (Penguin Books, London, 1994).
Grob, M., and Flaxco, K. W. Reading, writing and redesigning. *Nature* **388**, 419–420 (1997).
Grotewiel, M. S., Beck, C. D. O., Wu, K. H., Zhu, X-R., and Davis, R. L. Integrin-mediated short-term memory in *Drosophila*. *Nature* **391**, 455–460 (1998).
Groves, J. T. The importance of being selective. *Nature* **389**, 329–330 (1997).
Gu, W., and Roeder, R. G. Activation of p53 sequence-specific DNA binding by acetylation of the p53 C-terminal domain. *Cell* **90**, 595–606 (1997).
Guarnieri, F., Fliss, M., and Bancroft, C. Making DNA add. *Science* **273**, 220–223 (1996).
Gupta, R. S., and Golding, G. B. The origin of the eukaryotic cell. *TIBS* **21**, 166–171 (1996).
Gura, T. One molecule orchestrates amoebae. *Science* **277**, 182 (1997).
Hager, A. J., Pollard, J. D., and Szostak, J. W. Ribozymes: aiming at RNA replication and protein synthesis. *Chemistry & Biology*, **3**, 717–725 (1996).

Hagerman, P. J., and Amiri, K. M. A. Hammering away at RNA global structure. *Current Opinion in Structural Biology* **6**, 317–321 (1996).

Hagerman, P. J., and Tinoco, I. Nucleic acids from sequence to structure to function. *Current Opinion in Structural Biology* **6**, 277–280 (1996).

Hakomori, S-I., and Zhang, Y. Glycosphingolipid antigens and cancer therapy. *Chemistry & Biochemistry* **4**, 97–104 (1997).

Halder, G., Callaerts, P., and Gehring, W. J. Induction of ectopic eyes by targeted expression of the *eyeless* gene in *Drosophila*. *Science* **267**, 1788–1792 (1995).

Hammer, J., Valsasnini, P., Tolba, K., Bolin, D., Higelin, J., Takacs, B., and Sinigaglia, F. Promiscuous and allele-specific anchors in HLA-DR-binding peptides. *Cell* **74**, 197–203 (1993).

Hammer, M. The neural basis of associative reward learning in honeybees. *TINS* **20**, 245–252 (1997).

Hanawalt, P. C. Transcription-coupled repair and human disease. *Science* **266**, 1957–1958 (1994).

Hancock, J. M. Simple sequences and the expanding genome. *BioEssays* **18**, 421–425 (1996).

Harbury, P. B., Plecs, J. J., Tidor, B., Alber, T., and Kim, P. S. High-resolution protein design with backbone freedom. *Science* **282**, 1462–1467 (1998).

Harrison, L. G. On growth and form. *Nature* **375**, 745–746 (1995).

Hartenstein, V., Lee, A., and Toga, A. W. A graphic digital database of *Drosophila* embryogenesis. *TIG* **11**, 51–58 (1995).

Hartley, B. S. Evolution of enzyme structure, *Proc. R. Soc. Lond. B* **205**, 443–452 (1979).

Hartwell, L. A robust view of biochemical pathways. *Nature* **387**, 855–857 (1997).

Hatch, T. Chlamydia: old ideas crushed, new mysteries bared. *Science* **282**, 638–639 (1998).

Haupt, K., and Mosbach, K. Plastic antibodies: developments and applications. *TIBTECH* **16**, 468–475 (1998).

Hecht, J., and Concar, D. Earth oddities tell their tale. *New Scientist*, (August 1996), p. 7.

Henderson, R. Macromolecular structure and self-assembly. *The limits of reductionism in biology. Wiley, Chichester (Novartis Foundation Symposium 213)*, 36–55 (1998).

Henikoff, S., and Matzke, M. A. Exploring and explaining epigenetic effects. *TIG* **13**, 293–295 (1997).

Henikoff, S., Greene, E. A., Pietrokovski, S., Bork, P., Attwood, T. K., and Hood, L. Gene families: the taxonomy of protein paralogs and chimeras. *Science* **278**, 609–614 (1997).

Hensch, T. K., Fagiolini, M., Mataga, N., Stryker, M. P., Baekkeskov, S., and Kash, S. F. Local GABA circuit control of experience-dependent plasticity in developing visual cortex. *Science* **282**, 1504–1508 (1998).

Hentze, M. W., and Kulozik, A. E. A perfect message: RNA surveillance and nonsense-mediated decay. *Cell* **96**, 307–310 (1999).

Heringa, J., and Taylor, W. R. Three-dimensional domain duplication, swapping and stealing. *Current Opinion in Structural Biology* **7**, 416–421 (1997).

Herschlag, D. Ribozyme crevices and catalysis. *Nature* **395**, 548–549 (1998).

Hieter, P., and Boguski, M. Functional genomics: It's all how you read it. *Science* **278**, 601–602 (1997).

Hirao, I., and Ellington, A. D. Re-creating the RNA world. *Current Biology* **5**, 1017–1022 (1995).

Hjelmfelt, A., Weinberger, E. D., and Ross, J. Chemical implementation of neural networks and Turing machines. *Proc. Natl. Acad. Sci. USA* **88**, 10983–10987 (1991).

Hjelmfelt, A.,Weinberger, E. D., and Ross, J. Chemical implementation of finite-state machines. *Proc. Natl. Acad. Sci. USA* **89**, 383–387 (1992).

Hodgkin, J., and Herman, R. K. Changing styles in *C. elegans* genetics. *TIG* **14**, 352–357 (1998).

Hodgkin, J., Horvitz, H. R., Jasny, B. R., Kimble, J. *C. elegans*: Sequence to biology. *Science* **282**, 2011 (1998).

Höfer, T., and Maini, P.K. Turing patterns in fish skin? *Nature* **380**, 678 (1996).

Holder, N., and McMahon, A. Genes from zebrafish screens. *Nature* **384**, 515–516 (1996).

Holland, H. D. Evidence for life on earth more than 3850 million years ago. *Science* **275**, 38–39 (1997).

Holland, J. H. Genetic algorithms. *Scientific American*, 44–50 (July 1992).

Holland, J. H. *Hidden Order: How Adaptation Builds Complexity* (Addison-Wesley Publishing Company, Inc., Reading, MA, 1995).

Holland, P. W. H. Vertebrate evolution: Something fishy about *Hox* genes. *Current Biology* **7**, R570-R572 (1997).

Holland, P. W. H., and Garcia-Fernàndez, J. *Hox* genes and chordate evolution. *Developmental Biology* **173**, 382–395 (1996).

Holland, P. W. H., Garcia-Fernàndez, J., Williams, N. A., and Sidow, A. Gene duplications and the origins of vertebrate development. *Development* **Supplement**, 125–133 (1994).

Hölldobler, B., and Wilson, E. O. *Journey to the Ants: A Story of Scientific Exploration* (Harvard University Press, Cambridge, Massachusetts, 1995).

Holliday, R. A different kind of inheritance. *Scientific American*, 40–48 (June 1989).

Holm, L., and Sander, C. Mapping the protein universe. *Science* **273**, 595–602 (1996).

Holmgren, S. K., Bretscher, L. E., Taylor, K. M., and Raines, R. T. *Chemistry & Biology* **6**, 63–70 (1999).

Holstege, F. C. P., Jennings, E. G., Wyrick, J. J., Lee, T. I., Hengartner, C. J., Green, M. R., Golub, T. R., Lander, E. S., and Young, R. A. Dissecting the regulatory circuitry of a eukaryotic genome. *Cell* **95**, 717–728 (1998).

Hong, S., Candelone, J-P., Patterson, C. C., and Boutron, C. F. Greenland ice evidence of hemispheric lead pollution two millennia ago by Greek and Roman civilizations. *Science* **265**, 1841–1843 (1994).

Hopfield, J. J. Kinetic proofreading: A new mechanism for reducing errors in biosynthetic processes requiring high specificity. *Proc. Natl. Acad. Sci. U.S.A.* **71**, 4135–4139 (1974).

Horowitz, P. M. Ironing out the protein folding problem? *Nature Biotechnology* **17**, 136–137 (1999).

Horwich, A. L., and Weissman, J. S. Deadly conformations – protein misfolding in prion disease. *Cell* **89**, 499–510 (1997).

Hughes, A. L., and Hughes, M. K. Small genomes for better flyers. *Nature* **377**, 391 (1995).

Hunt, G. R. Manufacture and use of hook-tools by New Caledonian crows. *Nature* **379**, 249–251 (1996).

Hurst, L. D. The silence of the genes. *Current Biology* **5**, 459–461 (1995).

Hutchison, R. Through the clouds of obscurity. *Nature* **365**, 704–706 (1993).

Huynen, M. A. Exploring phenotype space through neutral evolution. *J. Mol. Evol.* **43**, 165–169 (1996).

Huynen, M. A., Diaz-Lazcoz, Y, and Bork, P. Differential genome display. *TIG* **13**, 389–390 (1997).

Inagaki, Y., Ehara, M., Watanabe, K. I., Hayashi-Ishimaru, Y., and Ohama, T. Directionally evolving genetic code: the UGA codon from stop to tryptophan in mitochondria. *J. Mol. Evol.* **47**, 378–384 (1998).

Issa, J-P. J., and Baylin, S. B. Epigenetics and human disease. *Nature Medicine* **2**, 281–282 (1996).

Iverson, B. L. Betas are brought into the fold. *Nature* **385**, 113–115 (1997).

Iyer, V. R., Eisen, M. B., Ross, D. T., Schuler, G., Moore, T., Lee, J. C. F., Trent, J. M., Staudt, L. M., Hudson, J., Boguski, M. S., Lashkari, D., Shalon, D., Botstein, D., and Brown, P. O. The transcriptional program in the response of human fibroblasts to serum. *Science* **283**, 83–87 (1999).

Jablonka, E., and Lamb, M. J. *Epigenetic Inheritance and Evolution: The Lamarckian Dimension* (Oxford University Press, Oxford, 1995).

Jacobsen, S. E., and Meyerowitz, E. M. Hypermethylated *SUPERMAN* epigenetic alleles in *Arabidopsis*. *Science* **277**, 1100–1103 (1997).

Jaeger, L. The new world of ribozymes. *Current Opinion in Structural Biology* **7**, 324–335 (1997).

Jaenicke, R. Folding and association of proteins. *Prog. Biophys. Molec. Biol.* **49**, 117–237 (1987).

Jaenisch, R. DNA methylation and imprinting: why bother? *TIG* **13**, 323–329 (1997).

Jan, Y. N., and Jan, L. Y. Asymmetric cell division. *Nature* **392**, 775–778 (1998).

Janner, A. De nive sexangula stellata. *Acta Cryst.* **A53**, 615–631 (1997).

Jayeraman, K. S. Indian researchers press for stricter rules to regulate 'gene-hunting'. *Nature* **379**, 381- (1996).

Jeffery, C. J. Moonlighting proteins. *TIBS* **24**, 8–11 (1999).

Jeffrey, P. D., Russo, A. A., Polyak, K., Gibbs, E., Hurwitz, J., Massagué, J., and Pavletich, N. P. Mechanism of CDK activation revealed by the structure of a cyclinA-CDK2 complex. *Nature* **376**, 313–320 (1995).

Jermann, T. M., Opitz, J. G., Stackhouse, J., and Benner, S. A. Reconstructing the evolutionary history of the artiodactyl ribonuclease superfamily. *Nature* **374**, 57–59 (1995).

Jiang, F., Kumar, R. A., Jones, R. A., and Patel, D. J. Structural basis of RNA folding and recognition in an AMP-RNA aptamer complex. *Nature* **382**, 183–186 (1996).

Jiménez-Sánchez, A. On the origin and evolution of the genetic code. *J. Mol. Evol.* **41**, 712–716 (1995).

Johnson-Laird, P. *The Computer and the Mind: An Introduction to Cognitive Science* (Fontana Press, London, 1993).

Johnston, M. Gene chips: Array of hope for understanding gene regulation. *Current Biology* **8**, R171-R174 (1998).

Johnston, M. Genome sequencing: The complete code for a eukaryotic cell. *Current Biology* **6**, 500–503 (1996).

Johnston, M. Towards a complete understanding of how a simple eukaryotic cell works. *TIG* **12**, 242–243 (1996).

Johnston, M., *et al.* Complete nucleotide sequence of *Saccharomyces cerevisiae* chromosome VIII. *Science* **265**, 2077–2082 (1994).

Jones, P. A., and Gonzalgo, M. L. Altered DNA methylation and genome instability: A new pathway to cancer? *Proc. Natl. Acad. Sci. USA* **94**, 2103–2105 (1997).

Joyce, G. F. Directed molecular evolution. *Scientific American*, 48–55 (1992).

Joyce, G. F. Evolutionary chemistry: getting there from here. *Science* **276**, 1658–1659 (1997).

Joyce, G. F. Ribozymes: Building the RNA world. *Current Biology* **6**, 965–967 (1996).

Joyce, G. F. RNA evolution and the origins of life. *Nature* **338**, 217–223 (1989).

Junge, W., Lill, H., and Engelbrecht, S. ATP synthase: an electrochemical transducer with rotatory mechanics. *TIBS* **22**, 420–423 (1997).

Kahn, P. Zebrafish hit the big time. *Science* **264**, 904–905 (1994).

Kaiser, J. Enivornment institute lays plans for gene hunt. *Science* **278**, 569–570 (1997).

Kaku, M. *Hyperspace: A Scientific Odyssey Through the 10th Dimension* (Oxford University Press, Oxford, 1995).

Kamtekar, S., Schiffer, J. M., Xiong, H., Babik, J. M., and Hecht, M. H. Protein design by binary patterning of polar and nonpolar amino acids. *Science* **262**, 1680–1685 (1993).

Kanaar, R., and Hoeijmakers, J. H. J. From competition to collaboration. *Nature* **391**, 335–337 (1998).

Kardar, M. Which came first, protein sequence or structure? *Science* **273**, 610 (1996).

Karp, P. D. Metabolic databases. *TIBS* **23**, 114–116 (1998).

Kass, S. U., Pruss, D., and Wolffe, A. P. How does DNA methylation repress transcription? *TIG* **13**, 444–449 (1997).

Kato, C., Inoue, A., and Horikoshi, K. Isolating and characterizing deep-sea marine microorganisms. *TIBTECH* **14**, 6–11 (1996).

Katz, L. C., and Shatz, C. J. Synaptic activity and the construction of cortical circuits. *Science* **274**, 1133–1138 (1996).

Kauffman, S. A. *The Origins of Order* (Oxford University Press, Oxford, 1993).

Kauffman, S. A., Weinberger, E. D., and Perelson, A. S. Maturation of the immune response via adaptive walks on affinity landscapes. *Theoretical Immunology, Part One, SFI Studies in the Sciences Complexity*, Perelson, A. S. (Ed.) (Addison-Wesley Publishing Company, 1988) pp. 349–382.

Kauffman, S. *At Home in the Universe: The Search for Laws of Complexity* (Viking, Great Britain, 1995).

Kauffman, S. Even peptides do it. *Nature* **382**, 496–497 (1996).

Kauffman, S. Evolving evolvability. *Nature* **382**, 309–310 (1996).

Keeling, P. J. A kingdom's progress: Archezoa and the origin of eukaryotes. *BioEssays* **20**, 87–95 (1998).

Keene, J. D. RNA surfaces as functional mimetics of proteins. *Chemistry & Biology* **3**, 505–513 (1996).

Keese, P., and Gibbs, A. Plant viruses: master explorers of evolutionary space. *Current Opinion in Genetics and Development* **3**, 873–877 (1993).

Kerr, R. A. Ancient life on Mars? *Science* **273**, 864–866 (1996).

Kerr, R. A. Animal oddballs brought into the ancestral fold? *Science* **270**, 580–581 (1995).

Kerr, R. A. Did Darwin get it all right? *Science* **267**, 1421–1422 (1995).

Kerr, R. A. Life goes to extremes in the deep earth – and elsewhere? *Science* **276**, 703–704 (1997).

Kerr, R. A. Requiem for life on Mars? Support for microbes fades. *Science* **282**, 1398–1399 (1998).

Kestenbaum, D. New math speeds the search for protein structures. *Science* **282**, 30–31 (1998).

Kiernan, V., Hecht, J., Cohen, P., and Concar, D. Did martians land in Antarctica? *New Scientist*, 4–5 (August 1996).

King, J. Refolding with a piece of the ring. *Nature Biotechnology* **15**, 514–515 (1997).

Kirkwood, A., Rioult, M. G., and Bear, M. F. Experience-dependent modification of synaptic plasticity in visual cortex. *Nature* **381**, 526–528 (1996).

Klenk, H-P. *et al.* The complete genome sequence of the hyperthermophilic, sulphate-reducing archaeon *Archaeoglobus fulgidus*. *Nature* **390**, 364–370 (1997).

Knight, R. D., and Landweber, L. F. Rhyme or reason: RNA-arginine interactions and the genetic code. *Chemistry & Biology* **5**, R215-R220 (1998).

Knoll, A. H. Breathing room for early animals. *Nature* **382**, 111–112 (1996).

Knoppers, B. M., and Chadwick, R. The human genome project: Under an international ethical microscope. *Science* **265**, 2035–2036 (1994).

Koch, C., and Laurent, G. Complexity and the nervous system. *Science* **284**, 96–98 (1999).

Koepp, M. J., Gunn, R. N., Lawrence, A. D., Cunningham, V. J., Dagher, A., Jones, T., Brooks, D. J., Bench, C. J., and Grasby, P. M. Evidence for striatal dopamine release during a video game. *Nature* **393**, 266–268 (1998).

Komiyama, N. H., Miyazaki, G., Taime, J. and Nagai, K. Transplanting a unique allosteric effect from crocodile into human haemoglobin. *Nature* **373**, 244–246 (1995).

Koonin, E. V., Mushegian, A. R., and Rudd, K. E. Sequencing and analysis of bacterial genomes. *Current Biology* **6**, 404–416 (1996).

Koonin, E. V., Tatusov, R. L., and Rudd, K. E. Sequence similarity of *Escherichia coli* proteins: Functional and evolutionary implications. *Proc. Natl. Acad. Sci. USA* **92**, 11921–11925 (1995).

Koonin, E. Why genome analysis? *TIG* **15**, 131 (1999).

Kortemme, T., Ramírez-Alvarado, M., and Serrano, L. Design of a 20-amino acid, three-stranded β-sheet protein. *Science* **281**, 253–256 (1998).

Koshi, J. M., and Goldstein, R. A. Mutation matrices and physical-chemical properties: Correlations and implications. *PROTEINS: Structure, Function, and Genetics* **27**, 336–344 (1997).

Koshland, D. E. Molecule of the year: The DNA repair enzyme. *Science* **266**, 1925 (1994).

Krings, M., Stone, A., Schmitz, R. W., Krainitzki, H., Stoneking, M., and Pääbo, S. Neandertal DNA sequences and the origin of modern humans. *Cell* **90**, 19–30 (1997).

Kunst, F. *et al.* The complete genome sequence of the Gram-positive bacterium *Bacillus subilis*. *Nature* **390**, 249–256 (1997).

Küppers, B-O. *Information and the Origin of Life* (The MIT Press, Cambridge, Massachusetts, 1990).

Kuwabara, P. E. Worming your way through the genome. *TIG* **13**, 455–460 (1997).

Kuwabara, T., Warashina, M., Orita, M., Koseki, S., Ohkawa, J., and Taira, K. Formation of a catalytically active dimer by tRNA ^{Val}-driven short ribozymes. *Nature Biotechnology* **16**, 961–965 (1998).

Kuwabara, T., Warashina, M., Tanabe, T., Tani, K., Asano, S., and Taira, K. A novel allosterically *trans*-activated ribozyme, the maxizyme, with exceptional specificity in vitro and in vivo. *Molecular Cell* **2**, 617–627 (1998).

Lamour, V., Quevillon, S., Diriong, S., N'Guyen, V. C., Lipinski, M., and Mirande, M. Evolution of the Glx-tRNA synthetase family: The glutaminyl enzyme as a case of horizontal gene transfer. *Proc. Natl. Acad. Sci. USA* **91**, 8670–8674 (1994).

Langton, C. G. (Ed) *Artificial Life III: Proceedings of the Workshop on Artificial Life Held June 1992, in Santa Fe, New Mexico* (Addison-Wesley Publishing Company, Reading, MA, 1994).

Langton, C. G., Taylor, C., Farmer, J. D., and Rasmussen, S. (Eds) *Artificial Life II: Proceedings of the Workshop on Artificial Life Held February 1990, in Santa Fe, New Mexico* (Addison-Wesley Publishing Company, Reading, MA, 1992).

Laufer, E., and Marigo, V. Evolution in developmental biology: of morphology and molecules. *TIG* **10**, 261–263 (1994).

Lawler, A. Ames tackles the riddle of life. *Science* **279**, 1840–1841 (1998).

Lawler, A. Finding puts mars exploration on front burner. *Science* **273**, 865 (1996).

Lawler, A. Mars meteorite quest goes global. *Science* **273**, 1653–1654 (1996).

Lawrence, P. A. *The Making of a Fly: The Genetics of Animal Design* (Blackwell Science, Oxford, 1995).

Lawrence, P. A., and Struhl, G. Morphogens, compartments, and pattern: Lessons from drosophila? *Cell* **85**, 951–961 (1996).

Lazcano, A. Biotic survivors. *Science* **277**, 46 (1997).

Lazcano, A., and Miller, S. L. The origin and early evolution of life: Prebiotic chemistry, the pre-RNA world, and time. *Cell* **85**, 793–798 (1996).

Leakey, R. *The Origin of Humankind* (Weidenfeld & Nicolson, London, 1994).

Lebrun, A., and Lavery, R. Unusual DNA conformations. *Current Opinion in Structural Biology* **7**, 348–354 (1997).

LeClerc, J. E., Li, B., Payne, W. L., and Cebula, T. A. High mutation frequencies among *Escherichia coli* and Salmonella pathogens. *Science* **274**, 1208–1211 (1996).

Lee, D. H., Granja, J. R., Martinez, J. A., Severin, K., and Ghadiri, M. R. A self-replicating peptide. *Nature* **382**, 525–528 (1996).

Lee, D. H., Severin, K., Yokobayashi, Y., and Ghadiri, M. R. Emergence of symbiosis in peptide self-replication through a hypercyclic network. *Nature* **390**, 591–594 (1997).

Lee, K-J., McCormick, W. D., Pearson, J. E., and Swinney, H. L. Experimental observation of self-replicating spots in a reaction-diffusion system. *Nature* **369**, 215–218 (1994).

Lehmann, A. R. Dual functions of DNA repair genes: molecular, cellular, and clinical implications. *BioEssays* **20**, 146–155 (1998).

Lengauer, C., Kinzler, K. W., and Vogelstein, B. Genetic instability in colorectal cancers. *Nature* **386**, 623–627 (1997).

Levinton, J. S. The big bang of animal evolution. *Scientific American*, 52–59 (November 1992).

Levy, S. *Artificial Life: The Quest for a New Creation* (Penguin Books, London, 1993).

Lewin, B. The mystique of epigenetics. *Cell* **93**, 301–303 (1998).

Lewis, E. B. Homeosis: the first 100 years. *TIG* **10**, 341–343 (1994).

Li, H., Helling, R., Tang, C., and Wingreen, N. Emergence of preferred structures in a simple model of protein folding. *Science* **273**, 666–669 (1996).

Li, T., and Nicolaou, K. C. Chemical self-replication of palindromic duplex DNA. *Nature* **369**, 218–221 (1994).

Li, W-H. *Molecular Evolution* (Sinauer Associates, Inc., Massachusetts, USA, 1997).

Lichten, M., and Goldman, A. S. H. Meiotic recombination hotspots. *Annu. Rev. Genetics* **29**, 423–444 (1995).

Lightowlers, R. N., Chinnery, P. F., Turnbull, D. M., and Howell, N. Mammalian mitochondrial genetics: heredity, heterplasmy and disease. *TIG* **13**, 450–455 (1997).

Lin, C. H., and Patel, D. J. Encapsulating an amino acid in a DNA fold. *Nature Structural Biology* **3**, 1046–1050 (1996).

Lin, K., Dorman, J. B., Rodan, A., and Kenyon, C. *daf-16*: An HNF-3/forkhead family member that can function to double the life-span of *Caenorhabditis elegans*. *Science* **278**, 1319–1322 (1997).

Lin, Y-J., Seroude, L., and Benzer, S. Extended life-span and stress resistance in the *drosophila* mutant *Methuselah*. *Science* **282**, 943–946 (1998).

Lindahl, T. Facts and artifacts of ancient DNA. *Cell* **90**, 1–3 (1997).

Lindahl, T. Instability and decay of the primary structure of DNA. *Nature* **362**, 709–715 (1993).

Lipton, R. J. DNA solution of hard computational problems. *Science* **268**, 542–548 (1995).

Lokey, R. S., and Iverson, B. L. Synthetic molecules that fold into a pleated secondary structure in solution. *Nature* **375**, 303–305 (1995).

Loomis, W. F., and Sternberg, P. W. Genetic networks. *Science* **269**, 649 (1995).

López-García, P., and Moreira, D. Metabolic symbiosis at the origin of eukaryotes. *TIBS* **24**, 88–93 (1999).

Lorimer, G. Folding with a two-stroke motor. *Nature* **388**, 720–723 (1997).

Louise-May, S., Auffinger, P., and Westhof, E. Calculations of nucleic acid conformations. *Current Opinion in Structural Biology* **6**, 289–298 (1996).

Lowe, C. J., and Wray, G. A. Radical alterations in the roles of homeobox genes during echinoderm evolution. *Nature* **389**, 718–721 (1997).

Luger, K., Mäder, A. W., Richmond, R. K., Sargent, D. F., and Richmond, T. J. Crystal structure of the nucleosome core particle at 2.8 Å resolution. *Nature* **389**, 251–260 (1997).

Luminet, J-P., Starkman, G. D., and Weeks, J. R. Is space finite? *Scientific American* , 68–75 (1999).

Luther, A., Brandsch, R., and von Kiedrowski, G. Surface-promoted replication and exponential amplification of DNA analogues. *Nature* **396**, 245–248 (1998).

Luzzatto, L., Bessler, M., and Rotoli, B. Somatic mutations in paroxysmal nocturnal hemoglobinuria: a blessing in disguise? *Cell* **88**, 1–4 (1997).

Lyubarev, A. E., and Kurganov, B. I. Biochemical organization and biochemical evolution. *Nanobiology* **4**, 47–54 (1996).

MacBeath, G., Kast, P., and Hilvert, D. Redesigning enzyme topology by directed evolution. *Science* **279**, 1958–1961 (1998).

Maddox, J. Polite row about models in biology. *Nature* **373**, 555 (1995).

Maeda, N., and Smithies, O. The evolution of multigene families: Human haptoglobin genes. *Ann. Rev. Genet.* **20**, 81–108 (1986).

Maley, L. E., and Marshall, C. R. The coming of age of molecular systematics. *Science* **279**, 505–506 (1998).

Manak, J. R., and Scott, M. P. A class act: conservation of homeodomain protein functions. *Development* , Supplement, 61–71 (1994).

Mandecki, W. The game of chess and searches in protein sequence space. *TIBTECH* **16**, 200 (1998).

Mann, S., and Ozin, G. A. Synthesis of inorganic materials with complex form. *Nature* **382**, 313–318 (1996).

Maoz, R., Matlis, S., DiMasi, E., Ocko, B. M., and Sagiv, J. Self-replicating amphiphilic monolayers. *Nature* **384**, 150–153 (1996).

Marchler-Bauer, A., Bryant, S. H. A measure of success in fold recognition. *TIBS* **22**, 236–240 (1997).

Marler, P. Unto the sweet bird's throat. *Nature* **382**, 592 (1996).

Marshall, E. NIH to produce a 'working draft' of the genome by 2001. *Science* **281**, 1774–1775 (1998).

Marshall, K. A., Robertson, M. P., and Ellington, A. D. A biopolymer by any other name would bind as well: a comparison of the ligand-binding pockets of nucleic acids and proteins. *Structure* **5**, 729–734 (1997).

Martin, W., and Muller, M. The hydrogen hypothesis for the first eukaryote. *Nature* **392**, 37–41 (1998).

Marx, J. DNA repair comes into its own. *Science* **266**, 728–730 (1994).

Masters, C. L., and Beyreuther, K. Tracking turncoat prion proteins. *Nature* **388**, 228–229 (1997).

Maynard Smith, J. *Did Darwin Get it Right?: Essays on Games, Sex and Evolution* (Penguin Books, London, 1993).

Maynard Smith, J. Natural selection and the concept of a protein space. *Nature* **225**, 563–564 (1970).

Maynard Smith, J., and Szathmáry, E. *The Major Transitions in Evolution* (W. H. Freeman and Company Limited, 1995).

Maynard Smith, J., and Szathmáry, E. *The Origins of Life: From the Birth of Life to the Origin of Language* (Oxford University Press, Oxford, 1999).

McAdams, H. H., and Shapiro, L. Circuit simulation of genetic networks. *Science* **269**, 650–658 (1995).

McEuen, P. L. Artificial atoms: New boxes for electrons. *Science* **278**, 1729–1730 (1997).

McGinnis, W., and Kuziora, M. The molecular architects of body design. *Scientific American*, 36–42 (February 1994).

McKay, D. S., Gibson, E. K., Thomas-Keprta, K. L., Vali, H., Romank, C. S., Clemett, S. J., Chillier, X. D. F., Maechling, C. R., and Zare, R. N. Search for past life on Mars: possible relic biogenic activity in martian meteorite ALH84001. *Science* **273**, 924–930 (1996).

McNaughton, B. Cognitive cartography. *Nature* **381**, 368–369 (1996).

Meinke, D. W., Cherry, J. M., Denn, C., Ronnsley, S. D., and Koornneef, M. *Arabidopsis thaliana*: A model plant for genome analysis. *Science* **282**, 662–682.

Mestel, R. Putting prions to the test. *Science* **273**, 184–189 (1996).

Meyer, A. Hox gene variation and evolution. *Nature* **391**, 225–227 (1998).

Meyer, E., and Duharcourt, S. Epigenetic programming of developmental genome rearrangements in ciliates. *Cell* **87**, 9–12 (1996).

Michel, F., and Westhof, E. Visualizing the logic behind RNA self-assembly. *Science* **273**, 1676–1677 (1996).

Miklos, G. L. G., and Rubin, G. M. The role of the genome project in determining gene function: Insights from model organisms. *Cell* **86**, 521–529 (1996).

Miller, D. P. Fluorescence studies of DNA and RNA and structure and dynamics. *Current Opinion in Structural Biology* **6**, 322–326 (1996).

Miller, R. V. Bacterial gene swapping in nature. *Scientific American*, 47–51 (January 1998).

Miller, S. L., and Lazcano, A. The origin of life – did it occur at high temperatures? *J. Mol. Evol.* **41**, 689–692 (1995).

Miller, S. L., and Orgel, L. E. *The Origins of Life on the Earth* (Prentice-Hall, Inc, New Jersey, 1974).

Minsky, M. Will robots inherit the earth? *Scientific American*, 87–91 (1994).

Mirkin, C. A., Letsinger, R. L., Mucic, R. C., Storhoff, J. J. A DNA-based method for rationally assembling nanoparticles into macroscopic materials. *Nature* **382**, 607–609 (1996).

Mlot, C. Microbes hint at a mechanism behind punctuated evolution. *Science* **272**, 1741 (1996).

Modrich, P. Mismatch repair, genetic stability, and cancer. *Science* **266**, 1959–1960 (1994).

Moerner, W. E. Those blinking single molecules. *Science* **277**, 1059–1061 (1997).

Mojzsis, S. J., Arrhenius, G., McKeegan, K. D., Harrison, T. M., Nutman, A. P., and Friend, C. R. L. Evidence for life on earth before 3,800 million years ago. *Nature* **384**, 55–59 (1996).

Monckton, D. G., Coolbaugh, M. I., Ashizawa, K. T., Siciliano, M. J., and Caskey, C. T. Hypermutable myotonic dystrophy CTG repeats in transgenic mice. *Nature Genetics* **15**, 193–196 (1997).

Monod, J. *Chance and Necessity: An Essay on the Natural Philosophy of Modern Biology* (Penguin Books, London, 1997).

Monroe, C., Meekhof, D. M., King, B. E., and Wineland, O. J. A "Schrödinger Cat" superposition state of an atom. *Science* **272**, 1131–1136.

Mooers, A. O., and Redfield, R. J. Digging up the roots of life. *Nature* **379**, 587–588 (1996).

Moore, J. C., and Arnold, F. H. Directed evolution of a para-nitrobenzyl esterase for aqueous-organic solvents. *Nature Biotechnology* **14**, 458–467 (1996).

Moore, M. J. Exploration by lamp light. *Nature* **374**, 766–767 (1995).

Moore, P. J., Reagan-Wallin, N. L., Haynes, K. F., and Moore, A. J. Odour conveys status on cockroaches. *Nature* **389**, 25 (1997).

Morell, V. How the malaria parasite manipulates its hosts. *Science* **278**, 223 (1997).

Morell, V. Life's last domain. *Science* **273**, 1043–1045 (1996).

Morell, V. Predator-free guppies take an evolutionary leap forward. *Science* **275**, 1880 (1997).

Morell, V. Proteins 'clock' the origins of all creatures – great and small. *Science* **271**, 448 (1996).

Morell, V. Starting species with third parties and sex wars. *Science* **273**, 1499–1502 (1996).

Morgan, D. O. Cyclin-dependent kinases: Engines, clocks, and microprocessors. *Annu. Rev. Cell Dev. Biol.* **13**, 261–291 (1997).

Morrell, V. Amazonian diversity: a river doesn't run through it. *Science* **273**, 1496–1497 (1996).

Morrell, V. The earliest art becomes older – and more common. *Science* **267**, 1908–1909 (1995).

Moxon, E. R., and Higgins, C. F. A blueprint for life. *Nature* **389**, 120–121 (1997).

Moxon, E. R., and Thaler, D. S. The tinkerer's evolving tool-box. *Nature* **387**, 659–662 (1997).

Mullins, J. Other worlds, other lives . . . *New Scientist*, 10 (August 1996).

Mullins, J., and Walker, G. The oldest meteorite from Mars. *New Scientist*, 10 (August 1996).

Murphy, G. G., and Glanzman, D. L. Mediation of classical conditioning in *Aplysia californica* by long-term potentiation of sensorimotor synapses. *Science* **278**, 467–469 (1997).

Murphy, M. P., and O'Neill, L. A. J. (Eds) *What is Life? The Next Fifty Years: Speculations on the Future of Biology* (Cambridge University Press, Cambridge, 1995).

Murray, A. W. How to compact DNA. *Science* **282**, 425–427 (1998).

Murzin, A. G. New protein folds. *Current Opinion in Structural Biology* **4**, 441–449 (1994).

Mushegian, A. R., and Koonin, E. V. A minimal gene set for cellular life derived by comparison of complete bacterial genomes. *Proc. Natl. Acad. Sci. U.S.A.* **93**, 10268–10273 (1996).

Myers, N. Mass extinction and evolution. *Science* **278**, 597–598 (1997).
Narlikar, G. J., and Herschlag, D. Mechanistic aspects of enzymatic catalysis: Lessons from comparison of RNA and protein enzymes. *Annu. Rev. Biochem.* **66**, 19–59 (1997).
Nash, H. A. Topological nuts and bolts. *Science* **279**, 1490–1491 (1998).
Naylor, G. J. P., and Brown, W. M. Structural biology and phylogenetic estimation. *Nature* **388**, 527–528 (1997).
Nédélec, F. J., Surrey, T., Maggs, A. C., and Leibler, S. Self-organization of microtubules and motors. *Nature* **389**, 305–308 (1997).
Nei, M., and Zhang, J. Molecular origin of species. *Science* **282**, 1428–1429 (1998).
Nelson, C. E., and Tabin, C. Footnote on limb evolution. *Nature* **375**, 630–631 (1995).
Nelson, J. C., Saven, J. G., Moore, J. S., and Wolynes, P. G. Solvophobically driven folding of nonbiological oligomers. *Science* **277**, 1793–1796 (1997).
Nelson, L. S., Rosoff, M. L., and Li, C. Disruption of a neuropeptide gene, *flp-1*, causes multiple behavioral defects in *Caenorhabditis elegans*. *Science* **281**, 1686–1690 (1998).
Netzer, W. J., and Hartl, F. U. Protein folding in the cytosol: chaperonin-dependent and -independent mechanisms. *TIBS* **23**, 68–73 (1998).
Netzer, W. J., and Hartl, F. U. Recombination of protein domains facilitated by co-translational folding in eukaryotes. *Nature* **388**, 343–349 (1997).
Nicolis, G., and Prigogine, I. *Exploring Complexity: An Introduction* (W.H. Freeman and Company, New York, 1989).
Nicoll, R. A., and Malenka, R. C. Long-distance long-term depression. *Nature* **388**, 427–429 (1997).
Nielsen, M. A., Knill, E., and Laflamme, R. Complete quantum teleportation using nuclear magnetic resonance. *Nature* **396**, 52–55 (1998).
Nigro, J. M., Cho, K. R., Fearon, E. R., Kern, S. E., Ruppert, J. M., Oliner, J. D., Kinzler, K. W., and Vogelstein, B. Scrambled exons. *Cell* **64**, 607–613 (1991).
Nisbet, E. G., and Fowler, C. M. R. Some liked it hot. *Nature* **382**, 404–405 (1996).
Nixon, A. E., Ostermeier, M., and Benkovic, S. J. Hybrid enzymes: manipulating enzyme design. *TIBTECH* **16**, 258–264 (1998).
Normile, D. Artificial life gets real as scientists meet in Japan. *Science* **272**, 1872–1873 (1996).
Normile, D. New views of the origins of mammals. *Science* **281**, 774–775 (1998).
Nottebohm, F. From bird song to neurogenesis. *Scientific American*, 56–61 (February 1989).
Nowak, M. A., Boerlijst, M. O., Cooke, J., and Maynard Smith, J. Evolution of genetic redundancy. *Nature* **388**, 167–171 (1997).
Nowak, R. Bacterial genome sequence bagged. *Science* **269**, 468–470 (1995).

Nowak, R. Mining treasures from 'junk DNA'. *Science* **263**, 608–610 (1994).
O'Brien, P. J., and Herschlag, D. Catalytic promiscuity and the evolution of new enzymatic activities. *Chemistry & Biology* **6**, R91-R105 (1999).
O'Neill, L., Murphy, M., and Gallagher, R. B. What are we? Where did we come from? Where are we going? *Science* **263**, 181–183 (1994).
O'Riain, M. J., Jarvis, J. U. M., and Faulkes, C. G. A dispersive morph in the naked mole-rat. *Nature* **380**, 619–621 (1996).
Oliver, S. A network approach to the systematic analysis of yeast gene function. *TIG* **12**, 241–242 (1996).
Oliver, S. G. From DNA sequence to biological function. *Nature* **379**, 597–600 (1996).
Oliver, S. G., Winson, M. K., Kell, D. B., and Baganz, F. Systematic functional analysis of the yeast genome. *TIBTECH* **16**, 373–378 (1998).
Olsen, G. J., and Woese, C. R. Lessons from an archaeal genome: what are we learning from *Methanococcus jannaschii*? *TIG* **12**, 377–379 (1996).
Orengo, C. Classification of protein folds. *Current Opinion in Structural Biology* **4**, 429–440 (1994).
Orgel, L. E. and Crick, F. H. C. Selfish DNA: the ultimate parasite. *Nature* **284**, 604–607 (1980).
Orgel, L. E. Molecular replication. *Nature* **358**, 203–209 (1992).
Orgel, L. E. Selection *in vitro*. *Proc. R. Soc. Lond. B* **205**, 435–442 (1979).
Orgel, L. E. The origin of life – a review of facts and speculations. *TIBS* **23**, 491–495 (1998).
Osborne, K. A., Robichon, A., Burgess, E., Butland, S., Shaw, R. A., Coulthard, A., Pereira, H. S., Greenspan, R. J., and Sokolowski, M. B. Natural behaviour polymorphism due to a cGMP-dependent protein kinase of *Drosophila*. *Science* **277**, 834–836 (1997).
Oster, G., and Wang, H. ATP synthase: two motors, two fuels. *Structure* **7**, R67-R72 (1999).
Palmer, J. D. Organelle Genomes: Going, going, gone! *Science* **275**, 790–791 (1997).
Palopoli, M. F., and Patel, N. H. Neo-Darwinian developmental evolution: can we bridge the gap between pattern and process? *Current Opinion in Genetics & Development* **6**, 502–508 (1996).
Paré, P. W., and Tumlinson, J. H. Biology's new Rosetta stone. *Nature* **385**, 29–31(1997).
Parikh, S. S., Mol, C. D., and Tainer, J. A. Base excision repair enzyme family portrait: integrating the structure and chemistry of an entire DNA repair pathway. *Structure* **5**, 1543–1550 (1997).
Patino, M. M., Liu, J-J., Glover, J. R., and Lindquist, S. Support for the prion hypothesis for inheritance of a phenotypic trait in yeast. *Science* **273**, 622–626 (1996).
Paulovich, A. G., Toczyski, D. P., and Hartwell, L. H. When checkpoints fail. *Cell* **88**, 315–321 (1997).
Pennisi, E. Chemical shackles for genes? *Science* **273**, 574–575 (1996).

Pennisi, E. Evolutionary and systematic biologists converge. *Science* **273**, 181–182 (1996).

Pennisi, E. Genome data shake tree of life. *Science* **280**, 672–674 (1998).

Pennisi, E. Genome links typhus bug to mitochondrion. *Science* **282**, 1243 (1998).

Pennisi, E. Genome reveals wiles and weak points of syphilis. *Science* **281**, 324–325 (1998).

Pennisi, E. Heat shock protein mutes genetic changes. *Science* **282**, 1796 (1998).

Pennisi, E. Is it time to uproot the tree of life? *Science* **284**, 1305–1307 (1999).

Pennisi, E. Laboratory workhouse decoded. *Science* **277**, 1432–1434 (1997).

Pennisi, E. NRC Oks long-delayed survey of human genome diversity. *Science* **278**, 568 (1997).

Pennisi, E. Polymer folds just like a protein. *Science* **277**, 1764 (1997).

Pennisi, E. Single gene controls fruit fly life-span. *Science* **282**, 856 (1998).

Pennisi, E. Tracing backbone evolution through a tunicate's lost tail. *Science* **274**, 1082–1083 (1996).

Pennisi, E. Worming secrets from the *C. elegans* genome. *Science* **282**, 1972–1974 (1998).

Pennisi, E., and Roush, W. Developing a new view of evolution. *Science* **277**, 34–37 (1997).

Penrose, R. *Shadows of the Mind: A Search for the Missing Science of Consciousness* (Oxford University Press, Oxford, 1994).

Penrose, R. *The Emperor's New Mind: Concerning Computers, Minds and the Laws of Physics* (Penguin Books, London, 1991).

Perrimon, N., and McMahon, A. P. Negative feedback mechanisms and their roles during pattern formation. *Cell* **97**, 13–16 (1999).

Petka, W. A., Harden, J. L., McGrath, K. P., Wirtz, D., and Tirrell, D. A. Reversible hydrogels from self-assembling artificial proteins. *Science* **281**, 389–392 (1998).

Piazza, A., Rendine, S., Minch, E., Menozzi, P. Mountain, J., and Cavalli-Sforza, L. L. Genetics and the origin of European languages. *Proc. Natl. Acad. Sci. USA* **92**, 5836–5840 (1995).

Piccirilli, J. A., Krauch, T., Moroney, S. E., and Benner, S. A. Enzymatic incorporation of a new base pair into DNA and RNA extends the genetic alphabet. *Nature* **343**, 33–37 (1990).

Plaxco, K. W., and Gro, M. The importance of being unfolded. *Nature* **386**, 657–659 (1997).

Plotkin, H. C. An evolutionary epistemological approach to the evolution of intelligence. In H. J. Jevison (ed.), *The Evolutionary Biology of Intelligence* (Springer-Verlag, Berlin and New York, 1988).

Porter, J. A., Young, K. E., and Beachy, P. A. Cholesterol modification of hedgehog signaling proteins in animal development. *Science* **274**, 255–259 (1996).

Prieur, D. Microbiology of deep-sea hydrothermal vents. *TIBTECH* **15**, 242–244 (1997).

Priscu, J. C., Fritsen, C. H., Adams, E. E., Giovannoni, S. J., Paerl, H. W., McKay, C. P., Doran, P. T., Gordon, D. A., Lanoil, B. D., and Pinckney, J. L. Perennial Antarctic lake ice: An oasis for life in a polar desert. *Science* **280**, 2095–2098 (1998).

Proudfoot, N. Ending the message is not so simple. *Cell* **87**, 779–781 (1996).

Psenner, R., and Sattler, B. Life at the freezing point. *Science* **280**, 2073–2074 (1998).

Pullum, G. K. Language that dare not speak its name. *Nature* **386**, 321–322 (1997).

Purugganan, M.D. The molecular evolution of development. *BioEssays* **20**, 700–711 (1998).

Purvis, A., and Bromham, L. Estimating the transition/transversion ratio from independent pairwise comparisons with an assumed phylogeny. *J. Mol. Evol.* **44**, 112–119 (1997).

Purvis, A., and Harvey, P. H. The right size for a mammal. *Nature* **386**, 332–333 (1997).

Radzicka, A., and Wolfenden, R. A proficient enzyme. *Science* **267**, 90–93 (1995).

Raff, M. C. Size control: The regulation of cell numbers in animal development. *Cell* **86**, 173–175 (1996).

Ramachandran, V. S., Armel, C., Foster, C., and Stoddard, R. Object recognition can drive motion perception. *Nature* **395**, 852–853 (1998).

Ramsay, G. DNA chips: State-of-the-art. *Nature Biotechnology* **16**, 40–44 (1998).

Rao, A. Sampling the universe of gene expression. *Nature Biotechnology* **16**, 1311–1312 (1998).

Rappaz, M., and Kurz, W. Dendrites solidified by computer. *Nature* **375**, 103–104 (1995).

Raymond, J. L., Lisberger, S. G., and Mauk, M. D. The cerebellum: A neuronal learning machine? *Science* **272**, 1126–1131 (1996).

Redinbo, M. R., Stewart, L., Kuhn, P, Champoux, J. J., and Hol, W. G. J. Crystal structures of human topoisomerase I in covalent and noncovalent complexes with DNA. *Science* **279**, 1504–1513 (1998).

Regan, L. Born to be beta. *Current Biology* **4**, 656 (1994).

Regan, L. Proteins to order? *Structure* **6**, 1–4 (1998).

Regan, L. The β-sheet-forming propensities of amino acids have been measured in a new model system. The origins of observed variations in the propensities are unclear, but the results provide a useful tool for protein design. *Current Biology* **4**, 656–658 (1994).

Reichard, P. The evolution of ribonucleotide reduction. *TIBS* **22**, 81–85 (1997).

Richardson, A., and Landry, S. J., and Georgopoulos, C. The ins and outs of a molecular chaperone machine. *TIBS* **23**, 138–143 (1998).

Riddihough, G. Homing in on the homeobox. *Nature* **357**, 643–644 (1992).

Riddle, D. S., Santiago, J. V., Bray-Hall, S. T., Doshi, N., Grantcharova, V. P., Yi, Q., and Baker, D. Functional rapidly folding proteins from simplified amino acid sequences. *Nature Structural Biology* **4**, 805–809 (1997).

Riezman, H. The ins and outs of protein translocation. *Science* **278**, 1728–1729 (1997).

Ringwald, M., Baldock, R., Bard, J., Kaufman, M., Eppig, J. T., Richardson, J. E., Nadeau, J. H., and Davidson, D. Database for mouse development. *Science* **265**, 2033–2034 (1994).

Roa, A. Sampling the universe of gene expression. *Nature Biotechnology* **16**, 1311–1313 (1998).

Robertson, D. L., and Joyce, G. F. Selection *in vitro* of an RNA enzyme that specifically cleaves single-stranded DNA. *Nature* **344**, 467–468 (1990).

Robertson, M. P. and Ellington, A. D. How to make a nucleotide. *Nature* **395**, 223–225 (1998).

Roemer, I., Reik, W., Dean, W., and Klose, J. Epigenetic inheritance in the mouse. *Current Biology* **7**, 277–280 (1997).

Rost, B. Marrying structure and genomics. *Structure* **6**, 259–263 (1998).

Rost, B. Protein structures sustain evolutionary drift. *Folding & Design* **2**, S19-S24 (1997).

Roush, W. "Smart" genes use many cues to set cell fate. *Science* **272**, 652–653 (1996).

Roush, W. A new embryo zoo. *Science* **274**, 1608–1609 (1996).

Roush, W. Corn: A lot of change from a little DNA. *Science* **272**, 1873 (1996).

Roush, W. Sizing up dung beetle evolution. *Science* **277**, 184 (1997).

Roush, W. Worm longevity gene cloned. *Science* **277**, 897–898 (1997).

Rucker, R. *Mind Tools: The Mathematics of Information* (Penguin Books, London, 1988).

Rutherford, S. L., and Lindquist, S. Hsp90 as a capacitor for morphological evolution. *Nature* **396**, 336–341 (1998).

Rutter, G. A., Kennedy, H. J., Wood, C. D., White, M. R. H., and Tavaré, J. M. Real-time imaging of gene expression in single living cells. *Chemistry & Biology* **5**, R285-R290 (1998).

Sabeti, P. C., Unrau, P. J., and Bartel, D. P. Accessing rare activities from random RNA sequences: the importance of the length of molecules in the starting pool. *Chemistry & Biology* **4**, 767–774 (1997).

Saks, M. E., Sampson, J. R., and Abelson, J. Evolution of a transfer RNA gene through a point mutation in the anticodon. *Science* **279**, 1665–1670 (1998).

Sali, A. 100,000 protein structures for the biologist. *Nature Structural Biology* **5**, 1029–1032 (1998).

Salser, S. J., and Kenyon, C. Patterning *C. elegans*: homeotic cluster genes, cell fates and cell migrations. *TIG* **10**, 159–163 (1994).

Sancar, A. DNA repair in humans. *Annu. Rev. Genetics* **29**, 69–105 (1995).

Sancar, A. Mechanisms of DNA excision repair. *Science* **266**, 1954–1956 (1994).

Santocanale, C., and Diffley, J. F. X. A Mec1- and Rad53- dependent checkpoint controls late-firing origins of DNA replication. *Nature* **395**, 615–618 (1998).

Sawaya, M. R., Prasad, R., Wilson, S. H., Kraut, J., and Pelletier, H. Crystal structures of human DNA polymerase β complexed with gapped and nicked DNA: Evidence for an induced fit mechanism. *Biochemistry* **36**, 11205–11215 (1997).

Schatz, G. Just follow the acid chain. *Nature* **388**, 121–122 (1997).

Schena, M., Hellar, R. A., Theriault, T. P., Konrad, K., Lachenmeier, E., and Davis, R. W. Microarrays: biotechnology's discovery platform for functional genomics. *TIBTECH* **16**, 301–306 (1998).

Schena, M., Shalon, D., Davis, R. W., and Brown, P. O. Quantitative monitoring of gene expression patterns with a complementary DNA microarray. *Science* **270**, 467–470 (1995).

Schena, M., Shalon, D., Heller, R., Chai, A., Brown, P. O., and Davis, R. W. Parallel human genome analysis: Microarray-based expression monitoring of 1000 genes. *Proc. Natl. Acad. Sci. USA* **93**, 10614–10619 (1996).

Schultz, D. W., and Yarus, M. On malleability in the genetic code. *J. Mol. Evol.* **42**, 597–601 (1996).

Schulz, H. N., Brinkhoff, T., Ferdelman, T. G., Hernández Mariné, M., Teske, A., and Jørgensen, B. B. Dense populations of a giant sulfur bacterium in Namibian shelf sediments. *Science* **284**, 493–495 (1999).

Schwartz, A. W. Did minerals perform prebiotic combinatorial chemistry? *Chemistry & Biology* **3**, 515–518 (1996).

Scott, M. P. Intimations of a creature. *Cell* **79**, 1121–1124 (1994).

Searle, J. R. Is the brain's mind a computer program? *Scientific American*, 20–25 (January 1990).

Seemüller, E., Lupas, A., and Baumeister, W. Autocatalytic processing of the 20S proteasome. *Nature* **382**, 468–470 (1996).

Ségalat, L. Elkes, D. A., and Kaplan, J. M. Modulation of serotonin-controlled behaviors by G_o in *Caenorhabditis elegans*. *Science* **267**, 1648–1651 (1995).

Service, R. F. Amino acid alchemy transmutes sheets to coils. *Science* **277**, 179 (1997).

Service, R. F. Can chip devices keep shrinking? *Science* **274**, 1834–1836 (1996).

Service, R. F. Coming soon: The pocket DNA sequencer. *Science* **282**, 399–401 (1998).

Service, R. F. DNA chips survey an entire genome. *Science* **281**, 1122 (1998).

Service, R. F. DNA ventures into the world of designer materials. *Science* **277**, 1036–1037 (1997).

Service, R. F. Microchip arrays put DNA on the spot. *Science* **282**, 396–399 (1998).

Service, R. F. Mimicking an enzyme in look and deed. *Science* **279**, 479–480 (1998).

Service, R. F. New probes open windows on gene expression, and more. *Science* **280**, 1010–1011 (1998).

Service, R. F. Prompting complex patterns to form themselves. *Science* **270**, 1299–1300 (1995).

Service, R. F. Researchers construct cell look-alikes. *Science* **275**, 31 (1997).

Service, R. F. Self-assembly comes together. *Science* **265**, 316–318 (1994).

Severin, K., Lee, D. H., Kennan, A. J., and Ghadiri, R. A synthetic peptide ligase. *Nature* **389**, 706–709 (1997).

Shakhnovich, E. I., and Gutin, A. M. Implications of thermodynamics of protein folding for evolution of primary sequences. *Nature* **346**, 773–775 (1990).

Shapiro, J. A. Adaptive mutation: who's really in the garden? *Science* **268**, 373–374 (1995).

Shapiro, L., and Lima, C. D. The argonne structural genomics workshop: Lamaze class for the birth of a new science. *Structure* **6**, 265–267 (1998).

Sharkey, M., Graba, Y., and Scott, M. P. *Hox* genes in evolution: protein surfaces and paralog groups. *TIG* **13**, 145–151 (1997).

Sharp, F. R., Massa, S. M., and Swanson, R. A. Heat-shock protein protection. *TINS* **22**, 97–99 (1999).

Sharp, P. M. In search of molecular darwinism. *Nature* **385**, 111–112 (1997).

Shawlot, W., and Behringer, R. R. Requirement for *Lim1* in head-organizer function. *Nature* **374**, 425–430 (1995).

Shear, J. B., Fishman, H. A., Allbritton, N. L., Garigan, D., Zare, R. N., and Scheller, R. H. Single cells as biosensors for chemical separations. *Science* **267**, 74–77 (1995).

Sherman, J. M., and Pillus, L. An uncertain silence. *TIG* **13**, 308–313 (1997).

Shoemaker, D. D., and Ross, K. G. Effects of social organization on gene flow in the fire ant *Solenopsis invicta*. *Nature* **383**, 613–616 (1996).

Short, N. Patterns of pattern formation. *Nature* **378**, 331 (1995).

Shubin, N. Evolutionary cut and paste. *Nature* **394**, 12–13 (1998).

Shubin, N., Tabin, C., and Carroll, S. Fossils, genes and the evolution of animal limbs. *Nature* **388**, 639–648 (1997).

Siegelmann, H. T. Computation beyond the Turing limit. *Science* **268**, 545–548 (1995).

Sievers, D., and von Kiedrowski, G. Self-replication of complementary nucleotide-based oligomers. *Nature* **369**, 221–224 (1994).

Silar, P., and Daboussi, M-J. Non-conventional infectious elements in filamentous fungi. *TIG* **15**, 141–145 (1999).

Simpson, M. L., Sayler, G. S., Applegate, B. M., Ripp, S., Nivens, D. E., Paulus, M. J., and Jellison, G. E. Bioluminescent-bioreporter integrated circuits form novel whole-cell biosensors. *TIBTECH* **16**, 332–338 (1998).

Singer, W. Development and plasticity of cortical processing architectures. *Science* **270**, 758–764 (1995).

Skandalis, A., Encell, L. P., and Loeb, L. A. Creating novel enzymes by applied molecular evolution. *Chemistry & Biology* **4**, 889–898 (1997).
Slack, J. New shapes from old genes. *Nature* **382**, 124–125 (1996).
Smith, C. Beyond flesh & blood. *New Scientist*, 28–29 (November 1995).
Smith, C. G. Computation without current. *Science* **284**, 274 (1999).
Smith, D. R. *et al.* Complete genome sequence of Methanobacterium thermoautotrophicum deltaH: functional analysis and comparative genomics. *J. Bacteriol.* **179**, 7135–7155 (1997).
Smith, J. How to tell a cell where it is. *Nature* **381**, 367–368 (1996).
Smith, J. T-box genes: What they do and how they do it. *TIG* **15**, 154–158 (1999).
Sniegowski, P. Evolution: setting the mutation rate. *Current Biology* **7**, R487-R488 (1997).
Spector, M. S., and Schnur, J. M. DNA ordering on a lipid membrane. *Science* **275**, 791–792 (1997).
Spencer, D. M. Creating conditional mutations in mammals. *TIG* **12**, 181–187 (1996).
Spitzer, N. C., and Sejnowski, T. J. Biological information processing: bits of progress. *Science* **277**, 1060–1061 (1997).
Stein, W., and Varela, F. J. (Eds) *Thinking About Biology: An Invitation to Current Theoretical Biology* (Adison-Wesley Publishing Company, Reading, MA, 1993).
Steller, H. Mechanisms and genes of cellular suicide. *Science* **267**, 1445–1449 (1995).
Stephens, R. S., Kalman, S., Lammel, C., Fan, J., Marathe, R., Aravind, L., Mitchell, W., Olinger, L., Tatusov, R. L., Zhao, Q., Koonin, E. V., and Davis, R. W. Genome sequence of an obligate intracellular pathogen of humans: *Chlamydia trachomatis*. *Science* **282**, 754–759 (1998).
Stewart, C-B. Active ancestral molecules. *Nature* **374**, 12–13 (1995).
Stewart, I. Emergent macrosimplicity. *Nature* **379**, 33 (1996).
Stewart, I. *Life's Other Secret: The New Mathematics of the Living World* (Penguin Books, London, 1998).
Stewart, I., and Golubitsky, M. *Fearful Symmetry: Is God a Geometer?* (Penguin Books, London, 1993).
Stix, G. Waiting for breakthroughs. *Scientific American*, 78–83 (April 1996).
Stokstad, E. The bare bones of catalysis. *Science* **279**, 1852 (1998).
Stone, R. Putting a human face on a new breed of robot. *Science* **274**, 182 (1996).
Strauss, E. New ways to probe the molecules of life. *Science* **282**, 1406–1407 (1998).
Strobel, S. A., and Doudna, J. A. RNA seeing double: close-packing of helices in RNA tertiary structure. *TIBS* **22**, 262–266 (1997).
Strong, D. R. Fear no weevil? *Science* **277**, 1058–1059 (1997).
Struthers, M. D., Cheng, R. P., and Imperiali, B. Design of a monomeric 23-residue polypeptide with defined tertiary structure. *Science* **271**, 342–345 (1996).

Stupp, S. I., LeBonheur, V., Walker, K., Li, L. S., Huggins, K. E., Keser, M., and Amstutz, A. Supramolecular materials: Self-organized nanostructures. *Science* **276**, 384–389 (1997).

Sullivan, D. T. DNA excision repair and transcription: implications for genome evolution. *Current Opinion in Genetics & Development* **5**, 786–791 (1995).

Sun, F-L., Dean, W. L., Kelsey, G., Allen, N. D., and Reik, W. Transactivation of *Igf2* in a mouse model of Beckwith-Wiedemann syndrome. *Nature* **389**, 809–815 (1997).

Sutherland, J. D., and Blackburn, J. M. Killing two birds with one stone: a chemically plausible scheme for linked nucleic acid replication and coded peptide synthesis. *Chemistry & Biology* **4**, 481–488 (1997).

Szathmáry, E. From RNA to language. *Current Biology* **6**, 764 (1996).

Szathmáry, E. The first two billion years. *Nature* **387**, 662–663 (1997).

Szathmáry, E. Towards the evolution of ribozymes. *Nature* **344**, 115 (1990).

Szathmáry, E. What is the optimum size for the genetic alphabet? *Proc. Natl. Acad. Sci. USA* **89**, 2614–2618 (1992).

Szathmáry, E., and Maynard Smith, J. The major evolutionary transitions. *Nature* **374**, 227–232 (1995).

Tang, C. M., Hood, D. W., and Moxon, E. R. *Haemophilus* influence: the impact of whole genome sequencing on microbiology. *TIG* **13**, 399–404 (1997).

Tarasow, T. M., Tarasow, S. L., and Eaton, B. E. RNA-catalysed carbon-carbon bond formation. *Nature* **389**, 54–57 (1997).

Tatusov, R. L., Koonin, E. V., and Lipman, D. J. A genomic perspective on protein families. *Science* **278**, 631–637 (1997).

Taubes, G. All together for quantum computing. *Science* **273**, 1164 (1996).

Taubes, G. Computer design meets Darwin. *Science* **277**, 1931–1932 (1997).

Tautz, D. Debatable homologies. *Nature* **395**, 17–19 (1998).

Tautz, D. Selector genes, polymorphisms, and evolution. *Science* **271**, 160–161 (1996).

Taylor, C. B. Damage control. *The Plant Cell* **9**, 111–113 (1997).

Taylor, K. L., Cheng, N., Williams, R. W., Steven, A. C., and Wickner, R. B. Prion domain initiation of amyloid formation in vitro from native Ure2p. *Science* **283**, 1339–1343 (1999).

Thaler, D. S. The evolution of genetic intelligence. *Science* **264**, 224–225 (1994).

Thomas, C. F., and White, J. G. Four-dimensional imaging: the exploration of space and time. *TIBTECH* **16**, 175–182 (1998).

Thomas, G. Invisible circuits. *Nature* **389**, 907–908 (1997).

Thompson, D. W. *On Growth and Form* (Cambridge University Press, Cambridge, 1994).

Thompson, E. M., Adenot, P. Tsuji, F. I., and Renard, J-P. Real time imaging of transcriptional activity in live mouse preimplantation embryos using a secreted luciferase. *Proc. Natl. Acad. Sci. U.S.A.* **92**, 1317–1321 (1995).

Tilghman, S. M. The sins of the fathers and mothers: Genomic imprinting in mammalian development. *Cell* **96**, 185–193 (1999).

Ting, C-T., Tsaur, S-C., Wu, M-L., and Wu, C-I. A rapidly evolving homeobox at the site of a hybrid sterility gene. *Science* **282**, 1501–1504 (1998).

Tomb, J-F. *et al.* The complete genome sequence of the gastric pathogen *Helicobacter pylori*. *Nature* **388**, 539–547 (1997).

Tracy, R. B., Chédin, F., and Kowalczykowski, S. C. The recombination hot spot chi is embedded within islands of preferred DNA pairing sequences in the E. coli genome. *Cell* **90**, 205–206 (1997).

Trivedi, B. Modeling the oddities of biology. *Nature Biotechnology* **16**, 1316–1317 (1998).

Uhlenbeck, O. C., Pardi, A., and Feigon, J. RNA structure comes of age. *Cell* **90**, 833–840 (1997).

Ullmann, G. M., Hauswald, M., Jensen, A., Kostić, N. M., and Knapp, E-W. Comparison of the physiologically equivalent proteins cytochrome c_6 and plastocyanin on the basis of their electrostatic potentials. Tryptophan 63 in cytochrome c_6 may be isofunctional with tyrosine 83 in plastocyanin. *Biochemistry* **36**, 16187–16196 (1997).

Ulrich, H. D., Mundorff, E., Santarsiero, B. D., Driggers, E. M., Stevens, R. C., and Schultz, P. G. The interplay between binding energy and catalysis in the evolution of a catalytic antibody. *Nature* **389**, 271–275 (1997).

Umbanhowar, P. Patterns in the sand. *Nature* **389**, 541–542 (1997).

Unrau, P. J., and Bartel, D. P. RNA-catalysed nucleotide synthesis. *Nature* **395**, 260–263 (1998).

Uphoff, K. W., Bell, S. D., and Ellington, A. D. *In vitro* selection of aptamers: the dearth of pure reason. *Current Opinion in Structural Biology* **6**, 281–288 (1996).

Valentine J. W., Erwin, D. H., and Jablonski, D. Developmental evolution of metazoan bodyplans: The fossil evidence. *Developmental Biology* **173**, 373–381 (1996).

van der Oost, J., de Vos, W. M., and Antranikian, G. Extremophiles. *TIBTECH* **14**, 415–417 (1996).

Vauclair, J. Mental states in animals: cognitive ethology. *Trends in Cognitive Sciences* **1**, 35–39 (1997).

Verdine, G. L., and Bruner, S. D. How do DNA repair proteins locate damaged bases in the genome? *Chemistry & Biology* **4**, 329–334 (1997).

Vermeij, G. J. Animal origins. *Science* **274**, 525–526 (1996).

Vidal, G. The oldest eukaryotic cells. *Scientific American*, **250**, 48–57 (1984).

Vieille, C., and Zeikus, G. Thermozymes: identifying molecular determinants of protein structural and functional stability. *TIBTECH* **14**, 183–190 (1996).

Vogel, G. A sulfurous start for protein synthesis? *Science* **281**, 627–629 (1998).

Vogel, G. A two-piece protein assembles itself. *Science* **281**, 763–764 (1998).

Vogel, G. Searching for living relics of the cell's early days. *Science* **277**, 1604 (1997).

von Kiedrowski, G. Primordial soup or crepes? *Nature* **381**, 20–21 (1996).

von Neumann, J. *The Computer and the Brain* (Yale University Press, New Haven and London, 1986).

Vulić, M., Dionisio, F., Taddei, F., and Radman, M. Molecular keys to speciation: DNA polymorphism and the control of genetic exchange in enterobacteria. *Proc. Natl. Acad. Sci. USA* **94**, 9763–9767 (1997).

Wagner, S. D., Milstein, C., and Neuberger, M. S. Codon bias targets mutation. *Nature* **376**, 732 (1995).

Wales, D. J., Miller, M. A., and Walsh, T. R. Archetypal energy landscapes. *Nature* **394**, 758–760 (1998).

Walker, H. Still alive after all these years? *New Scientist*, 7 (August 1996).

Walter, M. Old fossils could be fractal frauds. *Nature* **383**, 385–386 (1996).

Warburton, P. E., and Kipling, D. Providing a little stability. *Nature* **386**, 553–555 (1997).

Ward, R., and Stringer, C. A molecular handle on the Neanderthals. *Nature* **388**, 225–226 (1997).

Watson, A. Microbiologists explore life's rich, hidden kingdoms. *Science* **275**, 1740–1742 (1997).

Watson, A. Why can't a computer be more like a brain? *Science* **277**, 1934–1936 (1997).

Weber, B. H., Depew, D. J., and Smith, J. D. (Eds) *Entropy, Information, and Evolution: New Perspectives on Physical and Biological Evolution* (The MIT Press, Cambridge, Massachusetts, 1990).

Wedel, A. B. Fishing the best pool for novel ribozymes. *TIBTECH* **14**, 459–465 (1996).

Wedemayer, G. J., Patten, P. A., Wang, L. H., Schultz, P. G., and Stevens, R. C. Structural insights into the evolution of an antibody combining site. *Science* **276**, 1665–1669 (1997).

Weeks, K. M. Protein-facilitated RNA folding. *Current Opinion in Structural Biology* **7**, 336–342 (1997).

Weinacht, T. C., Ahn, J., and Bucksbaum, P.H. Controlling the shape of a quantum wavefunction. *Nature* **397**, 233–235 (1999).

Weinberg, S. *The First Three Minutes: A Modern View of the Origin of the Universe* (Basic Books, New York, 1993).

Weinert, T. A DNA damage checkpoint meets the cell cycle engine. *Science* **277**, 1450–1451 (1997).

Wen, X., Fuhrman, S., Michaels, G. S., Carr, D. B., Smith, S., Barker, J. L., and Somogyi, R. Large-scale temporal gene expression mapping of central nervous system development. *Proc. Natl. Acad. Sci. U.S.A.* **95**, 334–339 (1998).

West, G. B., Brown, J. H., and Enquist, B. J. A general model for the origin of allometric scaling laws in biology. *Science* **276**, 122–126 (1997).

Westhof, E., and Michel, F. Ribozyme architectural diversity made visible. *Science* **282**, 251–252 (1998).

Wickner, R. B. [URE3] as an altered *URE2* protein: Evidence for a prion analog in *Saccharomyces cerevisiae*. *Science* **264**, 566–569 (1994).
Williams, G. C. *Natural Selection: Domains, Levels and Challenges* (Oxford University Press, Oxford, 1992).
Williams, J. C., Zeelen, J. P., Neubauer, G., Vriend, G., Backmann, J., Michels, P. A. M., Lambier, A-M., and Wierenga, R. K. Structural and mutagenesis studies of leishmania triosephosphate isomerase: a point mutation can convert a mesophilic enzyme into a superstable enzyme without losing catalytic power. *Protein Engineering* **12**, 243–250 (1999).
Williams, N. Gram-positive bacterium sequenced. *Science* **277**, 478 (1997).
Williamson, J. R. Aptly named aptamers display their aptitude. *Nature* **382**, 112–113 (1996).
Wilson, C., and Szostak, J. W. *In vitro* evolution of a self-alkylating ribozyme. *Nature* **374**, 777–782 (1995).
Wilson, C., and Szostak, J. W. Isolation of a fluorophore-specific DNA aptamer with weak redox activity. *Chemistry & Biology* **5**, 609–617 (1998).
Winson, M. K., and Kell, D. B. Going places: forced and natural molecular evolution. *TIBTECH* **14**, 323–325 (1996).
Wittgenstein, L. *Tractatus Logio-Philosophicus* (Routledge & Kegan Paul, London, 1988).
Wodicka, L., Dong, H., Mittmann, M., Ho, M-H., and Lockhart, D. J. Genome-wide expression monitoring in *Saccharomyces cerevisiae*. *Nature Biotechnology* **15**, 1359–1367 (1997).
Woese, C. R. Phylogenetic trees: Whither microbiology? *Current Biology* **6**, 1060–1063 (1996).
Wolpert, L. The evolutionary origin of development: cycles, patterning, privilege and continuity. *Development*, Supplement, 79–84 (1994).
Wolynes, P. G., Luthey-Schulten, Z., and Onuchic, J. N. Fast-folding experiments and the topography of protein folding energy landscapes. *Chemistry & Biology* **3**, 425–432 (1996).
Wolynes, P. G., Onuchic, J. N., and Thirumalai, D. Navigating the folding routes. *Science* **267**, 1619–1620 (1995).
Wong, J. T-F. Role of minimization of chemical distances between amino acids in the evolution of the genetic code. *Proc. Natl. Acad. Sci. USA* **77**, 1083–1086 (1980).
Wood, B. Human evolution. *BioEssays* **18**, 945–954 (1996).
Wood, R. D. DNA repair in eukaryotes. *Annu. Rev. Biochem.* **65**, 135–167 (1996).
Wood, R. D. DNA repair: Knockouts still mutating after first round. *Current Biology* **8**, R757-R760 (1998).
Wootton, J. C. Sequences with 'unusual' amino acid compositions. *Current Opinion in Structural Biology* **4**, 413–421 (1994).
Wray, G. A. Promoter logic. *Science* **279**, 1871–1872 (1998).
Wu, C-I. Now blows the east wind. *Nature* **380**, 105–106 (1996).

Wu, C-I., and Johnson, N. A. Endless forms, several powers. *Nature* **382**, 298–299 (1996).

Wu, C-I., and Li, W-H. Evidence for higher rates of nucleotide substitution in rodents than in man. *Proc. Natl. Acad. Sci. USA.* **82**, 1741–1745 (1985).

Wu, H., Yang, W-P., and Barbas, C. F. Building zinc fingers by selection: Toward a therapeutic application. *Proc. Natl. Acad. Sci. U.S.A.* **92**, 344–348 (1995).

Wuethrich, B. Why sex? Putting theory to the test. *Science* **281**, 1980–1982 (1998).

Wydro, R. M., Madira, W., Hiramatsu, T., Kogut, M., Kushner, D.J. Salt-sensitive *in vitro* protein synthesis by a moderately halophilic bacterium. *Nature* **269**, 824–825 (1977).

Yao, S., Ghosh, I., Zutshi, R., and Chmielewski, J. Selective amplification by auto- and cross-catalysis in a replicating peptide system. *Nature* **396**, 447–450 (1998).

Yélamos, J., Klix, N., Goyenechea, B., Lozano, F., Chui, Y. L., González Fernández, A., Pannell, R., Neuberger, M. S., and Milstein, C. Targeting of non-Ig sequences in place of the V segment by somatic hypermutation. *Nature* **376**, 225–229 (1995).

Yomo, T., Prijambada, I. D., Yamamoto, K., Shima, Y., Negoro, S., and Urabe, I. Properties of artificial proteins with random sequences. *Ann. NY Acad. Sci.* **864**, 131–135 (1998).

Young, D. B. Blueprint for the white plague. *Nature* **393**, 515–516 (1998).

Yuh, C-H., Bolouri, H., and Davidson, E. H. Genomic cis-regulatory logic: experimental and computational analysis of a sea urchin gene. *Science* **279**, 1896–1902 (1998).

Zeyl, C., and Bell, G. The advantage of sex in evolving yeast populations. *Nature* **388**, 465–468 (1997).

Zhang, B., and Cech, T. R. Peptide bond formations by *in vitro* selected ribozymes. *Nature* **390**, 96–100 (1997).

Zhang, C-T. Relations of the numbers of protein sequences, families and folds. *Protein Engineering* **10**, 757–761 (1997).